Proceedings of the
Steklov Institute of Mathematics
Volume 197 Issue 1 of 4 1993

Some Questions in the Theory of Oscillations and the Theory of Optimal Control

Collection of Papers

Edited by
R. V. Gamkrelidze
Corresponding Member of the Academy of Sciences of the USSR

A Translation of

ТРУДЫ
ОРДЕНА ЛЕНИНА
И ОРДЕНА ОКТЯБРЬСКОЙ РЕВОЛЮЦИИ
МАТЕМАТИЧЕСКОГО ИНСТИТУТА
имени В. А. СТЕКЛОВА
ТОМ 197 1991

ABSTRACT: This issue consists of two fundamental papers. The first, *Asymptotic theory of relaxation oscillations*, is by E. F. Mishchenko and A. Yu. Kolesov and is in essence a short monograph devoted to the theory of periodic motions in singularly perturbed systems. The second paper, *Regimes with increasingly more frequent switchings in optimal control problems*, is by M. I. Zelikin and V. F. Borisov and deals with structural properties of the solutions of a system having infinitely many switchings on a finite time interval to Hamiltonian systems with discontinuous right-hand side.

1991 *Mathematics Subject Classification*. Primary 34C23, 34D05, 34E15, 49K15, 58F14, 93B50, 93C15; Secondary 34C20, 34C45, 34D30, 34E05, 58F05, 58F21, 93C85.

Library of Congress Cataloging-in-Publication Data
Nekotorye voprosy teorii kolebaniĭ i teorii optimal′nogo upravleniia. English.
Some questions in the theory of oscillations and the theory of optimal control: collection of papers/edited by R. V. Gamkrelidze.
p. cm.—(Proceedings of the Steklov Institute of Mathematics; v. 197, issue 1 of 4, 1993)
ISBN 0-8218-3148-8 (acid-free paper)
1. Oscillations. 2. Control theory. 3. Mathematical optimization.
I. Gamkrelidze, R. V. II. Title. III. Series: Trudy Matematicheskogo instituta imeni V. A. Steklova. English; 1993, issue 1.
QA1.A413 1993, Issue 1
[QA865]
510 s–dc20 93-7068
[515′.352] CIP

This nonjournal serial does not always appear in strict sequential volume order in the original. As a consequence the four 1993 issues in translation are derived from: (Russian Tom) **197**(1991), **200**(1991), **193**(1992), **194**(1992).

АКАДЕМИЯ НАУК

СОЮЗА СОВЕТСКИХ СОЦИАЛИСТИЧЕСКИХ РЕСПУБЛИК

ТРУДЫ

ОРДЕНА ЛЕНИНА
И ОРДЕНА ОКТЯБРЬСКОЙ РЕВОЛЮЦИИ
МАТЕМАТИЧЕСКОГО ИНСТИТУТА
имени В. А. СТЕКЛОВА

CXCVII

НЕКОТОРЫЕ ВОПРОСЫ ТЕОРИИ КОЛЕБАНИЙ И ТЕОРИИ ОПТИМАЛЬНОГО УПРАВЛЕНИЯ

Под редакцией
члена-корреспондента АН СССР
Р. В. ГАМКРЕЛИДЗЕ

Москва «Наука» 1991

Correspondence between *Trudy Mat. Inst. Steklov* and *Proc. Steklov Inst. Math.*

Russian imprint	Vol.	English imprint
1966	74	1967
1965	75	1967
1965	76	1967
1965	77	1967
1965	78	1967
1965	79	1966
1965	80	1968
1966	81	1968
1966	82	1967
1965	83	1967
1965	84	1968
1966	85	1967
1965	86	1967
1966	87	1967
1967	88	1969
1967	89	1968
1967	90	1969
1967	91	1969
1966	92	1968
1967	93	1970
1968	94	1969
1968	95	1971
1968	96	1970
1968	97	1969
1968	98	1971
1967	99	1968
1971	100	1974
1972	101	1975
1967	102	1970
1968	103	1970
1968	104	1971
1969	105	1971
1969	106	1972
1971	107	1976
1968	108	1971

Russian imprint	Vol.	English imprint
1971	109	1974
1970	110	1972
1970	111	1972
1971	112	1973
1970	113	1972
1970	114	1974
1971	115	1974
1971	116	1973
1972	117	1974
1972	118	1976
1973	119	1976
1974	120	1976
1972	121	1974
1973	122	1975
1973	123	1975
1976	124	1978
1973	125	1975
1973	126	1975
1975	127	1977
1972	128	1974
1973	129	1976
1978	130	1979
1974	131	1975
1973	132	1975
1973	133	1977
1975	134	1977
1975	135	1978
1975	136	1978
1976	137	1978
1975	138	1977
1980	139	1982
1976	140	1979
1976	141	1979
1976	142	1979
1977	143	1980

Russian imprint	Vol.	English Imprint	Issue
1979	144	1980	2
1980	145	1981	1
1978	146	1980	3
1980	147	1981	2
1978	148	1980	4
1978	149	1981	3
1979	150	1981	4
1980	151	1982	2
1980	152	1982	3
1981	153	1982	4
1983	154	1984	4
1981	155	1983	1
1980	156	1983	2
1981	157	1983	3
1981	158	1983	4
1983	159	1984	2
1982	160	1984	1
1983	161	1984	3
1983	162	1985	1
1984	163	1985	4
1983	164	1985	2
1984	165	1985	3
1984	166	1986	1
1985	167	1986	2
1984	168	1986	3
1985	169	1986	4
1984	170	1987	1
1985	171	1987	2
1985	172	1987	3
1986	173	1987	4
1986	174	1988	1
1986	175	1988	2
1987	176	1988	3
1986	177	1988	4
1986	178	1989	1

Russian imprint	Vol.	English Imprint	Issue
1988	179	1989	2
1987	180	1989	3
1988	181	1989	4
1988	182	1990	1
1990	183	1991	4
1990	184	1991	2
1988	185	1990	2
1989	186	1991	1
1989	187	1990	3
1990	188	1991	3
1989	189	1990	4
1989	190	1992	1
1989	191	1992	2
1990	192	1992	3
1991	196	1992	4
1991	197	1993	1
1991	200	1993	2
1992	193	1993	3
1992	194	1993	4

This nonjournal serial does not always appear in strict sequential volume order in the original. Accordingly, the translation will occasionally reflect any gaps resulting thereby, as happens for example with the English imprints of 1984, 1985 and 1990.

CONTENTS

RUSSIAN CONTENTS*

*The American Mathematical Society scheme for transliteration of Cyrillic may be found at the end of index issues of *Mathematical Reviews*.

Труды Матем. Инст. Стеклова
197 (1991)

Proceedings of the Steklov
Institute of Mathematics
1993, Issue 1

ASYMPTOTIC THEORY OF RELAXATION OSCILLATIONS

UDC 517.926

E. F. MISHCHENKO AND A. YU. KOLESOV

The articles [1]–[4] laid the foundations for the general theory of relaxation oscillations in systems of ordinary differential equations. They contain, in particular, a construction of the asymptotics for periodic relaxation solutions up to terms of order ε (ε is a small parameter), along with formulations of conditions ensuring the existence of a relaxation cycle. For systems on the plane this theory was later put in a complete form by Rozov and was summarized in the first part of the monograph [5]. However, for almost thirty years the question of uniqueness of a relaxation cycle with fixed zero approximation remained open for multidimensional systems, along with the question of the complete asymptotics of trajectories in a neighborhood of a breakaway point (точка срыва).

In the first chapter of this paper a relaxation system is regarded as a certain kind of special small perturbation of its corresponding relay system. This approach is new only on the mathematical level, since it was actually used in the monograph [6] on a descriptive level. The theorem obtained in this chapter on C^1-closeness of the solutions of the relay system and the relaxation system enables us not only to prove the uniqueness of a multidimensional relaxation cycle and to construct the complete asymptotics of the trajectories in a neighborhood of a breakaway point, but also to consider other applications. In particular, with its help we prove a theorem on the existence under certain conditions of countably many dichotomous cycles for a three-dimensional relaxation system with two slow variables.

In the second chapter we introduce for the first time the concept of a parabolic relaxation system, and formulate for it a theorem analogous to the theorem in the first chapter on C^1-closeness to a parabolic relay system. Here, however, we give a criterion for stability of a spatially homogeneous cycle, and then apply it to some problems in biophysics.

In the third chapter we use the Krylov-Bogolyubov-Mitropol'skiĭ asymptotic method to construct the normal form for a mapping in a neighborhood of a fixed point. The results are then applied to the problem of bifurcation of spatially nonhomogeneous periodic solutions from a homogeneous relaxation cycle.

The contents of the first three chapters make up the basic aspects of the theory of relaxation oscillations. The fourth chapter, which is devoted entirely to duck trajectories of relaxation systems, should be regarded as an application of the first three chapters. For example, in it we give a great amount of attention to the construction of stable duck cycles which arise when a stable equilibrium state passes through a breakaway manifold. Up to the present time the methods of nonstandard analysis have been used to construct the theory of such cycles only for systems on the plane (see [7], where a detailed bibliography is given). In the present article we obtain

---

1991 *Mathematics Subject Classification*. Primary 34D05, 34C23, 34E15, 58F14; Secondary 58F21, 34E05, 34D30, 34C20, 34C45.

significantly more complete results for multidimensional relaxation systems with one slow variable with the help of the usual techniques of relaxation oscillations (see [5] and [8]). For a larger number of slow variables we prove a theorem on the existence and instability of duck cycles under a certain generality of the situation.

Markedly more complicated is a new problem on duck trajectories of multidimensional relaxation systems arising when there is a general violation of one of the basic restrictions of the first chapter—the normal switching condition. The assertions obtained here are applied to a van der Pol system with a harmonic action on the slow variable: an intuitive mechanism for destruction of its orbitally exponentially stable invariant torus is described.

Thus, the main problems arising in the analysis of the asymptotic behavior of periodic solutions of singularly perturbed ordinary differential equations and parabolic relaxation boundary value problems are solved in full scope.

This article is dedicated to the memory of L. S. Pontryagin.

Chapter 1. A theorem on C^1-closeness of the solutions of the relaxation system and the relay system. Asymptotics of relaxation oscillations

In this chapter, which contains our main results, we formulate and prove a theorem on C^1-closeness [9], which we use at once to give a solution of the problem posed in [5] on uniqueness of a multidimensional relaxation cycle. Then, following [10], we reduce the original relaxation system in a neighborhood of a breakaway point to a system whose right-hand sides depend regularly on a small parameter. A similar reduction facilitates in a very essential way the construction of the complete asymptotics of the solutions of the relaxation system in a neighborhood of a breakaway point [11]. Here the question of "matching" different asymptotic formulas and remainders is solved by qualitative methods based on the theorem on C^1-closeness.

As an immediate application of the theorem on C^1-closeness we study the situation when a three-dimensional relaxation system with two slow variables has countably many dichotomous cycles. The corresponding conditions are formulated in terms of the limit relay system [12].

At the end of the chapter we present the solution of a problem of Pontryagin [13] connected with tori of relaxation systems and generalizing the well-known Pontryagin-Rodygin theorem. On the basis of it we then generalize the Pontryagin-Rodygin theorem in another direction [14]: conditions are formulated for the existence of relaxation tori, i.e., tori such that in moving along them the trajectory "jumps" infinitely many times from one stable surface of slow motions to another.

§1. Statement of the problem, preliminary facts, and heuristic considerations

1.1. **Relay systems.** It is assumed below that the equation $f(x, y) = 0$, $x \in R^k$, $y \in R^m$, with smooth left-hand side determines in R^n $(n = k + m)$ a smooth m-dimensional surface Γ that decomposes into disjoint pieces [5]: $\Gamma = \Gamma_{1-} + \Gamma_{10} + \Gamma_{+} + \Gamma_{20} + \Gamma_{2-}$. Let $A(x, y) = f'_x$. By definition: $(x, y) \in \Gamma_{1-} + \Gamma_{2-}$ if all the eigenvalues of the matrix A have negative real parts; $(x, y) \in \Gamma_{10} + \Gamma_{20}$ if it has only a zero eigenvalue, while all the remaining eigenvalues are in the left complex half-plane as before; finally, $(x, y) \in \Gamma_{+}$ if at least one eigenvalue of A lies in the right half-plane. Moreover, we assume that $x = \Phi_1(y)$ or $x = \Phi_2(y)$ (Φ_1 and Φ_2 are smooth vector-valued functions) for (x, y) in Γ_{1-} or Γ_{2-}, respectively.

Suppose that $g(x, y)$ is a smooth R^m-valued function of the variables $x \in R^k$

and $y \in R^m$. We consider in R^m the differential equation with relay nonlinearity

$$\dot{y} = g(\Phi(y), y), \qquad \Phi(y) = \begin{cases} \Phi_1(y) \\ \Phi_2(y). \end{cases} \tag{1.1}$$

As usual, it is most simple to give a constructive definition of the concept of a solution of an equation of this type. Let us say that the vector y_0 is such that $(\Phi_1(y_0), y_0) \in \Gamma_{1-}$. Under this condition the solution of (1.1) first coincides with the solution of the equation

$$\dot{y} = g(\Phi_1(y), y) \tag{1.2}$$

with initial condition $y|_{t=0} = y_0$. Assume that as t increases, the point $(\Phi_1(y(t)), y(t))$ hits the set Γ_{10} at some first moment of time t_0. Then a switching takes place, and at the subsequent moments of time $y(t)$ is the solution of the equation

$$\dot{y} = g(\Phi_2(y), y) \tag{1.3}$$

with initial condition $y|_{t=t_0} = y(t_0)$. Assume that as t increases further, the point $(\Phi_2(y(t)), y(t))$ hits the set Γ_{20} at some first moment of time $t_1 > t_0$. Then again there is a switching, and so on.

Suppose that the relay system (1.1) has a periodic solution $y_0(t)$ with two or more switchings on time intervals with length of the period T_0, and assume for definiteness that $(\Phi_1(y_0(0)), y_0(0)) \in \Gamma_{1-}$. The technique developed in [15] makes it easy for us to investigate the question of stability of the solution under the following additional restriction.

Condition 1. Suppose that the increasing positive numbers $t_0, t_1, \dots$ are the successive switching times for the periodic solution $y_0(t)$, with $(\Phi_1(y_0(t_{2s})), y_0(t_{2s})) \in \Gamma_{10}$ and $(\Phi_2(y_0(t_{2s+1})), y_0(t_{2s+1})) \in \Gamma_{20}$, where $s = 0, 1, \dots$. We introduce the tangent planes to Γ at these points, and we then take them to the origin of coordinates. After this we consider their intersections with R^m. The vectors $g(\Phi_1(y_0(t_{2s})), y_0(t_{2s}))$ and $g(\Phi_2(y_0(t_{2s+1})), y_0(t_{2s+1}))$ are also taken to the origin. We refer to a *normal switching* if each of these vectors does not belong to the corresponding plane in R^m introduced above.

In the plane passing through the point $y_0(0)$ and orthogonal to the vector $y_0(0)$ we consider the Poincaré operator Π_0 of return along trajectories of the relay system (1.1). In view of Condition 1 this operator is defined in a neighborhood of the point $y_0(0)$ and is continuous (it does not have this property in the general case when Condition 1 fails). The investigation of its smoothness is somewhat complicated by the fact that the derivatives of the functions Φ_1 and Φ_2 increase unboundedly in norm as the vectors $(\Phi_1(y), y)$ and $(\Phi_2(y), y)$ approach Γ_{10} and Γ_{20}, respectively. However, in view of the restrictions imposed on A the "unboundedness" of the derivatives of Φ_1 and Φ_2 bears a one-dimensional character, and this enables us to establish the continuous differentiability of the operator Π_0. (A rigorous proof involves introducing in neighborhoods of switching points the special coordinates proposed in [5], which reduces the problem to the consideration of a scalar differential equation for which the analysis is trivial; we do not dwell in greater detail on this question, because this coordinate system is described below.) Accordingly, we have the following assertion.

Lemma 1.1. *Suppose that Condition 1 holds, and all the eigenvalues of the matrix Π_0', computed at the point $y_0(0)$, are less than 1 in modulus. Then the periodic solution $y_0(t)$ of the relay system (1.1) is orbitally exponentially stable.*

An important point below is an algorithm for investigating the stability properties of the periodic solution $y_0(t)$ of the relay system (1.1). Obviously, formal linearization of the system yields an equation in R^m with δ-shaped coefficients, and there are several ways to introduce the concept of a solution of such an equation. In our case its solutions between successive switching times are the solutions of the equation

$$\dot{h} = [D(t) - C(t)A^{-1}(t)B(t)]h. \tag{1.4}$$

Here $D(t) = g'_y$, $C(t) = g'_x$, $A(t) = f'_x$, and $B(t) = f'_y$, and the derivatives are taken at $y = y_0(t)$ and $x = \Phi_1(y_0(t))$ or $\Phi_2(y_0(t))$. A solution $h(t)$ of (1.4) has discontinuities of the first kind as t passes the switching times t_p, with

$$h(t_p + 0) = h(t_p - 0) + a(b, h(t_p - 0)), \qquad p = 0, 1, \dots, \tag{1.5}$$

where $a = \dot{y}_0(t_p - 0) - \dot{y}_0(t_p + 0)$, $b = \operatorname{grad} T(y)|_{y=y_0(t_p)}$, and $T(y)$ is the time taken to hit Γ_{10} or Γ_{20}, respectively, in moving from the point y along a trajectory of the relay system (1.1). It follows from the aforementioned properties of the derivatives of Φ_1 and Φ_2 that the matrix on the right-hand side of (1.4) is not bounded in norm as $t \to t_{2s} - 0$ and $t \to t_{2s+1} - 0$, respectively. However, the singularities arising here are integrable, as follows from the one-dimensional character of the singularity of the derivatives of Φ_1 and Φ_2 (see the remark before Lemma 1.1).

We dwell in greater detail on the relations (1.5). Denote by $\Pi_1(t, y)$ and $\Pi_2(t, y)$ the respective semigroups generated by equations (1.2) and (1.3). Obviously, the phase flow determined by the relay system (1.1) coincides with $\Pi_1(t, y)$ if $0 \le t < T(y)$ and with $\Pi_2(t - T(y), \Pi_1(T(y), y))$ if t is a little larger than $T(y)$. Using these formulas, we linearize this phase flow on the periodic solution $y_0(t)$ in a neighborhood of the switching point t_0. In this way we get that

$$h(t_0 + 0) - h(t_0 - 0) = (-\dot{y}_0(t_0 + 0) + \dot{y}_0(t_0 - 0))(\operatorname{grad} T|_{y=y_0(0)}, h(0)).$$

Here $h(t)$ is a trajectory of the linearized phase flow. After this we differentiate the obvious representation $T(y) = t_0 - \delta + T(y(t_0 - \delta))$, and then let δ go to 0. As a result,

$$(\operatorname{grad} T|_{y=y_0(0)}, h(0)) = (\operatorname{grad} T|_{y=y_0(t_0)}, h(t_0 - 0)).$$

By virtue of the previous formula we have arrived at condition (1.5) for the switching time t_0 under consideration.

In what follows we need an explicit formula for the vector b. Consider the matrices $A = A(t_p - 0)$ and $B = B(t_p - 0)$. Denote by α and β eigenvectors of A and A^*, respectively, corresponding to the zero eigenvalue, with it assumed that $(\alpha, \beta) = 1$. It turns out that

$$b = -B^*\beta/(\dot{y}_0(t_p - 0), B^*\beta), \tag{1.6}$$

where the denominator is nonzero by Condition 1. To prove this it suffices to introduce the coordinates indicated in [5] in a neighborhood of switching points.

Thus we have described a way of linearizing the relay system (1.1) that enables us to construct the monodromy matrix of the linear periodic problem (1.4), (1.5). We remark that in view of (1.6) it has a discontinuous periodic solution $\dot{y}_0(t)$, i.e., the monodromy matrix has 1 as an eigenvalue. The rest of its eigenvalues coincide with the eigenvalues of the matrix Π'_0. It follows from this and Lemma 1.1 that the well-known Andronov-Witt theorem has been extended to the case of relay systems.

In conclusion we remark that the system (1.4) with the relations (1.5) is obtained upon linearization of the relay system (1.1) on any trajectory $y_0(t)$ of it (not necessarily periodic).

1.2. **Properties of a hybrid dynamical system.** We proceed to an analysis of the hybrid dynamical system

$$x = \Phi(y), \qquad \dot{y} = g(x, y). \tag{1.7}$$

Suppose for definiteness that the initial condition $(x(0), y(0))$ of its solution $(x(t), y(t))$, which is located near Γ_{1-}, is chosen so that we hit Γ_{1-} upon projection parallel to R^k. Assuming that the "jump" takes place instantaneously, we relate the curve $(x(t), y(0))$ to the trajectory of the solution, where $0 \leq t < \infty$, $x(\infty) = \Phi_1(y(0))$, and $x(t)$ is the solution of the following differential equation in R^k:

$$\dot{x} = f(x, y) \tag{1.8}$$

for $y = y(0)$. Then $y(t)$ is determined from (1.2), and $x(t) = \Phi_1(y(t))$. A new time again arises when the point $(x(t_0), y(t_0))$ falls in the set Γ_{10}. It is connected with the differential equation (1.8) for $y = y(t_0)$. Suppose that it has a solution $x(t)$ such that $x(-\infty) = x(t_0)$ and $x(\infty) = \Phi_2(y(t_0))$. The curve $(x(t), y(t_0))$ $(-\infty < t < \infty)$ we also relate to the trajectory of the solution of system (1.7), assuming that here too the motion along it takes place instantaneously. The trajectories of the solutions of (1.7) are defined similarly at subsequent moments of time. As in the preceding subsection, we fix a trajectory $(x_0(t), y_0(t))$ of the hybrid system (according to the foregoing, $y_0(t)$ is the solution of (1.1) with initial condition $y|_{t=0} = y_0(0)$).

Condition 2. For $y = y_0(t_0)$ equation (1.8) has an exponentially stable one-dimensional integral manifold in a neighborhood of the equilibrium state $x_0 = \Phi_1(y_0(t_0))$. Suppose that on this manifold the right-hand side of the equation begins with a quadratic term. Assume that the analogous condition holds for $y = y_0(t_1)$, and so on.

It is important that the "outgoing tendrils" leaving the equilibrium states considered for equation (1.8) are unique. Interestingly, here there can be a continuum of incoming tendrils: a model example is supplied by the system $\dot{x}_1 = x_1^2$, $\dot{x}_2 = -x_2$, for which these tendrils are given by the equality $x_2 = c\exp(1/x_1)$, where $x_1 < 0$ and c is an arbitrary constant.

Condition 3. We consider in R^n balls of sufficiently small radius with centers on the unstable "tendrils" of equation (1.8) introduced in Condition 2. It is assumed that when the values of $x(0)$ and y are in these balls the solutions of (1.8) approach Γ_{1-} or Γ_{2-} asymptotically as $t \to \infty$.

The geometric meaning of Condition 3 is trivial, but the consequences of Condition 2 are relatively complex. Together, they ensure the existence and uniqueness of the pieces of trajectories of the solutions of the hybrid system (1.7) that are connected with equation (1.8). Moreover, these conditions imply the existence for the hybrid system (1.7) of a cycle that is a "lifting" of a cycle of the relay system (1.1) from R^m to R^n if the trajectory $y_0(t)$ of the relay system is closed. Thus, we have arrived at the following assertion.

Lemma 1.2. *Suppose that in addition to Lemma 1.1 Conditions 2 and 3 hold on the closed trajectory $y_0(t)$. Then the cycle Ц_0 of the hybrid system (1.7) is orbitally exponentially stable.*

The proof follows from Lemma 1.1, since the pieces of the trajectories of the hybrid system (1.7) connected with (1.8) are traversed instantaneously, and this reduces the problem of the stability of the cycle Ц$_0$ to the geometric closeness of certain curves, and the latter is ensured by Conditions 2 and 3.

It should be remarked that Conditions 1–3 are encountered in other terms in several places in the monograph [5]. Here we have formulated them in such a way as to render obvious the stability of the cycle of the hybrid system, which fact plays a central role in the proof of the existence of a relaxation cycle. Also important is the fact that we have an algorithm for investigating the stability of the cycle Ц$_0$, thanks to the connection between the relay system and the hybrid system.

1.3. **The connection between the hybrid system and the relaxation system.** We consider the relaxation system

$$\varepsilon\dot{x} = f(x, y), \qquad \dot{y} = g(x, y), \qquad 0 < \varepsilon \ll 1, \tag{1.9}$$

corresponding to the hybrid system (1.7). Let $(x_0(t, \varepsilon), y_0(t, \varepsilon))$ be its trajectory with initial condition $x_0(0, \varepsilon) = x_0$, $y_0(0, \varepsilon) = y_0$ such that the point x_0 lies in the domain of attraction of the equilibrium state $\Phi_1(y_0)$ of system (1.8) for $y = y_0$ and $(\Phi_1(y_0), y_0) \in \Gamma_1$. Next, we consider the linear system

$$\varepsilon\dot{h}_1 = A_{11}(t, \varepsilon)h_1 + A_{12}(t, \varepsilon)h_2, \qquad \dot{h}_2 = A_{21}(t, \varepsilon)h_1 + A_{22}(t, \varepsilon)h_2 \tag{1.10}$$

obtained upon linearization of system (1.9) on the indicated trajectory, and we denote by $h_1(t, \varepsilon)$, $h_2(t, \varepsilon)$ the solution of it with initial conditions $h_1(0, \varepsilon) = h_1^0$ and $h_2(0, \varepsilon) = h_2^0$ independent of ε. Denote by $(x_0(t), y_0(t))$ the solution of the hybrid system (1.7) with the same initial conditions $x_0(0) = x_0$ and $y_0(0) = y_0$, and with switching times $t_0 < t_1 < \cdots$. Suppose that Conditions 1–3 hold on the trajectory $y_0(t)$ of the relay system (1.1). Then the following assertion is valid.

Theorem 1.1. *The limit relations*

$$\lim_{\varepsilon\to 0}(x_0(t, \varepsilon), y_0(t, \varepsilon)) = (x_0(t), y_0(t)), \tag{1.11}$$

$$\lim_{\varepsilon\to 0}(h_1(t, \varepsilon), h_2(t, \varepsilon)) = (h_{10}(t), h_{20}(t)) \tag{1.12}$$

hold uniformly with respect to initial conditions (x_0, y_0) *in any closed bounded set, with respect to vectors* h_1^0 *and* h_2^0 *in the unit ball about zero, and with respect to* t *in a finite interval with arbitrarily small fixed neighborhoods of the points* $t = 0, t_0, t_1, \ldots$ *deleted, where* $h_{20}(t)$ *is the solution of system* (1.4), (1.5) *with initial condition* $h_{20}(0) = h_2^0$, *and*

$$h_{10}(t) = (\Phi_1'|_{y=y_0(t)})h_{20}(t) \quad \text{or} \quad (\Phi_2'|_{y=y_0(t)})h_{20}(t).$$

We underscore that the closeness of the solutions of the hybrid system and the relaxation system in the C-metric follows from results contained in [5]. Therefore, our problem is to prove the limit equality (1.12). We note further that Theorem 1.1 enables us to establish a theorem on the correspondence between the structurally stable stationary regimes of the relay system (1.1) and the relaxation system (1.9). For example, we have the following assertion.

Theorem 1.2. *Under the conditions of Lemma* 1.2 *the relaxation system* (1.9) *has a unique orbitally exponentially stable cycle* Ц(ε) *of period* $T(\varepsilon)$ *in some neighborhood of the cycle* Ц$_0$ *independent of* ε, *and* Ц$(\varepsilon) \to$ Ц$_0$ *and* $T(\varepsilon) \to T_0$ *as* $\varepsilon \to 0$.

For a proof we construct a plane nonparallel to R^k at the point $(\Phi_1(y_0(0)), y_0(0))$ by completing the secant from §1.1. It follows from results obtained in [5] that the Poincaré operator of return along trajectories of the system (1.9) transforms into itself some convex subset of this plane. Moreover, it follows from [5] that up to a quantity of order ε all the cycles $\amalg(\varepsilon)$ of (1.9) located in a neighborhood of $\amalg_0$ have the same asymptotics. This circumstance is of decisive significance in the proof of the theorem. Indeed, linearizing on the cycles indicated above, we arrive at linear $T(\varepsilon)$-periodic systems of the form (1.10) with monodromy matrices denoted by $W(\varepsilon)$. The limit equality (1.12) enables us to conclude that

$$\lim_{\varepsilon\to 0} W(\varepsilon) = \begin{pmatrix} 0 & -A^{-1}(T_0)B(T_0)V \\ 0 & V \end{pmatrix},$$

where $A(t)$ and $B(t)$ are the same as in (1.4), and V is the monodromy matrix of the periodic problem (1.4), (1.5). Theorem 1.2 is proved.

We remark that an assertion analogous to Theorem 1.2 is contained in [16], where it is proved that the derivative of the Poincaré return operator of system (1.9) converges as $\varepsilon \to 0$. In the solution of that problem (for which [5] served as a source) system (1.9) was reduced in [16] to a certain canonical form in which there remains only one fast variable, which is taken as the new time. Moreover, to make the exposition clearer only two slow variables were considered. The technique used in [16] is more specialized and differs essentially from ours. In [16] attention is not focused on the algorithmic part; for example, there is nothing analogous to the periodic problem (1.4), (1.5) which we use to formulate an effective criterion for uniqueness and stability of the cycle.

Let us return to Theorem 1.1. The proof of the limit equality (1.12) is technically cumbersome. To make the subsequent presentation clearer we describe the general character of variation in time of the functions $h_1(t, \varepsilon)$ and $h_2(t, \varepsilon)$, which change in a very nontrivial way upon passing through neighborhoods of breakaway points. We first consider the values of t when the motion along the trajectory $(x_0(t, \varepsilon), y_0(t, \varepsilon))$ of the relaxation system (1.9) is slow. For these t the eigenvalues of the matrix A_{11} in (1.10) lie to the left of the imaginary axis, and this allows us [8] to pass from system (1.10) to equation (1.4). It follows from the whole complex of subsequent constructions that the norm of $h_1(t, \varepsilon)$ increases upon passing through a neighborhood of a breakaway point, becoming of order ε^{-1}. Moving along the trajectory $(x_0(t, \varepsilon), y_0(t, \varepsilon))$, we go far beyond the breakaway point in a time of order ε. On this whole interval the norm of $h_1(t, \varepsilon)$ remains a quantity of order ε^{-1}. However, when in moving along the trajectory we get close to a drop point the norm of $h_1(t, \varepsilon)$ falls to values of the order of 1. Here the norm of $h_2(t, \varepsilon)$ remains bounded, a fact having to do with the fact that the L-norm of $h_1(t, \varepsilon)$ is bounded uniformly with respect to ε. The main difficulty on a mathematical level is connected with the proof that $h_2(t, \varepsilon)$ is bounded, say for $t_0 - \delta \le t \le t_0 + \delta$, and with the computation of the quantities

$$h_{20}(t_0 \pm 0) = \lim_{\delta\to 0}\lim_{\varepsilon\to 0} h_2(t_0 \pm \delta, \varepsilon). \tag{1.13}$$

After this it is fairly simple to prove (1.5).

1.4. **Some auxiliary facts.** First of all we describe the general properties of the solutions of system (1.9) and the linearized system (1.10) in a neighborhood of a breakaway point. With this goal we take the zero of the space R^n to the point $(\Phi_1(y_0(t_0)), y_0(t_0))$ and introduce a special coordinate system in a neighborhood of it. In R^k we consider a plane orthogonal to the vector β appearing on the right-hand side of (1.6), and we choose $k-1$ linearly independent basis directions. We supplement them by a one-dimensional curvilinear direction that is an invariant integral manifold of equation (1.8) for $y = y_0(t)$. In R^m we take as a basis the vector $\dot y_0(t_0 - 0)$ and $m - 1$ linearly independent directions lying in the plane orthogonal to the vector (1.6). The new coordinates are denoted by $\xi = (\xi_1, \dots, \xi_k)$

and $\eta = (\eta_1, \dots, \eta_m)$, respectively, with ξ_1 measuring the distance along the one-dimensional tendril of equation (1.8), and η_1 corresponding to the direction $\dot{y}_0(t_0 - 0)$.

After similarity transformatioms and normalization of the time, in the new variables system (1.9) takes the form [5]

$$\varepsilon\dot{\xi} = F(\xi, \eta), \qquad \dot{\eta} = G(\xi, \eta), \tag{1.14}$$

where the first component $F_1 = \xi_1^2 + \eta_1 + \cdots$ of the vector-valued function F does not contain terms linear in $\eta_2, \dots, \eta_m$, and $F_j(\xi_1, 0, \dots, 0) \equiv 0$ for $j = 2, \dots, k$.

These properties of F are obvious consequences of our way of passing to new coordinates, in which the zero approximation of the trajectory of (1.9) is most simply written analytically. Indeed, it follows from [5] that for $-q \le \xi_1 \le q$, where $q > 0$ is sufficiently small,

$$\eta_1(\xi_1, \varepsilon) = \eta_1^0(\xi_1) + O(\varepsilon^{2/3}), \qquad \eta_j(\xi_1, \varepsilon) = \eta_j^0(\xi_1) + O(\varepsilon \ln(1/\varepsilon)), \tag{1.15}$$

$$\xi_l(\xi_1, \varepsilon) = \xi_l^0(\xi_1) + O(\varepsilon^{2/3}), \qquad j = 2, \dots, m, \quad l = 2, \dots, k. \tag{1.16}$$

Here $\eta_1^0(\xi_1) = -\xi_1^2 + \cdots$ for $\xi_1 \le 0$ and $\eta_1^0(\xi_1) \equiv 0$ for $\xi_1 \ge 0$; the functions $\eta_j^0(\xi_1)$ and $\xi_l^0(\xi_1)$ are also identically equal to zero for positive ξ_1, while for $\xi_1 \le 0$ the expansions of the functions η_j^0 begin with cubic terms and the expansions of the functions ξ_l^0 begin with quadratic terms (in [5] the functions $\xi_l^0(\xi_1)$ do not vanish for $\xi_1 \ge 0$).

Everywhere below we assume that system (1.10) is the result of linearization of system (1.14) on the corresponding time intervals, and $\xi_1 = -q$ corresponds to the time $t = 0$ on the trajectory.

The coordinates we have introduced enable us to investigate the numerous integrals arising in the analysis of the properties of $h_1(t, \varepsilon)$ and $h_2(t, \varepsilon)$ by using the substitution

$$dt = \varepsilon d\xi_1 / F_1(\xi, \eta), \tag{1.17}$$

which is naturally connected with the trajectory of system (1.9) and whose invertibility follows from the fact that the variable ξ_1 is monotonically increasing along the trajectory [5]. Here it is essential that the estimate

$$1/F_1(\xi, \eta) \le N/\xi_1^2, \qquad \Delta_1 \varepsilon^{1/3} \le \xi_1 \le q, \tag{1.18}$$

follows from the structure of F_1 and from (1.15) and (1.16), where $\Delta_1 > 0$ is sufficiently large and $N > 0$ (here and below, the single letter N stands for positive constants whose exact value is inessential).

Below, $\xi_1^{(t)}$ is the coordinate ξ_1 of a point on the trajectory of system (1.14); in particular, $\xi_1^{(0)} = -q$. Denote by τ_2 the time when $\xi_1^{(\tau_2)} = q$. It is important that for $0 \le t \le \tau_2$ the matrix $A_{11}(t, \varepsilon)$ in (1.10) has a unique eigenvalue $\lambda(t, \varepsilon)$ changing sign, while its remaining eigenvalues lie to the left of the imaginary axis for these t. The leading eigenvalue of $A_{11}(t, \varepsilon)$ has the asymptotic expression

$$\lambda(t, \varepsilon) = \lambda_0(\xi_1^{(t)}) + O(\varepsilon^{2/3}), \qquad -q \le \xi_1^{(t)} \le q, \tag{1.19}$$

where $\lambda_0(\xi_1) = 2\xi_1 + O(\xi_1^2)$. We explain one way to determine the function $\lambda_0(\xi_1)$. Denote by $\lambda_0(\xi, \eta)$ the leading eigenvalue of the matrix F'_ξ. Substituting in $\lambda_0(\xi, \eta)$ the zero approximations from (1.15) and (1.16), we get $\lambda_0(\xi_1)$. The asymptotic expression (1.19) now follows in an elementary way from (1.15) and (1.16) and the structure of the vector-valued function F.

We conclude with the help of (1.19) that $\lambda(t, \varepsilon)$ has only one zero τ_1 for $0 \le t \le \tau_2$, and $\xi_1^{(\tau_1)} = O(\varepsilon^{(2/3)})$. The computation of the breakaway time carried out in [5] for the trajectory of system (1.4) leads to the equalities $\tau_1 = t_0 + O(\varepsilon^{2/3})$ and $\tau_2 - \tau_1 = O(\varepsilon^{2/3})$. The latter of them means that the region of positivity of $\lambda(t, \varepsilon)$ is large in comparison to ε. Using (1.17) and (1.19) and computing F_1 with the help of results in [5] with terms of order $\varepsilon^{2/3}$ taken into account (these terms turn out to be essential here), we can verify in addition that

$$\int_{\tau_1}^{\tau_2} \lambda(t, \varepsilon)\, dt = \frac{2}{3}\varepsilon \ln \frac{1}{\varepsilon} + O(\varepsilon). \tag{1.20}$$

(We omit a detailed derivation of (1.20), since it has only heuristic value in our constructions.) In the next section it will be shown that the norm of $h_1(\tau_1, \varepsilon)$ has order $\varepsilon^{-1/3}$. The reason the norm of $h_1(t, \varepsilon)$ takes the order ε^{-1} on going through the region of positivity of $\lambda(t, \varepsilon)$ becomes intuitively clear from this and (1.20).

The proof of the following important statement clarifies somewhat the general considerations of the preceding paragraph.

Lemma 1.3. *For* $0 \le \tau < t \le \tau_2$ *the Cauchy matrix* $K(t, \tau)$ *of the system*

$$\varepsilon \dot{h} = A_{11}(t, \varepsilon) h \tag{1.21}$$

satisfies the inequality

$$|K(t, \tau)| \le N K_0(t, \tau), \qquad K_0 = \exp\left\{\frac{1}{\varepsilon}\int_\tau^t \lambda(s, \varepsilon)\, ds\right\}. \tag{1.22}$$

(*Here and below,* $|\cdot|$ *is a norm or modulus.*)

Proof. We introduce the matrix $U(t, \varepsilon) = U(\xi, \eta)_{\xi=\xi(t,\varepsilon),\, \eta=\eta(t,\varepsilon)}$, where $\xi(t, \varepsilon)$ and $\eta(t, \varepsilon)$ are the components of the trajectory of system (1.14) for $0 \le t \le \tau_2$, and $U(\xi, \eta)$ reduces the matrix F'_ξ to block diagonal form. Making the substitution $h = U(t, \varepsilon)v$ in (1.21), we get

$$\varepsilon \dot{v} = [C_1(t, \varepsilon) + \varepsilon C_2(t, \varepsilon)]v, \tag{1.23}$$

where $C_2(t, \varepsilon) = -U^{-1}\dot{U}$, $C_1(t, \varepsilon) = \operatorname{diag}\{\lambda(t, \varepsilon), \Lambda(t, \varepsilon)\}$, and all the eigenvalues of the $(k-1)\times(k-1)$ matrix $\Lambda(t, \varepsilon)$ lie to the left of the imaginary axis. Moreover, since q is small, the elements of $\Lambda(t, \varepsilon)$ are close to constants, and hence

$$|K_1(t, \tau)| \le N_1 \exp\left\{-\frac{N_2}{\varepsilon}(t - \tau)\right\},$$

where $K_1(t, \tau)$ is the Cauchy matrix of the equation $\varepsilon \dot{w} = \Lambda(t, \varepsilon) w$. It follows from this and the Gronwall-Bellman lemma that to prove our assertion it suffices to see that $C_2(t, \varepsilon)$ is bounded in the L_1-norm uniformly with respect to ε.

It follows from the choice of the matrix $U(t, \varepsilon)$ that

$$|C_2(t, \varepsilon)| \le N|\dot{\xi}(t, \varepsilon)|. \tag{1.24}$$

We remark that the integral of $|\dot{\xi}(t, \varepsilon)|$ over $[0, \tau_2]$ is equal to the variation of the vector-valued function $\xi(t, \varepsilon)$, or the length of the curve. But, as is geometrically obvious, the latter is bounded uniformly with respect to ε. Indeed, choosing ξ_1 as the parameter on the curve and differentiating the asymptotic formulas from [5] for the trajectory of system (1.14), we see that

$$\frac{d\xi_l(\xi_1, \varepsilon)}{d\xi_1} = \frac{d\xi_l^0(\xi_1)}{d\xi_1} + O(\varepsilon^{1/3}), \qquad -q \le \xi_1 \le q, \tag{1.25}$$

where $l = 2, \dots, k$. The required fact follows trivially from this.

We also prove the inequality

$$K_0(t, \tau) \leq N(\xi_1^{(t)}/\xi_1^{(\tau)})^2, \qquad \Delta_1 \varepsilon^{1/3} \leq \xi_1^{(\tau)} \leq q, \tag{1.26}$$

which is used often at various stages of the proof of Theorem 1.1. For the ξ_1 under consideration we have $1/F_1 \leq 1/(\xi_1^2 - N\varepsilon^{2/3})$ (by the properties of F_1 and formulas (1.15) and (1.16)). In the formula for K_0 we now use the asymptotic representation (1.19) and the substitution (1.17). As a result we arrive at the inequality

$$K_0(t, \tau) \leq N_1 \exp\left\{\int_{\xi_1^{(\tau)}}^{\xi_l^{(t)}} \frac{2\xi_1}{\xi_1^2 - N_2\varepsilon^{2/3}}\, d\xi_1\right\},$$

which is equivalent to (1.26).

The above facts, which are basic, enable us to prove the properties of $h_1(t, \varepsilon)$ and $h_2(t, \varepsilon)$ described in the preceding subsection, and the limit equality (1.12) follows in a relatively simple manner from these properties. It is sufficiently clear that the analysis of the asymptotic properties of the given functions is based on the computation and estimation of a large number of special integrals. Therefore, we devote a separate section to the question.

§2. Proof of the theorem on C^1-closeness

2.1. **Asymptotic properties of the functions $h_1(t, \varepsilon)$ and $h_2(t, \varepsilon)$ at the beginning of the breakaway region.** It follows from the asymptotic properties of the trajectory of the relaxation system [5] that the equalities

$$\begin{aligned} A_{11}(t, \varepsilon) &= A(t) + O(\varepsilon^{2/3}), & A_{12}(t, \varepsilon) &= B(t) + O(\varepsilon^{2/3}), \\ A_{21}(t, \varepsilon) &= C(t) + O(\varepsilon^{2/3}), & A_{22}(t, \varepsilon) &= D(t) + O(\varepsilon^{2/3}) \end{aligned} \tag{2.1}$$

hold uniformly with respect to t in any finite interval outside certain order-$\varepsilon^{2/3}$ neighborhoods of the switching times t_p of the relay system (1.1) to which the relaxation system under consideration converges as $\varepsilon \to 0$, where the matrices A, B, C, and D are the same as in (1.4). The matrices (2.1) are uniformly bounded on the whole interval of change of t (including the neighborhoods of the points t_p). Using (2.1) and applying to system (1.10) the asymptotic integration technique in [8], we see that on any interval $[\delta, t_0 - \delta]$ ($\delta > 0$ is fixed) the functions $h_1(t, \varepsilon)$ and $h_2(t, \varepsilon)$ converge uniformly as $\varepsilon \to 0$ to the functions

$$h_{10}(t) = -A^{-1}(t)B(t)h_{20}(t), \tag{2.2}$$

$$h_{20}(t) = h_2^0 + \int_0^t [D(\tau) - C(\tau)A^{-1}(\tau)B(\tau)]h_{20}(\tau)\, d\tau. \tag{2.3}$$

We remark that as $t \to t_0 - 0$ the function (2.2) has a singularity of the form $1/\sqrt{t_0 - t}$: according to (1.19), the norm of $A^{-1}(t)$ grows like $1/\varepsilon_1^{(t)}$ as $t \to t_0 - 0$, and $\xi_1 \sim -\sqrt{-\eta_1}$ and $\eta_1 \sim t - t_0$ in view of the formulas for the zero approximation of the trajectory of the relaxation system.

Theorem 2.1. *Uniformly with respect to $t \in [0, \tau_1]$*

$$\lim_{\varepsilon \to 0} h_2(t, \varepsilon) = h_{20}(t), \tag{2.4}$$

where $h_{20}(t)$ is defined by (2.3). An analogous limit equality holds for $h_1(t, \varepsilon)$ on any interval of the form $[\delta, t_0 - \delta]$, and at the point τ_1

$$|h_1(\tau_1, \varepsilon)| \leq N\varepsilon^{-1/3}. \tag{2.5}$$

Proof. By the first equation in (1.10),

$$h_1(t, \varepsilon) = K(t, 0)h_1^0 + \frac{1}{\varepsilon}\int_0^t K(t, \tau)A_{12}(\tau, \varepsilon)h_2(\tau, \varepsilon)\,d\tau. \tag{2.6}$$

Substituting (2.6) into the second equation in (1.10), we arrive at

(2.7)

$$h_2(t, \varepsilon) - \int_0^t \left[A_{22}(\tau, \varepsilon)h_2(\tau, \varepsilon) + \frac{1}{\varepsilon}A_{21}(\tau, \varepsilon)\int_0^\tau K(\tau, \theta)A_{12}(\theta, \varepsilon)h_2(\theta, \varepsilon)\,d\theta\right]d\tau$$
$$= h_2^0 + \int_0^t A_{21}(\tau, \varepsilon)K(\tau, 0)h_1^0\,d\tau.$$

Substituting $h_2(t, \varepsilon) = h_{20}(t) + h_{21}(t, \varepsilon)$ in (2.7), we conclude that $h_{21}(t, \varepsilon)$ is a solution of the integral equation

$$Lh_{21} = \psi_1(t) + \psi_2(t) + \psi_3(t), \tag{2.8}$$

where the Volterra operator L is defined by the left-hand side of (2.7), and

$$\psi_1(t) = \int_0^t [A_{22}(\tau, \varepsilon) - D(\tau)]h_{20}(\tau)\,d(\tau),$$

$$\psi_2(t) = \int_0^t A_{21}(\tau, \varepsilon)K(\tau, 0)h_1^0\,d\tau,$$

$$\psi_3(t) = \int_0^t \left[\frac{1}{\varepsilon}A_{21}(\tau, \varepsilon)\int_0^\tau K(\tau, \theta)A_{12}(\theta, \varepsilon)h_{20}(\theta)\,d\theta + C(\tau)A^{-1}(\tau)B(\tau)h_{20}(\tau)\right]d\tau.$$

We prove that the norm of the operator L^{-1} in the space C of continuous functions on $[0, C_1]$ is bounded uniformly with respect to ε. With this goal we consider the nonhomogeneous equation $Lh - f(t)$. Let $N_1 = \max_{0\le t\le \tau_1}|f(t)|$. Then by Lemma 1.3, the function $v(t) = |h(t)|$ satisfies

$$v(t) \le N_1 + N_1\int_0^t v(\tau)\,d\tau + N_3\frac{1}{\varepsilon}\int_0^t \left(\int_0^\tau K_0(\tau, \theta)v(\theta)\,d\theta\right)d\tau. \tag{2.9}$$

Changing the order of integration in the last term on the right-hand side of (2.9), we get

$$\int_0^t \left(\int_\theta^t K_0(\tau, \theta)\,d\tau\right)v(\theta)\,d\theta \le \int_0^{\tau_1}\left(\int_\theta^{\tau_1} K_0(\tau, \theta)\,d\tau\right)v(\theta)\,d\theta. \tag{2.10}$$

Using (2.10) in (2.9) and using the Gronwall-Bellman lemma, we see that to prove the needed fact it is necessary to verify the boundedness of the integral

$$\frac{1}{\varepsilon}\int_0^{\tau_1}\left(\int_\theta^{\tau_1} K_0(\tau, \theta)\,d\tau\right)d\theta. \tag{2.11}$$

We prove that this integral (denote it by J) has the limit $-\int_0^{t_0}\lambda_0^{-1}(\theta)\,d\theta$ as $\varepsilon \to 0$, where $\lambda_0(\theta)$ is the zero approximation of $\lambda(t, \varepsilon)$. With this goal we break up (2.11) into the integrals over the intervals $[0, \tau_1 - \tau_{10}\varepsilon^{2/3}]$ and $[\tau_1 - \tau_{10}\varepsilon^{2/3}, \tau_1]$, where $\tau_{10} > 0$ is arbitrarily fixed. In the second of these the interval of integration is asymptotically small, and that enables us to estimate it by the quantity $N\varepsilon^{1/3}$. And we write the first integral in the form $J_1 + J_2$, breaking up the inside integral in it by the point $\tau_1 - \tau_{10}\varepsilon^{2/3}$. For J_2 we have the estimate

$$J_2 = \frac{1}{\varepsilon}\int_{\tau_1 - \tau_{10}\varepsilon^{2/3}}^{t_1} K_0(\tau, \tau_1 - \tau_{10}\varepsilon^{2/3})\,d\tau \int_0^{\tau_1 - \tau_{10}\varepsilon^{2/3}} K_0(\tau_1 - \tau_{10}\varepsilon^{2/3}, \theta)\,d\theta \le N\varepsilon^{1/3},$$

which becomes obvious if in the second factor we carry K_0 under the differential sign, i.e., write it in the form of a Stieltjes integral, take into account that $\lambda(\tau_1 - \tau_{10}\varepsilon^{2/3}, \varepsilon) = O(\varepsilon^{1/3})$ in view of (1.19), and in the first factor use the inequality $K_0 \le 1$. Analogously, we write J_1 in terms of a Stieltjes integral and then pass to the limit in it. To justify this we deviate, as usual, from the "singularity" $\tau_1 - \tau_{10}\varepsilon^{2/3}$ by a small quantity $\delta > 0$ and use the inequality

$$\lambda(t, \varepsilon) \le -N\sqrt{\tau_1 - \tau_{10}\varepsilon^{2/3} - t}, \qquad 0 \le t \le \tau_1 - \tau_{10}\varepsilon^{2/3}$$

(which follows from (1.19); see the remark after (2.3)), on the corresponding regions. The last inequality leads to the following estimates:

$$\begin{aligned}
&\frac{1}{\varepsilon}\int_{\tau_1-\tau_{10}\varepsilon^{2/3}-\delta}^{\tau_1-\tau_{10}\varepsilon^{2/3}} \left(\int_{\theta}^{\tau_1-\tau_{10}\varepsilon^{2/3}} K_0(\tau, \theta)\, d\tau\right) d\theta \\
&\qquad \le N_1\varepsilon^{1/3}\int_{-N_2\delta/\varepsilon^{2/3}}^{0} \exp(-|u|^{3/2}) \int_u^0 \exp(|v|^{3/2})\, dv\, du \le N_3\sqrt{\delta}, \\
&\frac{1}{\varepsilon}\int_0^{\tau_1-\tau_{10}\varepsilon^{2/3}-\delta} \left(\int_{\tau_1-\tau_{10}\varepsilon^{2/3}-\delta}^{\tau_1-\tau_{10}\varepsilon^{2/3}} K_0(\tau, \theta)\, d\tau\right) d\theta \\
&\qquad \le N_1\varepsilon^{1/3}\int_{-\infty}^{-\delta N_2/\varepsilon^{2/3}} \exp(-|u|^{3/2})\, du \int_{-\delta N_2/\varepsilon^{2/3}}^{0} \exp(|\tau|^{3/2})\, dv \le N_3\varepsilon^{1/3},
\end{aligned}$$

and these justify passing to the limit.

Thus the quantity (2.11) is bounded. To prove the limit equality (2.4) it remains to show that the right-hand side of (2.8) tends to zero as $\varepsilon \to 0$, uniformly with respect to $0 \le t \le \tau_1$. For $\psi_1(t)$ this follows from (2.1). Writing $\psi_2(t)$ in Stieltjes form, we see that it has order ε. Finally, proceeding as in the case of the integral (2.11), we conclude that $\psi_3(t) \to 0$ as $\varepsilon \to 0$, uniformly with respect to $0 \le t \le \tau_1$. Thus, (2.4) is proved. Taking this into account in (2.6), we get the analogous equality for $h_1(t, \varepsilon)$ on any interval of the form $[\delta, t_0 - \delta]$.

Let us proceed to the proof of (2.5). Using Lemma 1.3 and the boundedness established above for $h_2(t, \varepsilon)$ on $[0, \tau_1]$, we deduce from (2.6) that

$$|h_1(\tau_1, \varepsilon)| \le N_1 + N_2\frac{1}{\varepsilon}\int_0^{\tau_1} K_0(\tau_1, \tau)\, d\tau. \tag{2.12}$$

We break up the interval of integration in the integral on the right-hand side of (2.12) by the point $\tau_1 - \tau_{10}\varepsilon^{2/3}$. The needed estimate is obtained if we rewrite the first integral in Stieltjes form and recall that $\lambda(\tau_1 - \tau_{10}\varepsilon^{2/3}, \varepsilon) = O(\varepsilon^{1/3})$, and in the second integral use the inequality $K_0 \le 1$. Theorem 2.1 is proved.

It follows from the asymptotic properties of the trajectory $(x_0(t, \varepsilon), y_0(t, \varepsilon))$ of system (1.9) [5] that $(x_0(\tau_1, \varepsilon), y_0(\tau_1, \varepsilon)) \to (\Phi_1(y_0(t_0)), y_0(t_0))$ as $\varepsilon \to 0$, i.e., Theorem 2.1 really is connected with the beginning of the region of fast motion. An analogous statement can be established also for the end of the region of fast motion. We explain what we have in mind. Let $\tau_3 = \tau_2 + \Delta_2\varepsilon\ln(1/\varepsilon)$, where $\Delta_2 > 0$ is sufficiently large. By results in [5], $(x_0(\tau_3, \varepsilon), y_0(\tau_3, \varepsilon)) \to (\Phi_2(y_0(t_0)), y_0(t_0))$ as $\varepsilon \to 0$, i.e., for suitable $t \ge \tau_3$ we again have an analogue of Theorem 2.1, provided, of course, that the boundedness of $h_1(\tau_3, \varepsilon)$ and $h_2(\tau_3, \varepsilon)$ as $\varepsilon \to 0$ is first established. Therefore, it follows from Theorem 2.1 and the corresponding assertion for $t \ge \tau_3$ that the quantities (1.13) are determined from the limit equalities

$$h_{20}(t_0 - 0) = \lim_{\varepsilon\to 0} h_2(\tau_1, \varepsilon), \qquad h_{20}(t_0 + 0) = \lim_{\varepsilon\to 0} h_2(\tau_3, \varepsilon).$$

2.2. **Boundedness of $h_2(t, \varepsilon)$ for $\tau_1 \le t \le \tau_3$.** We proceed to the proof that $h_2(t, \varepsilon)$ is bounded for $\tau_1 \le t \le \tau_3$, which, as mentioned in the preceding section, is a central point in the proof of the limit equality (1.12). The importance of the boundedness of $h_2(t, \varepsilon)$ is clear also from the formula

$$
\begin{aligned}
h_2(\tau_3, \varepsilon) - h_2(\tau_1, \varepsilon) &= \frac{1}{\varepsilon}\int_{\tau_1}^{\tau_3} A_{21}(\tau, \varepsilon)\int_0^{\tau} K(\tau, \theta)A_{12}(\theta, \varepsilon)h_2(\theta, \varepsilon)\,d\theta\,d\tau \\
&\quad + \int_{\tau_1}^{\tau_3} [A_{22}(\tau, \varepsilon)h_2(\tau, \varepsilon) + A_{21}(\tau, \varepsilon)K(\tau, 0)h_1^0]\,d\tau,
\end{aligned} \tag{2.13}
$$

which follows from (2.7) and which we later use to compute the limit of $h_2(\tau_3, \varepsilon)$ as $\varepsilon \to 0$. Indeed, if $h_2(t, \varepsilon)$ is bounded on $[\tau_1, \tau_3]$, then the second term on the right-hand side of (2.13) can be estimated in norm by a quantity of order $\varepsilon^{2/3}$, i.e., only the first term will be needed in our analysis.

Let us prove first that $h_2(t, \varepsilon)$ is bounded on $[\tau_1, \tau_2]$. Rewriting the integral equation (2.7) for the initial time τ_1 and arguing as in the proof of Theorem 2.1, we conclude that the necessary fact follows from the boundedness of the quantities

$$
|h_1(\tau_1, \varepsilon)|\int_{\tau_1}^{\tau_2} K_0(\tau, \tau_1)\,d\tau, \qquad \frac{1}{\varepsilon}\int_{\tau_1}^{\tau_2}\left(\int_{\theta}^{\tau_2} K_0(\tau, \theta)\,d\tau\right)d\theta. \tag{2.14}
$$

We break up these integrals into special sums and study each sum separately. Here a decisive role is played by the substitution (1.17) and the inequalities (1.18) and (1.26).

By (1.22), estimates of the integrals in (2.14) are closely connected with the nature of the behavior of the function $\lambda(t, \varepsilon)$. Employing (1.19), we distinguish the time intervals on which it has various asymptotics, and which are convenient to use in working with the integrals in (2.14). Let $\xi_1^{(t_*)} = \Delta_1\varepsilon^{1/3}$ and $\xi_1^{(t_{**})} = \delta_1\varepsilon^{1/3}\ln(1/\varepsilon)$, where $\delta_1 > 0$ is small, and Δ_1 is the same as in (1.18) and (1.26). By (1.19),

$$
0 \le \lambda(t, \varepsilon) \le N\varepsilon^{1/3}, \qquad \tau_1 \le t \le t_*, \tag{2.15}
$$

$$
0 \le \lambda(t, \varepsilon) \le \delta_2\varepsilon^{1/3}\ln(1/\varepsilon), \qquad \tau_1 \le t \le t_{**}, \tag{2.16}
$$

where $\delta_2 > 0$ is small.

We break up the first integral in (2.14) into two integrals by the point t_*. Inequality (2.15) allows us to estimate the first of them from above by the quantity $N\varepsilon^{2/3}$, and to obtain the inequality

$$
\int_{t_*}^{\tau_2} K_0(\tau, \tau_1)\,d\tau = \int_{t_*}^{\tau_2} K_0(\tau, t_*)\,d\tau K_0(t_*, \tau_1) \le N\int_{t_*}^{\tau_2} K_0(\tau, t_*)\,d\tau \tag{2.17}
$$

for the second. For the integral on the right-hand side of (2.17) the substitution (1.17) and the inequalities (1.18) and (1.26) can be used to establish the estimate

$$
\int_{t_*}^{\tau_2} K_0(\tau, t_*)\,d\tau \le N_1\varepsilon\int_{\xi_1^{(t_*)}}^{\xi_1^{(\tau_2)}}\left(\frac{\xi_1^{(\tau)}}{\xi_1^{(t_*)}}\right)^2 \frac{d\xi_1^{(\tau)}}{(\xi_1^{(\tau)})^2} \le N_2\varepsilon^{1/3}. \tag{2.18}
$$

From (2.5) and (2.18) we conclude that the first quantity in (2.14) is bounded.

Breaking up the interval of integration in the second integral in (2.14) by the point t_{**}, we denote by I_1 and I_2 the terms of the resulting sum. Making the substitution (1.17) in I_2 and using (1.18) and (1.26), we conclude that

$$
I_2 \le N_1\varepsilon\int_{\xi_1^{(t_{**})}}^{\xi_1^{(\tau_2)}}\left(\int_{\xi_1^{(\theta)}}^{\xi_1^{(\tau_2)}}\left(\frac{\xi_1^{(\tau)}}{\xi_1^{(\theta)}}\right)^2\frac{d\xi_1^{(\tau)}}{(\xi_1^{(\tau)})^2}\right)\frac{d\xi_1^{(\theta)}}{(\xi_1^{(0)})^2} \le N_2\Big/\ln^3\frac{1}{\varepsilon}.
$$

Again using the point t_{**}, we break up the inside integral in I_1, writing it in the form $I_{11} + I_{12}$. By (2.16),

$$(2.19)\quad I_{11} \le \frac{1}{\varepsilon}\int_{\tau_2}^{t_{**}}\left(\int_{\theta}^{t_{**}} \exp\left\{\delta_2\varepsilon^{-2/3}\ln\frac{1}{\varepsilon}(\tau-\theta)\right\}d\tau\right)d\theta \le N\varepsilon^{1/3-\delta_3}\Big/\ln^2\frac{1}{\varepsilon},$$

where $\delta_3 > 0$ is small. Then we write the first term in the representation

$$(2.20)\qquad \varepsilon I_{12} = \int_{\tau_1}^{t_{**}} K_0(t_{**}, \theta)\,d\theta \int_{t_{**}}^{\tau_2} K_0(\tau, t_{**})\,d\tau$$

in the form

$$(2.21)\qquad \int_{\tau_1}^{t_*} K_0(t_*, \theta)\,d\theta K_0(t_{**}, t_*) + \int_{t_*}^{t_{**}} K_0(t_{**}, \theta)\,d\theta.$$

Using (2.15) and (1.26), we get

$$(2.22)\qquad \int_{\tau_1}^{t_*} K_0(t_*, \theta)\,d\theta \le N\varepsilon^{2/3}, \qquad K_0(t_{**}, t_*) \le N\ln^2\frac{1}{\varepsilon}.$$

The estimates of the second term in (2.21) and the second factor in (2.20) are based on the substitution (1.17), which leads to the inequalities

$$(2.23)\qquad \int_{t_*}^{t_{**}} K_0(t_{**}, \theta)\,d\theta \le N_1\varepsilon\int_{\xi_1^{(t_*)}}^{\xi_1^{(t_{**})}}\left(\frac{\xi_1^{(t_{**})}}{\xi_1^{(\theta)}}\right)^2\frac{d\xi_1^{(\theta)}}{(\xi_1^{(\theta)})^2} \le N_2\varepsilon^{2/3}\ln^2\frac{1}{\varepsilon},$$

$$(2.24)\qquad \int_{t_{**}}^{\tau_2} K_0(\tau, t_{**})\,d\tau \le N_1\varepsilon\int_{\xi_1^{(t_{**})}}^{\xi_1^{(\tau_2)}}\left(\frac{\xi_1^{(\tau)}}{\xi_1^{(t_{**})}}\right)^2\frac{d\xi_1^{(\tau)}}{(\xi_1^{(\tau)})^2} \le N_2\varepsilon^{1/3}\Big/\ln^2\frac{1}{\varepsilon}.$$

Combination of (2.19)–(2.24) now implies the boundedness of I_1.

Thus we have proved that $h_2(t, \varepsilon)$ is bounded on $[\tau_1, \tau_2]$, i.e., the main difficulty has been overcome. Indeed, the concluding step is now simple. Let $\tau_2' = \tau_2 + \Delta_3\varepsilon$, where $\Delta_3 > 0$ is sufficiently large, and consider the interval $[\tau_2, \tau_2']$. After the substitution $t \to \varepsilon t$ the dependence on ε in (1.10) becomes regular, and it must be integrated on a finite time interval. Therefore,

$$(2.25)\qquad |h_1(t, \varepsilon)| \le N|h_1(\tau_2, \varepsilon)|, \qquad \tau_2 \le t \le \tau_2'.$$

As it turns out,

$$(2.26)\qquad |h_1(\tau_2, \varepsilon)| \le N/\varepsilon.$$

Inequality (2.26) can be deduced from the estimate

$$|h_1(\tau_2, \varepsilon)| \le N_1 + N_2\frac{1}{\varepsilon}\int_0^{\tau_2} K_0(\tau_2, \tau)\,d\tau,$$

which follows from (2.6) (with Lemma 1.3 taken into account) as follows: the integral on the right-hand side is broken up by the point t_* into two integrals; to estimate the first one we use the representation $K_0(\tau_2, \tau) = K_0(\tau_2, t_*)K_0(t_*, \tau)$, and then (1.26) and (2.15); in the case of the second integral,

$$\frac{1}{\varepsilon}\int_{t_*}^{\tau_2} K_0(\tau_2, \tau)\,d\tau \le N_1\int_{\xi_1^{(t_*)}}^{\xi_1^{(\tau_2)}}\left(\frac{\xi_1^{(\tau_2)}}{\xi_1^{(\tau)}}\right)^2\frac{d\xi_1^{(\tau)}}{(\xi_1^{(\tau)})^2} \le \frac{N_2}{\varepsilon}.$$

Inequality (2.26) enables us to conclude that $h_2(t, \varepsilon)$ is bounded for the values of t being considered. It remains to take the interval $\tau_2' \le t \le \tau_3$. For these t the points of the trajectory of the relaxation system (1.9) lie by the choice of τ_2' in a small neighborhood of the drop point that is independent of ε. Therefore, the elements of the matrix $A_{11}(t, \varepsilon)$ are close to constants, and its eigenvalues lie to the left of the imaginary axis. It follows from this that for $\tau_2' \leq \tau \leq t \leq \tau_3$

$$(2.27) \qquad |K(t, \tau)| \leq N_1 \exp\{-N_2/\varepsilon(t-\tau)\},$$

where, recall, $K(t, \tau)$ is the Cauchy matrix of equation (1.21). The inequality (2.27) and the estimate (2.25) for $t = \tau_2'$ ensure that $h_2(t, \varepsilon)$ is bounded on the whole interval $[\tau_1, \tau_3]$.

At the end of the preceding subsection we mentioned that for the validity of the analogue of Theorem 2.1 for suitable $t \geq \tau_3$ it is still necessary to establish that $h_1(\tau_3, \varepsilon)$ is bounded. This fact can be proved as follows: the integral equation (2.6) is rewritten for the initial time τ_2'; then to estimate its right-hand side we use the already-proved boundedness of $h_2(t, \varepsilon)$ and inequalities (2.25)–(2.27); this chain leads to the estimate

$$|h_1(\tau_3, \varepsilon)| \leq N_1 \frac{1}{\varepsilon} \exp\left\{-\frac{N_2}{\varepsilon}(\tau_3 - \tau_2')\right\} + N_3 \frac{1}{\varepsilon} \int_{\tau_2'}^{\tau_3} \exp\left\{-\frac{N_4}{\varepsilon}(\tau_3 - \tau)\right\} d\tau,$$

which concludes the proof.

It follows from (2.6) that $h_1(t, \varepsilon)$ is bounded outside arbitrarily small neighborhoods of the points t_p. This and the second equation in (1.10) imply that $h_2(t, \varepsilon)$ converges to solutions of (1.4) uniformly with respect to t outside intervals of the form $[\tau_1, \tau_3]$, i.e., regions of fast motion on the trajectory of the relaxation system. Thus, to find $h_{20}(t)$ for $t_0 < t < t_1$ ($h_{20}(t)$ is known for $0 \leq t < t_0$ in view of Theorem 2.1) we must compute $h_{20}(t_0 + 0) = \lim_{\varepsilon \to 0} h_2(\tau_3, \varepsilon)$, and so on.

2.3. **Computation of the limit of $h_2(\tau_3, \varepsilon)$ as $\varepsilon \to 0$.** As already mentioned, the problem reduces to an analysis of the first term on the right-hand side of (2.13), which we denote by S. We first of all single out the parts of S tending to zero as $\varepsilon \to 0$. Here we must improve somewhat the analytic apparatus being used, because for negative ξ_1 inequality (1.18) ceases to be valid. In view of (1.19) and Lemma 1.3 only the case of asymptotically small values of ξ_1 is really of interest. It will be shown below that the interval $\tau_1 - \delta(\varepsilon) \leq t \leq \tau_1$ with $\delta(\varepsilon) = \varepsilon^{2/3}\ln^2(1/\varepsilon)$ is of basic significance. We remark that, since $\xi_1 = O(\sqrt{-\eta_1})$ for $\eta_1 \leq 0$ and $\eta_1(\tau_1 - \delta(\varepsilon), \varepsilon) = -\delta(\varepsilon) + O(\varepsilon^{2/3})$, we have

$$(2.28) \qquad \xi_1^{(\tau_1 - \delta(\varepsilon))} = O(\varepsilon^{1/3}\ln(1/\varepsilon)).$$

Further, for $-N_1\varepsilon^{1/3}\ln(1/\varepsilon) \leq \xi_1 \leq N_2\varepsilon^{1/3}$ we have

$$(2.29) \qquad F_1(\xi, \eta) = \xi_1^2 + \varepsilon^{2/3} v_0(\xi_1/\varepsilon^{1/3}) + O(\varepsilon \ln^3(1/\varepsilon)),$$

where the function v_0 is defined in [5]. Formula (2.29) gives a constructive character to the substitution (1.17).

Let $S = S_1 + S_2$, with the inside integral broken up by the point $\tau_1 - \delta(\varepsilon)$. By (2.13),

$$(2.30) \qquad |S_1| \leq N \frac{1}{\varepsilon} \int_{\tau_1}^{\tau_3} \left(\int_0^{\tau_1 - \delta(\varepsilon)} |K(\tau, \theta)|\, d\theta \right) d\tau.$$

Using the representation $K(\tau, \theta) = K(\tau, \tau_1)K(\tau_1, \tau_1 - \delta(\varepsilon))K(\tau_1 - \delta(\varepsilon), \theta)$ and Lemma 1.3, we pass from (2.30) to the estimate

$$(2.31) \quad |S_1| \leq N \frac{1}{\varepsilon} \int_{\tau_1}^{\tau_3} |K(\tau, \tau_1)|\, d\tau \int_0^{\tau_1 - \delta(\varepsilon)} K_0(\tau_1 - \delta(\varepsilon), \theta)\, d\theta K_0(\tau_1, \tau_1 - \delta(\varepsilon)).$$

We represent the first factor on the right-hand side of (2.31) as the sum of integrals over $[\tau_1, \tau_2]$ and $[\tau_2, \tau_3]$. Lemma 1.3 is used to estimate the first integral. Then the resulting integral of K_0 is broken up into two integrals by the point t_*. Inequality (2.15) is used in the first integral, and the substitution (1.17) in the second, with (1.18) and (1.26) taken into account. As a result,

$$\int_{\tau_1}^{\tau_2} |K(\tau, \tau_1)|\,d\tau \le N_1 \int_{\tau_1}^{\tau_2} K_0(\tau, \tau_1)\,d\tau \le N_2\varepsilon^{1/3}, \tag{2.32}$$

and for the second integral

$$\begin{aligned} \int_{\tau_2}^{\tau_3} |K(\tau, \tau_1)|\,d\tau &\le N_1 \int_{\tau_2}^{\tau_3} |K(\tau, \tau_2)|\,d\tau K_0(\tau_2, \tau_1) \\ &\le N_2 \int_{\tau_2}^{\tau_3} |K(\tau, \tau_2)|\,d\tau K_0(\tau_2, t_*) \le N_3\varepsilon^{-2/3} \int_{\tau_2}^{\tau_3} |K(\tau, \tau_2)|\,d\tau. \end{aligned} \tag{2.33}$$

The first inequality in (2.33) is a consequence of Lemma 1.3, the second follows from (2.15), and the third from (1.26). The last integral in (2.33) can be estimated from above by a quantity of order ε in view of (2.27) and the estimate $|K(\tau, \tau_2)| \le N$, $\tau_2 \le \tau \le \tau_2'$, because $\tau_2' - \tau_2 = O(\varepsilon)$. From this and (2.32), the first factor in (2.31) satisfies the inequality

$$\int_{\tau_1}^{\tau_3} |K(\tau, \tau_1)|\,d\tau \le N\varepsilon^{1/3}. \tag{2.34}$$

We write the second factor in (2.31) in terms of a Stieltjes integral. Then we use the asymptotic equality

$$\lambda(\tau_1 - \delta(\varepsilon), \varepsilon) = O(\varepsilon^{1/3}\ln(1/\varepsilon)),$$

which follows from (1.19) and (2.28). In summary, we see that

$$\begin{aligned} \int_0^{\tau_1-\delta(\varepsilon)} K_0(\tau_1 - \delta(\varepsilon), \theta)\,d\theta &\le N_1\varepsilon^{2/3}\left(\ln\frac{1}{\varepsilon}\right)^{-1} \int_0^{\tau_1-\delta(\varepsilon)} d_\theta K_0(\tau_1 - \delta(\varepsilon), \theta) \\ &\le N_2\varepsilon^{2/3} \Big/ \ln\frac{1}{\varepsilon}. \end{aligned} \tag{2.35}$$

It remains to estimate the last factor in (2.31). The substitution (1.17) and formula (2.29) lead to the inequality

$$\begin{aligned} K_0(\tau_1, \tau_1 - \delta(\varepsilon)) &\le N_1 \exp\left\{\int_{-N_2\ln(1/\varepsilon)}^{0} \frac{2u}{u^2 + v_0(u)}\,du\right\} \\ &\le N_3 \exp\left\{-N_4 \ln^3\frac{1}{\varepsilon}\right\}. \end{aligned} \tag{2.36}$$

Combining (2.34)–(2.36), we finally get

$$|S_1| \le N_1 \exp\left\{-N_2\ln^3\frac{1}{\varepsilon}\right\} \Big/ \ln(1/\varepsilon).$$

Let $S_2 = S_{21} + S_{22}$, where the interval of integration in the inside integral is broken up by the point τ_1. After this we represent S_{22} in the form of a sum $S_{22}^1 + S_{22}^2$ of integrals over $[\tau_1, t_{**}]$ and $[t_{**}, \tau_3]$. Using Lemma 1.3 and (2.16), we conclude that

$$|S_{22}^1| \le N_1\frac{1}{\varepsilon}\int_{\tau_1}^{t_{**}} \left(\int_{\tau_1}^{\tau} K_0(\tau, \theta)\,d\theta\right) d\tau \le N_2\varepsilon^{1/3-\delta_4} \Big/ \ln^2\frac{1}{\varepsilon},$$

where $\delta_4 > 0$ is small. Breaking up the inside integral in S_{22}^2 by the point t_{**}, we write it as the sum $S_{22}^{21} + S_{22}^{22}$. The substitution (1.17) and the inequalities (1.18), (1.26), and (2.27) lead to the conclusion that $S_{22}^{22} \to 0$ as $\varepsilon \to 0$. Indeed, breaking up the inside and outside integrals in the inequality

$$|S_{22}^{22}| \le N\frac{1}{\varepsilon}\int_{t_{**}}^{\tau_3}\left(\int_{t_{**}}^{\tau}|K(\tau, \theta)|\,d\theta\right)d\tau$$

by the point τ_2, we represent its right-hand side as a sum. In the integrals obtained for $t_{**} \le \tau$ and $\theta \le \tau_2$ we make the substitution (1.17) and use the inequalities (1.18) and (1.26), while for $\tau_2 \le \tau$, $\theta \le \tau_2'$ and $\tau_2' \le \tau$, $\theta \le \tau_3$ we use the already-mentioned estimate $|K(\tau, \theta)| \le N$ and inequality (2.27), respectively. In this way we get that $|S_{22}^{22}| \le N/\ln^2(1/\varepsilon)$.

Accordingly, the integrals S_{21} and S_{22}^{21} make an essential contribution to S as $\varepsilon \to 0$:

$$(2.37)\quad S_{21} = \frac{1}{\varepsilon}\int_{\tau_1}^{\tau_3} A_{21}(\tau, \varepsilon)K(\tau, \tau_1)\,d\tau\int_{\tau_1-\delta(\varepsilon)}^{\tau_1} K(\tau_1, \theta)A_{12}(\theta, \varepsilon)h_2(\theta, \varepsilon)\,d\theta,$$

$$(2.38)\quad S_{22}^{21} = \frac{1}{\varepsilon}\int_{t_{**}}^{\tau_3} A_{21}(\tau, \varepsilon)K(\tau, t_{**})\,d\tau\int_{\tau_1}^{t_{**}} K(t_{**}, \theta)A_{12}(\theta, \varepsilon)h_2(\theta, \varepsilon)\,d\theta.$$

We prove that (2.37) and (2.38) are bounded. From (2.37) we get

$$|S_{21}| \le N\frac{1}{\varepsilon}\int_{\tau_1}^{\tau_3}|K(\tau, \tau_1)|\,d\tau\int_{\tau_1-\delta(\varepsilon)}^{\tau_1} K_0(\tau_1, \theta)\,d\theta.$$

The first integral on the right-hand side has already been encountered in (2.31), where it was shown not to exceed $N\varepsilon^{1/3}$. And it is easy to see by breaking up the second by the point $\tau_1 - \tau_{10}\varepsilon^{2/3}$ that it does not exceed $N\varepsilon^{2/3}$. Indeed, the second of the integrals obtained in this way can be estimated from above by the quantity $N\varepsilon^{2/3}$, because $K_0 \le 1$, while for the first integral the needed estimate can be obtained after writing the integral in Stieltjes form and using the equality $\lambda(\tau_1-\tau_{10}\varepsilon^{2/3}, \varepsilon) = O(\varepsilon^{1/3})$ encountered above. It follows from (2.38) that

$$|S_{22}^{21}| \le N\frac{1}{\varepsilon}\int_{\tau_1}^{t_{**}} K_0(t_{**}, \theta)\,d\theta\int_{t_{**}}^{\tau_3}|K(\tau, t_{**})|\,d\tau.$$

Comparing with (2.20), we conclude on replacing τ_3 by τ_2 that the analogous product of integrals has already been studied. Therefore, the boundedness of $|S_{22}^{21}|$ is a consequence of the chain of inequalities

$$\begin{aligned}\int_{\tau_2}^{\tau_3}|K(\tau, t_{**})|\,d\tau &\le N_1K_0(\tau_2, t_{**})\int_{\tau_2}^{\tau_3}|K(\tau, \tau_2)|\,d\tau\\ &\le N_2\varepsilon^{-2/3}\left(\ln^2\frac{1}{\varepsilon}\right)^{-1}\int_{\tau_2}^{\tau_3}|K(\tau, \tau_2)|\,d\tau.\end{aligned}$$

Here the first step is implemented with the use of the representation $K(\tau, t_{**}) = K(\tau, \tau_2)K(\tau_2, t_{**})$ and Lemma 1.3, and the second is based on (1.26). It remains to observe that the last integral was already encountered in (2.33).

Direct computations of the limits of S_{21} and S_{22}^{21} as $\varepsilon \to 0$ is difficult. Below we indicate a way of computing the limit of the sum S_{21} and S_{22}^{21} that uses only the fact that it converges as $\varepsilon \to 0$. In addition to Theorem 2.1, for this it is necessary to know the limit of $h_2(t, \varepsilon)$ as $\varepsilon \to 0$ on the interval $[\tau_1, t_{**}]$.

Theorem 2.2. *$h_{22}(t, \varepsilon) \to h_{20}(t_0 - 0)$ uniformly with respect to $\tau_1 \le t \le t_{**}$.*

Proof. From the integral equation (2.7), rewritten for the initial time τ_1, we derive the inequality

$$\max_{\tau_1 \le t \le t_{**}} |h_2(t, \varepsilon) - g_2(\tau_1, \varepsilon)| \le N_1 |h_1(\tau_1, \varepsilon)| \int_{\tau_1}^{t_{**}} K_0(\tau, \tau_1)\, d\tau + N_2 \frac{1}{\varepsilon} \int_{\tau_1}^{t_{**}} \left(\int_{\tau_1}^{\tau} K_0(\tau, \theta)\, d\theta \right) d\tau. \tag{2.39}$$

We start by estimating the first term on the right-hand side of (2.39). We have

$$\int_{\tau_1}^{t_{**}} K_0(\tau, \tau_1)\, d\tau = \int_{\tau_1}^{t_*} K_0(\tau, \tau_1)\, d\tau + K_0(t_*, \tau_1) \int_{t_*}^{t_{**}} K_0(\tau, t_*)\, d\tau.$$

Here the first term and the coefficient of the second are estimated with the help of (2.15), and in the second integral the substitution (1.17) is made with (1.18) and (1.26) taken into account. Finally, we get

$$\int_{\tau_1}^{t_{**}} K_0(\tau, \tau_1)\, d\tau \le N \varepsilon^{2/3} \ln \frac{1}{\varepsilon}.$$

Recalling (2.5), we see that the first term on the right-hand side of (2.39) does not exceed $N\varepsilon^{1/3} \ln(1/\varepsilon)$. Changing the order of integration in the second integral in (2.39), we note that we already considered it in §2.2, where we obtained the estimate (2.19) for it. Thus, to conclude the proof of the theorem it remains to take into account that, in view of Theorem 2.1, $h_2(\tau_1, \varepsilon) \to h_{20}(t_0 - 0)$ as $\varepsilon \to 0$.

Theorems 2.1 and 2.2 permit us to replace $h_2(\theta, \varepsilon)$ by $h_{20}(t_0 - 0)$ in (2.37) and (2.38). Therefore, as $\varepsilon \to 0$, (2.13) becomes

$$h_{20}(t_0 + 0) - h_{20}(t_0 - 0) = H h_{20}(t_0 - 0), \tag{2.40}$$

where H is some matrix whose structure is clear on an intuitive level (see Lemma 1.3 and the way of choosing the coordinates ξ, η indicated in §1.4). By (2.37) and (2.38), for a rigorous treatment of the question it is necessary to prove that $K(t, \tau)$ is "almost decomposable" for $\tau_1 - \delta(\varepsilon) \le \tau \le t \le t_{**}$.

We write equation (1.21) as a system

$$\begin{aligned} \varepsilon \dot{h}^1 &= A_{11}^{11}(t, \varepsilon) h^1 + A_{11}^{12}(t, \varepsilon) h^2, \\ \varepsilon \dot{h}^2 &= A_{11}^{21}(t, \varepsilon) h^1 + A_{11}^{22}(t, \varepsilon) h^2, \end{aligned} \tag{2.41}$$

setting $h = (h^1, h^2)$, where $h^1 \in R$ and $h^2 \in R^{k-1}$. On the time interval of interest to us $A_{11}(t, \varepsilon) = F'_\xi$, and hence the coefficients of system (2.41) have significant characteristics. First, the function $A_{11}^{11}(t, \varepsilon)$ has the same asymptotics as $\lambda(t, \varepsilon)$. Second, the eigenvalues of the matrix $A_{11}^{22}(t, \varepsilon)$ lie in the left complex half-plane, and its coefficients are close to constants, uniformly with respect to ε. Third, it follows from (2.28) and the way of choosing t_{**} that

$$|A_{11}^{12}(t, \varepsilon)| \le N_1 \varepsilon^{1/3} \ln(1/\varepsilon), \qquad |A_{11}^{21}(t, \varepsilon)| \le N_2 \varepsilon^{2/3} \ln^2(1/\varepsilon). \tag{2.42}$$

Theorem 2.3. *Suppose that*

$$K_{bc}(t, \tau) = \operatorname{diag}\{K_1(t, \tau), K_2(t, \tau)\},$$

where

$$K_1(t, \tau) = \exp\left\{ \frac{1}{\varepsilon} \int_\tau^t A_{11}^{11}(s, \varepsilon)\, ds \right\},$$

and $K_2(t, \tau)$ *is the Cauchy matrix of the system* $\varepsilon \dot{h}^2 = A_{11}^{22}(t, \varepsilon)h^2$. *Then*

$$K(t, \tau) = K_{bc}(t, \tau) + O(\varepsilon^{1/3}\ln^3(1/\varepsilon))$$

uniformly with respect to $\tau_1 - \delta(\varepsilon) \le \tau < t \le t_{**}$.

Proof. Denote by $h^1(t, \tau)$ and $h^2(t, \tau)$ the solutions of (2.41) with initial conditions $h^1(\tau, \tau) = h_0^1$ and $h^2(\tau, \tau) = h_0^2$ independent of ε. We have

$$h^1(t, \tau) = K_1(t, \tau)h_0^1 + \frac{1}{\varepsilon}\int_\tau^t K_1(t, s)A_{11}^{12}(s, \varepsilon)h^2(s, \tau)\, ds, \tag{2.43}$$

$$h^2(t, \tau) = K_2(t, \tau)h_0^2 + \frac{1}{\varepsilon}\int_\tau^t K_2(t, s)A_{11}^{21}(s, \varepsilon)h^1(s, \tau)\, ds. \tag{2.44}$$

In view of the inequality $|K_2(t, \tau)| \le N_1 \exp\{-N_2(t-\tau)/\varepsilon\}$ and the second inequality in (2.42), we deduce from (2.44) that

$$\sup_{t,\tau}|h^2(\tau, t)| \le N_1 + N_2\varepsilon^{2/3}\ln^2(1/\varepsilon)\sup_{t,\tau}|h^1(t, \tau)|, \tag{2.45}$$

and from the first inequality in (2.42) and (2.43) we get

$$\sup_{t,\tau}|h^1(t, \tau)| \le \left(N_1 + N_2\ln^3(1/\varepsilon)\sup_{t,\tau}|h^2(t, \tau)|\right)\sup_{t,\tau} K_1(t, \tau). \tag{2.46}$$

Since K_1 is asymptotically equivalent to K_0, we get

$$K_1(t, \tau) \le N_1 K_0(t, \tau) \le N_2 K_0(t_{**}, \tau_1) \le N_3\ln^2(1/\varepsilon)$$

for the t and τ under consideration. By (2.45) and (2.46),

$$\sup_{t,\tau}|h^1(t, \tau)| \le N\ln^5(1/\varepsilon). \tag{2.47}$$

Using (2.47), we derive from (2.44) that

$$h^2(t, \tau) = K_2(t, \tau)h_0^2 + O(\varepsilon^{2/3}\ln^7(1/\varepsilon)). \tag{2.48}$$

Finally, substituting (2.48) in (2.43), we get

$$h^1(t, \tau) = K_1(t, \tau)h_0^1 + O(\varepsilon^{1/3}\ln^3(1/\varepsilon)). \tag{2.49}$$

The theorem is obtained by combining the asymptotic representations (2.48) and (2.49).

We apply Theorem 2.3 to (2.37) and (2.38) while considering the properties of $A_{12}(\theta, \varepsilon)$ (in the first row the first element tends to 1 as $\varepsilon \to 0$ and the remaining ones tend to 0) that reflect the features of the function $F_1(\xi, \eta)$ (it does not contain terms of the form $c_j\eta_j$, where $j = 2, \dots, m$). In this way we arrive at the conclusion that the nonzero elements of the matrix H can be found only in the first column. Since here the equality (2.40) is satisfied by the function $\dot{y}_0(t)$, the first column of H is the expansion of the vector $\dot{y}_0(t_0+0) - \dot{y}_0(t_0-0)$ in the coordinates $(\eta_1, \dots, \eta_m)$. We have come to the condition (1.5) with $p = 0$.

Accordingly, the function $h_{20}(t)$ is the solution of system (1.4), (1.5) with the initial condition $h_{20}(0) = h_2^0$, and $h_{10}(t)$ is given by $h_{10}(t) = -A^{-1}(t)B(t)h_{20}(t)$. The proof of Theorem 1.1 is complete.

§3. The principle of reduction in a neighborhood of a breakaway point

3.1. **First step of the reduction principle.** In this subsection we pass from an arbitrary relaxation system in a neighborhood of a breakaway point to a system with one fast variable. Recall that in the special coordinates described in §1.4 the relaxation system has the following form in a neighborhood of a breakaway point:

$$\varepsilon\dot{\xi}_j = f_j(\xi_1, \xi_2, \eta_1, \eta_2), \qquad \dot{\eta}_j = g_j(\xi_1, \xi_2, \eta_1, \eta_2), \tag{3.1}$$

where $j = 1, 2$. Here the smooth functions f_j and g_j of the variables $\xi_1, \eta_1 \in R$, $\xi_2 \in R^{k-1}$, and $\eta_2 \in R^{m-1}$, $k, m \geq 1$, are defined in a sufficiently small neighborhood of zero, and

$$f_2(0, 0, 0, 0) = 0, \qquad g_2(0, 0, 0, 0) = g_1(0, 0, 0, 0) = 1,$$
$$\partial\xi_1 f_2(0, 0, 0, 0) = 0, \qquad f_1(\xi_1, \xi_2, \eta_1, \eta_2) = \xi_1^2 + \eta_1 + \cdots,$$

where the dots stand for quadratic terms other than ξ_1^2, along with terms of higher order of smallness. It is also assumed that all the eigenvalues of the matrix $\partial_{\xi_2} f_2(0, 0, 0, 0)$ lie to the left of the imaginary axis.

Lemma 3.1. *The system* (3.1) *has an exponentially stable* (*with exponent of order* ε^{-1}) *invariant integral manifold*

$$\xi_2 = V(\xi_1, \eta_1, \eta_2, \varepsilon), \tag{3.2}$$

where the function V *depends smoothly on all arguments.*

To prove this we note that after the substitution $t \to \varepsilon t$ we arrive at the thoroughly studied case of a center integral manifold [17].

On the integral manifold (3.2) we get the relaxation system

$$\varepsilon\dot{\xi} = F(\xi, \eta_1, \eta_2, \varepsilon), \qquad \dot{\eta}_j = G_j(\xi, \eta_1, \eta_2, \varepsilon) \qquad (j = 1, 2) \tag{3.3}$$

with the single fast variable $\xi = \xi_1$; the right-hand sides of this system have the previous properties for $\varepsilon = 0$.

3.2. **Second step of the reduction principle.** We first determine a smooth function $\eta_1 = \Phi(\xi, \eta_2)$ $(\Phi(0, 0) = 0)$ from the equation $F|_{\varepsilon=0} = 0$ in a neighborhood of zero. Then we take into account the equality $\eta_1 = \Phi(\xi, \eta_2)$ in the equation $\partial_\xi F|_{\varepsilon=0} = 0$, and then we find the smooth function $\xi = \varphi(\eta_2)$ $(\varphi(0) = 0)$ from it. Now let

$$H_0(\xi, \eta_2) = \begin{cases} \Phi(\xi, \eta_2) & \text{for } -q \leq \xi \leq \varphi(\eta_2), \\ \Phi(\varphi(\eta_2), \eta_2) & \text{for } \varphi(\eta_2) \leq \xi \leq q, \end{cases}$$

where the positive constant q is sufficiently small.

Theorem 3.1. *The system* (3.3) *has an invariant integral manifold*

$$\eta_1 = H(\xi, \eta_2, \varepsilon) \qquad (H(\xi, \eta_2, 0) = H_0(\xi, \eta_2)). \tag{3.4}$$

Here the function H *and its derivatives with respect to* ξ *and* η_2 *are jointly continuous for* $|\xi|, \|\eta_2\| \leq q$, $0 \leq \varepsilon \leq \varepsilon_0$, *where* ε_0 *is sufficiently small. Moreover, it is* G_1-*Lipschitz, i.e.,*

$$|H_{G_1+\Delta G_1} - H_{G_1}| \leq M\|\Delta G_1\|_C, \tag{3.5}$$

where M *is some constant, and* ΔG_1 *is measured in the metric of functions continuous with respect to* ξ, η_1, *and* η_2 *in the domain* $|\xi|, |\eta_1|, \|\eta_2\| \leq q$. *Further, the solutions of* (3.3) *exponentially approximate the integral manifold* (3.4) *with exponent of order* $\varepsilon^{\lambda-1}$, $0 < \lambda < 1/3$, *as long as* $-q \leq \xi \leq -\varepsilon^\lambda$, *and stay in a neighborhood of it for* $-\varepsilon^\lambda \leq \xi \leq q$.

For $-q \leq \xi \leq -q_1 < 0$ the theorem follows from results in [17]. Then we extend the integral manifold (3.4) along the trajectories of system (3.3) up to the point $\xi = q$

(this is possible, because $\dot{\xi} > 0$ and $\dot{\eta}_1 > 0$ on the trajectories of (3.3)). From this method of extension we see that it suffices to verify (3.5) on each trajectory of (3.3) separately. But the latter problem reduces to the integration of (1.10), with the term

$$\Delta G_1|_{\xi=\xi(t,\varepsilon),\,\eta_1=\eta_1(t,\varepsilon),\,\eta_2=\eta_2(t,\varepsilon)}$$

added to the right-hand side of the second equation, where $\xi(t, \varepsilon)$, $\eta_1(t, \varepsilon)$, and $\eta_2(t, \varepsilon)$ are the components of the trajectory of (3.3). The fact we need now follows trivially from this and from the results of the preceding sections.

Accordingly, the problem in [5] of constructing the asymptotics of the solutions of (3.1) in a neighborhood of a breakaway point reduces to the integration of the equation

$$\frac{d\eta_2}{d\xi} = \partial_\xi H \cdot G_2(\xi, H, \eta_2, \varepsilon)/(G_1 - \partial_{\eta_2} H \cdot G_2) \tag{3.6}$$

in R^{m-1}.

Lemma 3.2. *For* $0 \le \varepsilon \le \varepsilon_0$ *the right-hand side of equation* (3.6) *satisfies a Lipschitz condition with respect to the variables* ξ *and* η_2, *with some universal constant.*

From the asymptotic properties given in [5] for the trajectories of (3.3) we derive

$$H = \begin{cases} H_0 + O(\varepsilon/\xi) & \text{for } -q \le \xi \le -\varepsilon^\lambda, \\ H_0 + O(\varepsilon^{2/3}) & \text{for } -\varepsilon^\lambda \le \xi \le q, \end{cases} \tag{3.7}$$

$$\partial_\xi H = \begin{cases} \partial_\xi H_0 + O(\varepsilon/\xi^2) & \text{for } -q \le \xi \le -\varepsilon^\lambda, \\ \partial_\xi H_0 + O(\varepsilon^{1/3}) & \text{for } -\varepsilon^\lambda \le \xi \le q. \end{cases} \tag{3.8}$$

If we then divide the second and third equations of (3.3) by the first and investigate the variational system, we see that

$$\partial_{\eta_2} H = \begin{cases} \partial_{\eta_2} H_0 + O(\varepsilon/\xi) & \text{for } -q \le \xi \le -\varepsilon^\lambda, \\ \partial_{\eta_2} H_0 + O(\varepsilon^{2/3}) & \text{for } -\varepsilon^\lambda \le \xi \le q. \end{cases} \tag{3.9}$$

Writing now (3.6) starting directly from (3.3) (i.e., representing its right-hand side in the form $\varepsilon G_2/F$) and differentiating its right-hand side with respect to ξ or η_2, and then taking (3.7)–(3.9) into account, we get the lemma.

Theorem 3.1 enables us to answer some fundamental questions arising in the analysis of asymptotic expansions of the function H on different regions of change of ξ. Using results from [9] and the fact that H is G_1-Lipschitz, we say that a "matching" of the asymptotic formulas for $\xi = -\varepsilon^\lambda$ can be justified.

As is easy to see, the method of finding H from the equation

$$\partial_\xi H \cdot F + \varepsilon(\partial_{\eta_2} H, G_2) = \varepsilon G_1$$

is analogous in its general features to the construction of trajectories of a relaxation system on the plane. Computations carried out by A. Yu. Kolesov have shown that the asymptotics of the function H are most symmetric and simple when system (3.3) is transformed to the so-called normal form introduced in [16].

§4. Asymptotics of multidimensional relaxation oscillations in a neighborhood of a breakaway point

4.1. **The normal form.** According to §3.1, under standard restrictions an arbitrary relaxation system reduces in a neighborhood of a breakaway point to a system with one fast variable

$$\varepsilon\dot{\xi} = f(\xi, \eta_1, \eta_2), \qquad \dot{\eta}_j = g_j(\xi, \eta_1, \eta_2), \qquad j = 1, 2. \tag{4.1}$$

Here $0 < \varepsilon \ll 1$, and the functions f and g_j are smooth with respect to the variables $\xi, \eta_1 \in R$, and $\eta_2 \in R^{m-1}$, $m \geq 2$, in a neighborhood of the origin and are such that

$$g_1(0, 0, 0) = 1, \qquad g_2(0, 0, 0) = 0, \qquad f(\xi, \eta_1, \eta_2) = \xi^2 + \eta_1 + \cdots,$$

where the dots stand for quadratic terms other than ξ^2 along with terms of a higher order of smallness. The system (4.1) is further simplified as follows. From the equation $f'_\xi = 0$ we find the function $\xi = \varphi(\eta_1, \eta_2)$, and from the equation $f(\varphi, \eta_1, \eta_2) = 0$ we find the function $\eta_1 = \psi(\eta_2)$. Then successively we make in (4.1) the substitution

$$(\xi, \eta_1, \eta_2) \to (\xi - \varphi(\psi(\eta_2), \eta_2), \eta_1 - \psi(\eta_2), \eta_2),$$

which rectifies the $(m-1)$-dimensional breakaway manifold, and for fixed η_2 we make the substitutions described in [5] with respect to ξ and η_1, which carry the surface $f = 0$ into $\eta_1 = -\xi^2$. As a result we get the system

$$\varepsilon\dot{\xi} = \xi^2 + \eta_1 + \varepsilon\theta(\xi, \eta_1, \eta_2), \qquad \dot{\eta}_j = g_j(\xi, \eta_1, \eta_2), \qquad j = 1, 2, \tag{4.2}$$

where the smooth functions θ and g_j are such that $\theta(0, 0, 0) = 0$, $g_2(0, 0, 0) = 0$, and $g_1(0, 0, 0) = 1$. It can be assumed in addition that

$$g_2(0, 0, \eta_2) \equiv 0, \qquad g_1(0, 0, \eta_2) \equiv 1. \tag{4.3}$$

Indeed, the first of these equalities will be satisfied after the substitution $\eta_2 \to \eta_2 + \mathcal{G}(\eta_2)\eta_1$, where $\mathcal{G}(\eta_2) = -g_2(0, 0, \eta_2)/g_1(0, 0, \eta_2)$, and the second after the substitution $(\xi, \eta_1, \eta_2) \to (\gamma^{-1/3}\xi, \gamma^{-2/3}\eta_1, \eta_2)$, where $\gamma = g_1(0, 0, \eta_2)$, and division of the right-hand sides by $\gamma^{1/3}$. We remark further that the regular addition of order ε on the right-hand side of the first equation in (4.2) can be omitted, since (as will be clear from what follows) it does not influence the structure of the asymptotics of the solutions.

Without loss of generality we can confine ourselves to the case $m = 2$, since for $m > 2$ in all the asymptotic expansions encountered below the scalar coefficients are replaced by vectors, which is inessential. In view of what was remarked earlier, this leads us to the standard three-dimensional system

$$\varepsilon\dot{\xi} = \xi^2 + \eta_1, \qquad \dot{\eta}_1 = g_1(\xi, \eta_1, \eta_2), \qquad \dot{\eta}_2 = g_2(\xi, \eta_1, \eta_2), \tag{4.4}$$

where the smooth functions g_1 and g_2 satisfy (4.3); this system is called the *normal form* of the multidimensional relaxation system in a neighborhood of the breakaway point.

Recall (see Theorem 3.1) that system (4.4) has an invariant integral manifold

$$\eta_1 = H(\xi, \eta_2, \varepsilon) \qquad (H(\xi, \eta_2, 0) = H_0(\xi, \eta_2)), \tag{4.5}$$

where the function H and its derivatives with respect to ξ and η_2 are jointly continuous for $|\xi|, |\eta_2| \leq q$, $0 \leq \varepsilon \leq \varepsilon_0$, and for sufficiently small q, $\varepsilon_0 > 0$, and

$$H_0(\xi, \eta_2) = \begin{cases} -\xi^2 & \text{for } -q \leq \xi \leq 0, \\ 0 & \text{for } 0 \leq \xi \leq q. \end{cases}$$

Moreover, the function (4.5) is g_1-Lipschitz, i.e.,

$$|H_{g_1+\Delta g_1} - H_{g_1}| \leq M\|\Delta g_1\|_C,$$

where M is some positive constant, and Δg_1 is measured in the metric of the functions continuous with respect to ξ, η_1, and η_2 in the domain $|\xi|, |\eta_1|, |\eta_2| \leq q$. Finally, the solutions of (4.4) approximate exponentially the integral manifold (4.5) with exponent of order $\varepsilon^{\lambda-1}$, $0 < \lambda < 1/3$, as long as $-q \le \xi \le -\varepsilon^{\lambda}$, and remain in a neighborhood of it for $-\varepsilon^{\lambda} \le \xi \le q$.

Accordingly, two problems arise naturally: construction of the asymptotics of the integral manifold (4.5) and asymptotic integration of the equation

$$\frac{d\eta_2}{d\xi} = \frac{\partial H}{\partial \xi} g_2 \Big/ \left(g_1 - \frac{\partial H}{\partial \eta_2} g_2 \right) (= \varepsilon g_2/(\xi^2 + H)) \tag{4.6}$$

on the integral manifold. As shown in §3.2, its right-hand side satisfies a Lipschitz condition with respect to the variables ξ and η_2 for $0 \le \varepsilon \le \varepsilon_0$, with some universal constant, i.e., equation (4.6) depends regularly on ε.

4.2. **Asymptotics of the integral manifold at the beginning and in the middle of the breakaway region.** To find H from the equation

$$\frac{\partial H}{\partial \xi}(\xi^2 + H) + \varepsilon \frac{\partial H}{\partial \eta_2} g_2(\xi, H, \eta_2) = \varepsilon g_1(\xi, H, \eta_2), \tag{4.7}$$

we consider successively the following three regions: $-q \le \xi \le -\varepsilon^{\lambda_1}$, $-\varepsilon^{\lambda_1} \le \xi \le \varepsilon^{\lambda_2}$, and $\varepsilon^{\lambda_2} \le \xi \le q$, where $0 < \lambda_1, \lambda_2 < 1/3$. In this subsection we construct the asymptotics of H on the first two of them.

For $-q \le \xi \le -\varepsilon^{\lambda_1}$ we look for the integral manifold of system (4.4) in the form

$$H = -\xi^2 + \sum_{n=1}^{\infty} \varepsilon^n a_n(\xi, \eta_2). \tag{4.8}$$

Substituting (4.8) in (4.7), we find that $a_1 = -g_1(\xi - \xi^2, \eta_2)/(2\xi)$, and the functions a_n can be expressed in terms of $a_1, \dots, a_{n-1}$. As in [5], we can use induction to establish the asymptotic expansions

$$a_n(\xi, \eta_2) = \frac{1}{\xi^{3n-2}} \sum_{k=0}^{\infty} a_{k,n}(\eta_2) \xi^k \tag{4.9}$$

for $\xi \to -0$, where the smooth functions $a_{k,n}(\eta_2)$ are determined recursively. For example, for the leading coefficients $a_{0,n}$ of the series (4.9) we have the formulas

$$a_{0,1} = -\frac{1}{2}, \qquad a_{0,n} = -\frac{3n-4}{4} \sum_{k=1}^{n-1} a_{0,k} a_{0,n-k}, \qquad n \ge 2,$$

which coincide exactly with those given in [5].

It follows from (4.9) and the choice of λ_1 that, uniformly with respect to η_2,

$$\varepsilon^n a_n(-\varepsilon^{\lambda_1}, \eta_2) = O(\varepsilon^{n(1-3\lambda_1)+2\lambda_1}). \tag{4.10}$$

This enables us to clarify the asymptotic properties of the series (4.8) on the interval $-q \le \xi \le -\varepsilon^{\lambda_1}$. We employ a special device which will also be used in what follows. Denote by $H_N(\xi, \eta_2, \varepsilon)$ the Nth partial sum of the series (4.8) and let the function g_1 in (4.7) and (4.4) be perturbed by the addition

$$\Delta g_1 = \varepsilon^{-1} \left[\frac{\partial H_N}{\partial \xi}(\xi^2 + H_N) + \varepsilon \frac{\partial H_N}{\partial \eta_2} g_2(\xi, H_N, \eta_2) - \varepsilon g_1(\xi, H_N, \eta_2) \right]. \tag{4.11}$$

It follows from the way in which H_N is defined and (4.10) that, in the metric of the space of continuous functions, Δg_1 becomes arbitrarily small as N increases. It is important that the perturbed system (4.4) has the integral manifold $\eta_1 = H_N$. This circumstance enables us to use the g_1-Lipschitz property to get the following assertion.

Theorem 4.1. *The asymptotic formula*

$$H(\xi, \eta_2, \varepsilon) = H_{N-1}(\xi, \eta_2, \varepsilon) + O(\varepsilon^{N(1-3\lambda_1)+2\lambda_1}) \tag{4.12}$$

is valid for the function (4.5) *uniformly with respect to* $-q \le \eta_2 \le q$ *when* $-q \le \xi \le -\varepsilon^{\lambda_1}$, *where* N *is an arbitrary positive integer, and* $0 < \lambda_1 < 1/3$.

Let $-\varepsilon^{\lambda_1} \le \xi \le \varepsilon^{\lambda_2}$. Carrying out the normalizations $\eta_1 = \mu^2 v$, $\xi = \mu u$, and $\mu^3 = \varepsilon$ in (4.7), we get that

$$\frac{\partial v}{\partial u}(u^2 + v) + \mu^2 g_2(\mu u, \mu^2 v, \eta_2)\frac{\partial v}{\partial \eta_2} = g_1(\mu u, \mu^2 v, \eta_2). \tag{4.13}$$

Setting $\mu = 0$ in (4.13) and considering the property (4.3) of g_1, we see that the zero approximation of the function $v = v(u, \eta_2, \mu)$ is a solution of the Riccati differential equation $dv/du = 1/(u^2 + v)$. It was shown in [5] to have a unique solution $v_0(u)$ with the properties

$$\begin{aligned} v_0(u) &= -u^2 - 1/(2u) + O(1/u^4), \qquad && u \to -\infty, \\ v_0(u) &= \Omega_0 - 1/u + O(1/u^3), \qquad \Omega_0 > 0, && u \to +\infty. \end{aligned} \tag{4.14}$$

The equalities (4.14) mean that the solution $v_0(u)$ deviates least from the function

$$H_0(u) = \begin{cases} -u^2, & u \le 0, \\ 0, & u \ge 0, \end{cases}$$

that is the zero approximation of the integral manifold H in the new variables. Therefore, it is expedient to look for the function $v(u, \eta_2, \mu)$ in the form

$$v(u, \eta_2, \mu) = v_0(u) + \sum_{n=1}^{\infty} \mu^n v_n(u, \eta_2). \tag{4.15}$$

Substituting (4.15) in (4.13) and equating coefficients of like powers of μ, we have the sequence of differential equations

$$\frac{\partial v_n}{\partial u} + \frac{v_n}{[u^2 + v_0(u)]^2} = \Phi_n(u, \eta_2) \tag{4.16}$$

for determining the functions v_n, where the right-hand sides Φ_n depend only on $v_0, \dots, v_{n-1}$. Regarding (4.16) as an ordinary differential equation, we choose v_n to be a solution of it deviating least from a bounded solution. It is unique and can be written in terms of the Green function in the form

$$v_n = M(u)\int_{-\infty}^{u} \Phi_n(\theta, \eta_2)\frac{d\theta}{M(\theta)}, \qquad M(u) = \exp\int_u^{\infty} \frac{d\theta}{[\theta^2 + v_0(\theta)]^2}.$$

Let v_n^+ and v_n^- be the asymptotic expansions of v_n as $u \to +\infty$ and $-\infty$, respectively. As in [5], we use induction to establish the equalities

$$v_n^- = u^{n-1}\sum_{k=0}^{\infty} b_{k,n}(\eta_2)/u^{3k}, \qquad n \ge 1, \tag{4.17}$$

$$v_n^+ = u^{n-1}\sum_{\nu=0}^{n}\sum_{k=3\nu}^{\infty}{}^{*} \frac{b_{k,\nu,n}(\eta_2)}{u^k}\ln^{\nu} u + b_{0,\pi(n),n}(\eta_2)\ln^{\pi(n)} u, \qquad n \ge 0, \tag{4.18}$$

which are important for what follows. Here all the coefficients depend smoothly on η_2, the symbol $\sum^*$ stands for the sum $\sum_{k=m,k\ne m+1}^{\infty}$, and

$$\pi(n) = [n/3] + \begin{cases} 0 & \text{if } n \not\equiv 1 \pmod 3, \\ 1 & \text{if } n \equiv 1 \pmod 3. \end{cases}$$

We proceed to clarify the asymptotic properties of the series (4.15) for $-\varepsilon^{\lambda_1-1/3} \le u \le \varepsilon^{\lambda_2-1/3}$, and to match the formulas obtained here with (4.12) for $\xi = -\varepsilon^{\lambda_1}$, which is the central point in the construction of the asymptotics as a whole. It follows from (4.17) and (4.18) that, uniformly with respect to $-q \le \eta_2 \le q$,

$$\begin{aligned} \mu^{n+2}v_n(-\varepsilon^{\lambda_1-1/3}, \eta_2) &= O(\varepsilon^{1+\lambda_1(n-1)}), \\ \mu^{n+2}v_n(\varepsilon^{\lambda_2-1/3}, \eta_2) &= O(\varepsilon^{1+\lambda_2(n-1)}). \end{aligned} \tag{4.19}$$

We denote by $V_N(u, \eta_2, \mu)$ the Nth partial sum of (4.15) and perturb the function g_1 in (4.4) and (4.13) by the additional term Δg_1 defined by (4.11), where now $H_N = \mu^2 V_N(\xi/\mu, \eta_2, \mu)$. As above, (4.19) ensures that Δg_1 is arbitrarily small for sufficiently large N, uniformly with respect to $-2\varepsilon^{\lambda_1} \le \xi \le \varepsilon^{\lambda_2}$ and $|\eta_2| \le q$. Moreover, the perturbed system (4.4) has the integral manifold $\eta_1 = H_N$ for the ξ under consideration, while for $-2\varepsilon^{\lambda_1} \le \xi \le -\varepsilon^{\lambda_1}$ it has the integral manifold H and the asymptotics (4.12) by virtue of the g_1-Lipschitz property. We prove that H coincides up to ε^{α} with H_N for $\xi = -\varepsilon^{\lambda_1}$, where $\alpha > 0$ is arbitrary. Assuming the opposite, we take the initial condition of system (4.4) to be a point lying on the intersection of the manifold H_N with the plane $\xi = -\varepsilon^{\lambda_1}$ and not belonging to the manifold H. Let ε^{α_1}, $\alpha_1 > 0$, be the distance from this point to H. Integrating in the direction of decreasing time and using the facts that for $-2\varepsilon^{\lambda_1} \le \xi \le -\varepsilon^{\lambda_1}$ the manifold H is stable with exponent of order $\varepsilon^{\lambda_1-1}$, while the time of passage of the interval $-2\varepsilon^{\lambda_1} \le \xi \le -\varepsilon^{\lambda_1}$ has order $\varepsilon^{2\lambda_1}$ (this follows from the equation $\dot{\xi} = -g_1(\xi, -\xi^2, \eta_2)/(2\xi)$ for the zero approximation of the trajectory), we conclude that the solution under consideration certainly leaves the order-ε^{α_2} neighborhood of H for any $\alpha_2 > 0$. Indeed, at the point $\xi = -2\varepsilon^{\lambda_2}$ the deviation from the manifold (4.5) being computed from the linear approximation is a quantity of order $\varepsilon^{\alpha_1}\exp(M\varepsilon^{3\lambda_1-1})$, $M > 0$. On the other hand, the solution lies on the surface H_N, and hence for some $\alpha_2 > 0$ it is at a distance of order ε^{α_2} from H on the whole interval $-2\varepsilon^{\lambda_1} \le \xi \le -\varepsilon^{\lambda_1}$. This is a contradiction.

Accordingly, the integral manifold H_N of the perturbed system (4.4) is an extension of H to the interval $-\varepsilon^{\lambda_1} \le \xi \le \varepsilon^{\lambda_2}$. If we now use the g_1-Lipschitz property, we arrive at the following statement.

Theorem 4.2. *For $-\varepsilon^{\lambda_1} \le \xi \le 0$ the integral manifold* (4.5) *has the asymptotic formula*

$$H(\xi, \eta_2, \varepsilon) = \mu^2 V_N(\xi/\mu, \eta_2, \mu) + O(\varepsilon^{1+\lambda_1 N}) \tag{4.20}$$

uniformly with respect to $|\eta_2| \le q$, and for $0 \le \xi \le \varepsilon^{\lambda_2}$

$$H(\varepsilon, \eta_2, \varepsilon) = \mu^2 V_N(\xi/\mu, \eta_2, \mu) + O(\varepsilon^{1+\lambda_2 N}). \tag{4.21}$$

4.3. **Asymptotics of the integral manifold at the end of the breakaway region.** For $\varepsilon^{\lambda_2} \le \xi \le q$ we divide the second and third equation of (4.4) by the first. As a result we get the system

$$\frac{d\eta_1}{d\xi} = \frac{\mu^3 g_1(\xi, \eta_1, \eta_2)}{\xi^2 + \eta_1}, \qquad \frac{d\eta_2}{d\xi} = \frac{\mu^3 g_2(\xi, \eta_1, \eta_2)}{\xi^2 + \eta_1}, \tag{4.22}$$

with right-hand sides regular for the values of ξ under consideration. Letting the trajectories of system (4.22) go out from the points of the curve that is the intersection of H with the plane $\xi = \varepsilon^{\lambda_2}$, we get the integral manifold H for $\xi \ge \varepsilon^{\lambda_2}$. Of course, the solutions of (4.22) can be expanded in asymptotic series whose structures are determined by the initial conditions for $\xi = \varepsilon^{\lambda_2}$. Setting $\xi = \varepsilon^{\lambda_2}$ in (4.21) and taking the asymptotic equalities (4.18) into account, we conclude that for $\xi \geq \varepsilon^{\lambda_2}$ the solution of (4.7) should be sought in the form

$$H = \sum_{n=2}^{\infty} \mu^n h_n, \qquad h_n = \sum_{\nu=0}^{\pi(n-2)} W_{n,\nu}(\xi, \eta_2) \ln^\nu \frac{1}{\mu}. \tag{4.23}$$

Substituting (4.23) in (4.7) and equating the coefficients of the same powers of μ, we get

$$\frac{\partial h_n}{\partial \xi} = \sum_{\nu=0}^{\pi(n-2)-1} V_{n,\nu}(\xi, \eta_2) \ln^\nu \frac{1}{\mu}, \tag{4.24}$$

where the functions $V_{n,\nu}$ depend only on the $W_{m,k}$ with the indices $m \leq n-1$. For example,

$$\frac{\partial h_2}{\partial \xi} = 0, \qquad \frac{\partial h_3}{\partial \xi} = \frac{g_1(\xi, 0, \eta_2)}{\xi^2}, \qquad \frac{\partial h_4}{\partial \xi} = 0. \tag{4.25}$$

From (4.24) and (4.25) we get a series of equations with separated variables:

$$\frac{\partial W_{2,0}}{\partial \xi} = 0, \quad \frac{\partial W_{3,0}}{\partial \xi} = \frac{g_1(\xi, 0, \eta_2)}{\xi^2}, \quad \frac{\partial W_{3,1}}{\partial \xi} = 0, \quad \frac{\partial W_{4,0}}{\partial \xi} = 0,$$
$$\frac{\partial W_{n,\nu}}{\partial \xi} = V_{n,\nu}, \quad \frac{\partial W_{n,\pi(n-2)}}{\partial \xi} = 0, \qquad \nu = 0, 1, \ldots, \pi(n-2), \quad n \geq 5.$$

As in [5], we choose their particular solutions according to the following rule. Let

$$W_{2,0} = b_{0,0,0}, \qquad W_{3,1} = b_{0,1,1}, \qquad W_{4,0} = b_{0,0,2},$$
$$W_{3,0} = b_{0,0,1} + (\mathrm{H}) \int_0^\xi \frac{g_1(s, 0, \eta_2)}{s^2} \, ds,$$
$$W_{n,\nu} = b_{0,\nu,n-2} + (\mathrm{H}) \int_0^\xi V_{n,\nu}(s, \eta_2) \, ds, \quad W_{n,\pi(n-2)} = b_{0,\pi(n-2),n-2}, \quad n \geq 5.$$

Here $b_{0,\nu,m}$ is the same as in (4.18), and $(\mathrm{H}) \int$ is the Hadamard regularization. Recall [5] that for an equation with separated variables the Hadamard regularization is a choice of a particular solution without a constant term in its expansion as $\xi \to +0$. We remark that asymptotic expressions for the functions $W_{n,\nu}$ as $\xi \to +0$ are obtained if we set $u = \xi/\mu$ in (4.15) and use (4.18). In this way we get (cf. [5])

$$W_{n,\nu} = \sum_{\kappa=0}^{\pi(n-2)-\nu-1} \sum_{k=2-n+\kappa+\nu}^{\infty} d_{k,\kappa}^{n,\nu}(\eta_2) \xi^k \ln^\kappa \xi + d_{0,\pi(n-2)-\nu}^{n,\nu}(\eta_2) \ln^{\pi(n-2)-\nu} \xi, \tag{4.26}$$
$$\nu = 0, 1, \ldots, \pi(n-2), \quad n \geq 2,$$

where the coefficients are expressed in terms of the coefficients in (4.18) by

$$d_{k,\kappa}^{n,\nu} = C_\nu^{\nu+\kappa} b_{n+k-2,-k,\nu+\kappa},$$
$$n = 3, 5, 6, \ldots, \qquad \nu = 0, 1, \ldots, \pi(n-2) - 1,$$
$$\kappa = 0, 1, \ldots, \pi(n-2) - 1 - \nu, \qquad k \geq 2 - n + \nu + \kappa,$$
$$d_{0,\pi(n-2)-\nu}^{n,\nu} = C_\nu^{\pi(n-2)} b_{n-2,0,\pi(n-2)}, \qquad n \geq 2, \quad \nu = 0, 1, \ldots, \pi(n-2).$$

It follows from the regularity of system (4.22) that the partial sums of (4.23) are asymptotic approximations of the integral manifold H, i.e., we have the following statement.

Theorem 4.3. *For* $\varepsilon^{\lambda_2} \le \xi \le q$ *the function* H *has the asymptotic representation*

$$H(\xi, \eta_2, \varepsilon) = \sum_{n=2}^{N} \mu^n h_n + O(\varepsilon^{(N+1)/3-\lambda_2(N-1)}) \tag{4.27}$$

uniformly with respect to $|\eta_2| \le q$, *where* N *is an arbitrary positive integer.*

4.4. **The asymptotics of the trajectories at the beginning of the breakaway region.** We have reduced the problem of constructing the asymptotics of the trajectories of system (4.4) to the asymptotic integration of equation (4.6); this is connected with the multidimensionality of the system, because there is no analogue of this equation in the case of the plane. We confine ourselves to the construction of the asymptotics of the Cauchy problem for equation (4.6) with initial condition

$$\eta_2(-q, \varepsilon) = \eta_2^0, \qquad |\eta_2^0| \le q, \tag{4.28}$$

independent of ε. It follows from the regularity properties of equation (4.6) noted in §4.1 that on each of the three regions of change of ξ the asymptotics of the solution is dictated by the asymptotics of the right-hand side and the initial condition, and the question of proving the asymptotic formulas with remainder is trivial.

For $-q \le \xi \le -\varepsilon^{\lambda_1}$ we use the asymptotic equality (4.8) in (4.6). Then it is natural to look for the solution of the problem (4.6), (4.28) in the form

$$\eta_2 = \eta_{2,0}(\xi) + \sum_{n=1}^{\infty} \varepsilon^n \eta_{2,n}(\xi). \tag{4.29}$$

Substituting (4.29) in (4.6), we get the following Cauchy problem for determining $\eta_{2,0}(\xi)$:

$$\frac{d\eta_{2,0}}{d\xi} = -\frac{2\xi g_2(\xi, -\xi^2, \eta_{2,0})}{g_1(\xi, -\xi^2, \eta_{2,0})}, \qquad \eta_{2,0}(-q) = \eta_2^0. \tag{4.30}$$

After elementary but fairly unwieldy computations we deduce that

$$\frac{d\eta_{2,n}}{d\xi} = M_n, \qquad \eta_{2,n}(-q) = 0, \qquad n \ge 1, \tag{4.31}$$

where

$$M_n = \frac{\Lambda_n}{\gamma_1} + \sum_{l=1}^{n} \Lambda_{n-l} \sum_{\nu=1}^{l} \frac{(-1)^\nu}{\gamma_1^{\nu+1}} \sum_{\substack{i_1+\cdots+i_\nu=l+\nu \\ i_j \ge 2}} \gamma_{i_1} \cdots \gamma_{i_\nu}, \tag{4.32}$$

(4.33)

$$\gamma_n = a_n(\xi, \eta_{2,0}) + \sum_{m=1}^{n-1} \sum_{k=1}^{n-m} \frac{a_{m\eta_1}^{(k)}(\xi, \eta_{2,0})}{k!} \sum_{\substack{i_1+\cdots+i_k=n-m \\ i_j \ge 1}} \eta_{2,i_1} \cdots \eta_{2,i_k}, \qquad n \ge 1,$$

$$\Lambda_n = \sum_{0<k+m\le n} \frac{g_{2\eta_1\eta_2}^{(k)(m)}(\xi, -\xi^2, \eta_{2,0})}{k!m!} A_{k,m}^n, \qquad n \ge 1, \tag{4.34}$$

$$\Lambda_0 = g_2(\xi, -\xi^2, \eta_{2,0}), \qquad A_{k,m}^n = \sum_{s=0}^{n-k-m} v_s^k W_{n-k-m-s}^m, \tag{4.35}$$

$$
\begin{aligned}
(4.36)\qquad & v_\nu^k = \sum_{\substack{i_1+\cdots+i_k=k+\nu \\ i_j\geq 1}} \gamma_{i_1}\cdots\gamma_{i_k}, \qquad W_\nu^m = \sum_{\substack{i_1+\cdots+i_m=m+\nu \\ i_j\geq 1}} \eta_{2,i_1}\cdots\eta_{2,i_m}, \\
& v_0^0 = W_0^0 = 1, \qquad v_\nu^0 = W_\nu^0 = 0, \quad \nu = 1, \ldots.
\end{aligned}
$$

Here and below, the symbol $F^{(m)(k)\cdots}_{\eta_1\eta_2\cdots}$ denotes the derivative of F of the form $\partial^{m+k+\cdots}/\partial\eta_1^k\partial\eta_2^k\cdots$. It follows from the structure of formulas (4.32)–(4.36) that (4.31) can be written in the form

$$(4.37)\qquad d\eta_{2,n}/d\xi = \kappa(\xi)\eta_{2,n} + \varphi_n(\xi),$$

where $\kappa(\xi)$ is the derivative of the right-hand side of (4.30) with respect to $\eta_{2,0}$ at $\eta_{2,0} = \eta_{2,0}(\xi)$, and the functions $\varphi_n(\xi)$ depend only on the $\eta_{2,s}$ with indices $s \leq n-1$. Thus, the functions $\eta_{2,n}$ are determined recursively.

Using (4.32)–(4.36), we show that, as $\xi \to -0$,

$$(4.38)\qquad \eta_{2,0}(\xi) = \eta_{2,0}(0) + \sum_{n=3}^{\infty} \kappa_n \xi^n,$$

$$(4.39)\qquad \eta_{2,n}(\xi) = \frac{1}{\xi^{3n-3}} \sum_{k=0}^{n}\sum_{\nu=3k}^{\infty} c_{k,\nu,n}\xi^\nu \ln^k|\xi| + c_n \ln^n|\xi|, \qquad n \geq 1.$$

It is clear that in the proof we need only the last one, because (4.38) is a consequence of the regularity of problem (4.30). The proof of (4.39) is based on certain algebraic properties of its right-hand side. Denote by P_n an arbitrary polynomial of order n in the variable $\xi^3\ln|\xi|$ with coefficients that can be expanded in series of integer powers of ξ. Let

$$\widetilde{P}_n = \frac{1}{\xi^{3n-3}} P_n + c_n \ln^n|\xi|,$$

where c_n is some constant. In the new notation (4.39) takes the form

$$(4.40)\qquad \eta_{2,n} = \widetilde{P}_n.$$

It can be verified without difficulty that

$$(4.41)\qquad \widetilde{P}_n\widetilde{P}_m = \widetilde{P}_{n+m}, \qquad \widetilde{P}_n P_m = P_{n+m}/\xi^{3n}, \qquad \int \frac{P_n}{\xi^{3n-2}}\,d\xi = \widetilde{P}_n,$$

where different representatives of the classes of functions introduced are denoted by the same letters P_n and $\widetilde{P}_n$.

For $\eta_{2,1}$ the equality (4.40) follows from the equation

$$\frac{d\eta_{2,1}}{d\xi} = \kappa(\xi)\eta_{2,1} + g'_{2\eta_1}(\xi, -\xi^2, \eta_{2,0}) - \frac{g_2(\xi, -\xi^2, \eta_{2,0})}{[a_1(\xi, \eta_{2,0})]^2} a_2(\xi, \eta_{2,0})$$

with the use of the properties (4.9) of the functions a_1 and a_2. Assume that (4.40) is valid for all $\eta_{2,m}$ with indices $m \leq n-1$. Then by (4.41) and the properties (4.9) of the functions a_n we get successively that

$$
\begin{aligned}
\gamma_n &= P_{n-1}/\xi^{3n-2}, & v_\nu^k &= P_\nu/\xi^{3\nu+k}, & w_\nu^m &= \widetilde{P}_{m+\nu}, \\
A^n_{k,m} &= P_{n-k}/\xi^{3n-2k}, & \Lambda_n &= P_{n-1}/\xi^{3n-2}, & \varphi_n &= P_n/\xi^{3n-2},
\end{aligned}
$$

where functions P_m and $\widetilde{P}_n$ of the same structure are also denoted in the same way. In view of the last property in (4.41) it follows from these equalities and (4.37) that

$$\eta_{2,n} = \int_{-q}^{\xi} \exp\left\{\int_s^\xi \kappa(\sigma)\,d\sigma\right\}\varphi_n(s)\,ds = \int \frac{P_n}{s^{3n-2}}\,ds = \widetilde{P}_n.$$

The proof of (4.40) (and hence also (4.39)) is complete. It enables us to make the following statement.

Theorem 4.4. *For* $-q \leq \xi \leq -\varepsilon^{\lambda_1}$ *the asymptotic equality*

$$\eta_2 = \eta_{2,0}(\xi) + \sum_{n=1}^{N} \varepsilon^n \eta_{2,n}(\xi) + O(\varepsilon^{(1-3\lambda_1)N+1}) \tag{4.42}$$

is valid for the solution of the Cauchy problem (4.6), (4.28).

Using this equality in the representation (4.12), we get an asymptotic expression for $\eta_1(\xi, \varepsilon)$.

Proposition 4.1. *For* $-q \leq \xi \leq -\varepsilon^{\lambda_1}$ *the coordinate* η_1 *of system* (4.4) *has the asymptotic formula*

$$\eta_1 = -\xi^2 + \sum_{n=1}^{N} \varepsilon^n \gamma_n(\xi) + O(\varepsilon^{N(1-3\lambda_1)+1-\lambda_1}), \tag{4.43}$$

where N *is an arbitrary positive integer, and the functions* γ_n *defined by* (4.33) *have the asymptotic expansions*

$$\gamma_n(\xi) = \frac{1}{\xi^{3n-2}} \sum_{k=0}^{n-1} \sum_{\nu=3k}^{\infty} c^n_{k,\nu} \xi^\nu \ln^k |\xi|, \qquad n \geq 1, \tag{4.44}$$

as $\xi \to -0$.

4.5. **The asymptotics of the trajectories in the middle of the breakaway region.** For $-\varepsilon^{\lambda_1} \leq \xi \leq \varepsilon^{\lambda_2}$ we make the substitutions $\xi = \mu u$, $\eta_1 = \mu^2 v$, $\eta_2 = \eta_{2,0}(0) + \mu^3 \omega$, and $\varepsilon = \mu^3$ in (4.6). With the help of (4.15) we get

$$\frac{d\omega}{du} = \frac{1}{\mu} \frac{g_2(\mu u, \mu^2 v, \eta_{2,0}(0) + \mu^3 \omega)}{u^2 + v}, \tag{4.45}$$

the right-hand side of which can be expanded in an asymptotic series of integer powers of μ. We are interested in the solution of (4.45) with an initial condition dependent on ε and with asymptotics deducible from (4.38), (4.39), and (4.42) with $\xi = -\varepsilon^{\lambda_1}$. Analyzing the asymptotic expansions obtained here, we see that the necessary solution should be sought in the form

$$\omega = \sum_{n=0}^{\infty} \mu^n \omega_n(u, \mu), \qquad \omega_n = \sum_{k=0}^{[n/3]+1} \omega_{n,k}(u) \ln^k \mu. \tag{4.46}$$

After substituting (4.46) in (4.45) and equating the coefficients of the same powers of μ we arrive at the sequence of differential equations

$$d\omega_n/du = M_{n+1}, \qquad n \geq 0, \tag{4.47}$$

where

$$M_n = \frac{\Lambda_n}{u^2 + v_0(u)} + \sum_{l=1}^{n} \Lambda_{n-l} \sum_{\nu=1}^{l} \frac{(-1)^\nu}{[u^2 + v_0(u)]^{\nu+1}} \sum_{\substack{i_1+\cdots+i_\nu=l \\ i_j \geq 1}} \gamma_{i_1} \cdots \gamma_{i_\nu}, \qquad n \geq 1, \tag{4.48}$$

$$\Lambda_n = \sum_{\substack{0<m+2k+3\nu\leq n \\ m+k\neq 0}} \frac{g^{(m)(k)(\nu)}_{2\xi\ \eta_1\eta_2}(0, 0, \eta_{2,0}(0))}{m!k!\nu!} A^n_{m,k,\nu}, \qquad \Lambda_0 = 0, \tag{4.49}$$

$$\gamma_n = v_n(u, \eta_{2,0}(0))$$

(4.50) $$+\sum_{m=1}^{[n/3]}\sum_{k=1}^{m}\frac{v^{(k)}_{n-3m\,\eta_2}(u, \eta_{2,0}(0))}{k!}\sum_{\substack{i_1+\cdots+i_k=3m-3k\\ i_j\geq 0}}\omega_{i_1}\cdots\omega_{i_k}, \qquad n\geq 0,$$

(4.51) $$A^n_{m,k,\nu} = u^m \sum_{s=0}^{n-m-2k-3\nu} z^k_s \lambda^\nu_{n-m-2k-3\nu-s},$$

(4.52) $$z^k_r = \sum_{\substack{i_1+\cdots+i_k=r\\ i_j\geq 0}}\gamma_{i_1}\cdots\gamma_{i_k}, \qquad \lambda^\nu_r = \sum_{\substack{i_1+\cdots+i_\nu=r\\ i_j\geq 0}}\omega_{i_1}\cdots\omega_{i_\nu},$$
$$z^0_0 = \lambda^0_0 = 1, \qquad z^0_r = \lambda^0_r = 0, \quad r = 1, 2, \dots.$$

Estimating the maximal power of $\ln\mu$ in (4.48)–(4.52) with the help of the equality

$$\max_{\substack{i_1+\cdots+i_\nu=r\\ i_j\geq 0}}\{[i_1/3]+\cdots+[i_\nu/3]\} = [r/3],$$

we conclude that

(4.53) $$M_{n+1} = \sum_{k=0}^{[n/3]} M_{n,k}(u)\ln^k\mu,$$

where the $M_{n,k}$ depend on the ω_m with indices $m\leq n-3$. Therefore, the sequence of differential equations is recurrent and splits into series of equations with separated variables

$$d\omega_{n,k}/du = M_{n,k}(u), \qquad k = 0, 1, \dots, [n/3],$$
$$d\omega_{n,[n/3]+1}/du = 0, \qquad n\geq 0.$$

The solutions of these equations are chosen in such a way that for $\xi = -\varepsilon^{\lambda_1}$ all the terms in (4.42) and (4.46) independent of λ_2 coincide. Hence,

(4.54) $$\omega_{n,k} = \int_0^u M_{n,k}(s)\,ds + (\mathrm{H})\int_{-\infty}^0 M_{n,k}(s)\,ds + \beta(n)c_{k,n,[n/3]+1},$$
$$k = 0, 1, \dots, [n/3],$$

(4.55) $$\omega_{n,[n/3]+1} = \beta(n)c_{[n/3]+1}, \qquad \beta(n) = \begin{cases} 1 & \text{if } n\equiv 0 \pmod 3,\\ 0 & \text{if } n\not\equiv 0 \pmod 3,\end{cases}$$

where $c_{k,n,[n/3]+1}$ and $c_{[n/3]+1}$ are the same as in (4.39). On the basis of (4.54) and (4.55) we get, for example,

(4.56)
$$\omega_0(u) = g'_{2\xi}(0, 0, \eta_{2,0}(0))\left[\int_0^u \frac{s\,ds}{s^2+v_0(s)} + (\mathrm{H})\int_{-\infty}^0 \frac{s\,ds}{s^2+v_0(s)}\right] + c_{0,0,1} + c_1\ln\mu.$$

(We remark that the second and third terms are not taken into account in the formula analogous to (4.56) in [5], where the asymptotics of the trajectories are constructed up to quantities of order ε.) As already mentioned above, M_{n+1} depends on ω with $m\leq n-3$. Therefore, $M_{n,k}\equiv 0$ for $n = 0, \dots, 3k-1$. In view of (4.54) and (4.55), this implies that

$$\omega_{n,k}(u)\equiv 0, \qquad n = 0, 1, \dots, 3k-4,$$
$$\omega_{3k-3,k}\equiv c_k, \qquad \omega_{3k-2,k}\equiv\omega_{3k-1,k}\equiv 0.$$

The asymptotic expansions of $\omega_{n,k}$ as $u \to -\infty$ are simple to obtain: let $\xi = \mu u$ and $\eta_2 = \eta_{2,0}(0) + \mu^3\omega$ in (4.29), and regroup the terms of the asymptotic series. In this way we get

$$\omega^-_{n,0} = \kappa_{n+3}u^{n+3} + u^n \sum_{m=0}^{[n/3]} \sum_{\nu=m}^{\infty} \frac{c_{m,n,\nu}}{u^{3\nu-3}} \ln^m |u| + \beta(n)c_{[n/3]+1} \ln_{[n/3]+1} |u|, \tag{4.57}$$

$$\begin{aligned}\omega^-_{n,k} = u^n \sum_{m=k}^{[n/3]} \sum_{\nu=m}^{\infty} C^m_k \frac{c_{m,n,\nu}}{u^{3\nu-3}} \ln^{m-k} |u| \\ + \beta(n)c_{[n/3]+1} C_k^{[n/3]+1} \ln^{[n/3]+1-k} |u|, \qquad n \geq 3k,\end{aligned} \tag{4.58}$$

where κ_{n+3}, $c_{m,n}$, and $c_{[n/3]+1}$ are the same as in (4.38) and (4.39). If we now analyze (4.48)–(4.52) and use induction on n, we see that

$$\omega^+_{n,k} = u^{n-3k} \sum_{\nu=0}^{n-3k} \sum_{l=3\nu}^{\infty}{}^{*} g^{n,k}_{\nu,l} \frac{\ln^\nu u}{u^l} + g^{n,k}_{\pi(n+1)-k,0} \ln^{\pi(n+1)-k} u, \qquad n \geq 3k. \tag{4.59}$$

For example, we give a proof for $k = 0$. As in the proof of (4.39), we use the same symbol P_n to denote different functions in the class of polynomials of degree n in the variable $(\ln u)/u^3$ with coefficients that expand in series of integer powers of $1/u$ and do not contain first-order terms. In addition we introduce the class of functions $\widetilde{P}_n = u^n P_n + g \ln^{\pi(n+1)} u$, where g is an arbitrary number. Then for $k = 0$ the equality (4.59) takes the form

$$\omega^+_{n,0} = \widetilde{P}_n. \tag{4.60}$$

For $\omega^+_{0,0}$, (4.60) follows directly from (4.56) and the asymptotic properties of $v_0(u)$ (see (4.18) for $n = 0$). Assume that it holds for all indices not exceeding $n - 1$. Then by the easily verified equalities

$$P_n P_m = P_{n+m}, \qquad \widetilde{P}_{i_1} \cdots \widetilde{P}_{i_k} = u^{i_1+\cdots+i_k+3k} P_{i_1+\cdots+i_k}$$

we conclude from (4.48)–(4.52) that $M_{n,0} = u^{n-1}P_n$. But from this we find that

$$\omega^+_{n,0} = \int u^{n-1} P_n \, du = u^n P_n + g \ln^{\pi(n+1)} u = \widetilde{P}_n.$$

This proves (4.59) for $k = 0$. The proof for $k \geq 1$ is similar but somewhat more unwieldy.

The asymptotic representations (4.58) and (4.59) enable us to make the following statement.

Theorem 4.5. *For* $-\varepsilon^{\lambda_1} \leq \xi \leq 0$ *the solution of the Cauchy problem* (4.6), (4.28) *satisfies the asymptotic equality*

$$\eta_2 = \eta_{2,0}(0) + \mu^3 \sum_{n=0}^{N} \mu^n \omega_n(\xi/\mu, \mu) + O(\varepsilon^{\lambda_1(N+4)}), \tag{4.61}$$

and for $0 \leq \xi \leq \varepsilon^{\lambda_2}$ *it satisfies*

$$\eta_2 = \eta_{2,0}(0) + \mu^3 \sum_{n=0}^{N} \mu^n \omega_n(\xi/\mu, \mu) + O(\varepsilon^{\lambda_2(N+1)+1}). \tag{4.62}$$

We take (4.61) and (4.62) into account in the equalities (4.20) and (4.21) giving the asymptotics of the integral manifold H for the corresponding values of ξ. In summary we get the asymptotic representation

$$\eta_1 = \mu^2 \sum_{n=0}^{\infty} \mu^n \gamma_n \tag{4.63}$$

of the coordinate η_1 of (4.4). Here the functions $\gamma_n = \gamma_n(u, \mu)$ determined by (4.50) have the structure

$$\gamma_n = \sum_{k=0}^{[n/3]} \gamma_{n,k}(u) \ln^k \mu. \tag{4.64}$$

Note that $\gamma_{3k,k} \equiv 0$ for $k \geq 1$, since $v_0(u)$ does not depend on η_2, and the corresponding derivatives in (4.50) are equal to zero. The functions $\gamma_{n,k}$ have the following asymptotic representations as $u \to -\infty$:

$$\gamma_{0,0}^- = v_0(u)^- = -u^2 + \sum_{k=0}^{\infty} \frac{\alpha_k}{u^{3k+1}}, \tag{4.65}$$

$$\gamma_{n,k}^- = u^{n-1} \sum_{m=k}^{[n/3]} \sum_{\nu=m}^{\infty} \frac{\alpha_{m,\nu}^{n,k}}{u^{3\nu}} \ln^{m-k} |u|, \qquad n \geq 1, \tag{4.66}$$

which are obtained from (4.43) and (4.44) after the substitutions $\xi = \mu u$ and $\eta_1 = \mu^2 v$ and then expansion again in powers of μ. We deduce from (4.50) and (4.59) that as $u \to +\infty$

$$\gamma_{n,k}^+ = u^{n-1-3k} \sum_{\nu=0}^{n-3k} \sum_{l=3\nu}^{\infty}{}^{*} \frac{\beta_{\nu,l}^{n,k}}{u^l} \ln^\nu u + \beta_{\pi(n)-k,0}^{n,k} \ln^{\pi(n)-k} u, \qquad n \geq 0. \tag{4.67}$$

The formulas (4.63)–(4.67) lead to the following assertion.

Proposition 4.2. *For* $-\varepsilon^{\lambda_1} \leq \xi \leq 0$

$$\eta_1 = \mu^2 \sum_{n=0}^{N} \mu^n \gamma_n(\xi/\mu, \mu) + O(\varepsilon^{\lambda_1 N+1}), \tag{4.68}$$

and for $0 \leq \xi \leq \varepsilon^{\lambda_2}$

$$\eta_1 = \mu^2 \sum_{n=0}^{N} \mu^n \gamma_n(\xi/\mu, \mu) + O(\varepsilon^{\lambda_2 N+1}). \tag{4.69}$$

4.6. **The asymptotics of the trajectories at the end of the breakaway region.** For $\varepsilon^{\lambda_2} \leq \xi \leq q$ we consider the equation

$$\frac{d\eta_2}{d\xi} = \frac{\mu^3 g_2(\xi, \eta_1, \eta_2)}{\xi^2 + \eta_1}, \tag{4.70}$$

which is obtained from (4.6) by setting $\varepsilon = \mu^3$ in the latter and substituting the asymptotic series (4.23) in place of $\eta_1 = H$. We shall be interested in the solution of it with initial condition (4.69) when $\xi = \varepsilon^{\lambda_2}$. This solution is sought in the form

$$\eta_2 = \eta_{2,0}(0) + \sum_{n=3}^{\infty} \mu^n \eta_{2,n}, \qquad \eta_{2,n} = \sum_{k=0}^{\pi(n-2)} \eta_{2,n,k}(\xi) \ln^k \mu. \tag{4.71}$$

Substituting (4.71) in (4.70) and equating the coefficients of the same powers of μ, we get

$$d\eta_{2,n}/d\xi = B_{n-3}, \qquad n \geq 3, \tag{4.72}$$

where

$$B_n = \sum_{\kappa=0}^{n} \Lambda_{n-\kappa} \sum_{k=0}^{[\kappa/2]} \frac{(-1)^k}{\xi^{2k+2}} \lambda_{\kappa-2k}^k, \qquad n \geq 0, \tag{4.73}$$

$$\begin{gathered} \Lambda_n = \sum_{0<2k+3m\leq n} \frac{g_{2\eta_1\eta_2}^{(k)(m)}(\xi, 0, \eta_{2,0}(0))}{k!m!} A_{k,m}^n, \qquad n \geq 1, \\ \Lambda_0 = g_2(\xi, 0, \eta_{2,0}(0)), \end{gathered} \tag{4.74}$$

$$\begin{aligned} \gamma_n &= h_n(\xi, \eta_{2,0}(0)) \\ &+ \sum_{m=1}^{[n/3]} \sum_{k=1}^{m} \frac{h_{n-3m\eta_2}^{(k)}(\xi, \eta_{2,0}(0))}{k!} \sum_{\substack{i_1+\cdots+i_k=3m \\ i_j\geq 3}} \eta_{2,i_1} \cdots \eta_{2,i_k}, \qquad n \geq 2, \end{aligned} \tag{4.75}$$

$$A_{k,m}^n = \sum_{r=0}^{n-2k-3m} \lambda_r^k \delta_{n-2k-3m-r}^m, \tag{4.76}$$

$$\begin{gathered} \lambda_\nu^k = \sum_{i_1+\cdots+i_k=2k+\nu} \gamma_{i_1} \cdots \gamma_{i_k}, \qquad \delta_\nu^m = \sum_{\substack{i_1+\cdots+i_m=3m+\nu \\ i_j\geq 3}} \eta_{2,i_1} \cdots \eta_{2,i_m}, \\ \lambda_0^0 = \delta_0^0 = 1, \qquad \delta_\nu^0 = \lambda_\nu^0 = 0, \quad \nu \geq 1. \end{gathered} \tag{4.77}$$

Analyzing (4.72)–(4.77) and using the equalities

$$\begin{aligned} &\max_{\substack{i_1+\cdots+i_k=l \\ i_j\geq 2}} \{\pi(i_1-2) + \cdots + \pi(i_k-2)\} \\ &\quad = \max_{\substack{i_1+\cdots+i_k=l \\ i_j\geq 3}} \{\pi(i_1-2) + \cdots + \pi(i_k-2)\} = \pi(l-2), \end{aligned}$$

we conclude that

$$B_{n-3} = \sum_{\nu=0}^{\pi(n-2)-1} B_{n-3,\nu} \ln^\nu \mu,$$

where the $B_{n-3,\nu}$ depend on the $\eta_{2,m}$ with $m \leq n-3$. From this and (4.71) we get series of differential equations dependent on n and with separated variables

$$\begin{gathered} d\eta_{2,n,\nu}/d\xi = B_{n-3,\nu}, \qquad \nu = 0, 1, \ldots, \pi(n-2)-1, \\ d\eta_{2,n,\pi(n-2)}/d\xi = 0. \end{gathered} \tag{4.78}$$

As in the preceding subsection, we choose solutions of (4.78) in such a way that for $\xi = \varepsilon^{\lambda_2}$ all the terms in (4.71) and (4.62) independent of λ_2 coincide. In this way we see that

$$\begin{gathered} \eta_{2,n,\nu} = (\mathrm{H}) \int_0^\xi B_{n-3,\nu}(s)\, ds + \sum_{\{k\,:\, 2k+\nu\leq n-3,\, k\leq\nu\}} (-1)^{\nu-k} g_{\nu-k,n-3k-3}^{n-3,k}, \\ \nu = 0, 1, \ldots, \pi(n-2)-1, \end{gathered} \tag{4.79}$$

$$\eta_{2,n,\pi(n-2)} = \beta(n)c_{[n/3]} + \sum_{k=0}^{[n/3]-1} g^{n-3,k}_{\pi(n-2)-k,0}(-1)^{\pi(n-2)-k}, \tag{4.80}$$

where all the coefficients are taken from the expansions of the functions $\omega_{n,k}$ as $u \to +\infty$.

The asymptotic expansions of the functions (4.79) as $\xi \to +0$ are obtained from (4.46) and (4.59), where we set $u = \xi/\mu$ and expand again with respect to μ. After performing these operations, we conclude that

(4.81)

$$\eta_{2,n,s} = \frac{1}{\xi^{n-3}} \sum_{\kappa=0}^{\pi(n-2)-s-1} \sum_{p=\kappa+s}^{\infty} z^{n,s}_{p,\kappa}\xi^p \ln^\kappa \xi + z^{n,s}_{0,\pi(n-2)-s} \ln^{\pi(n-2)-s} \xi,$$

$$z^{n,s}_{p,\kappa} = \sum_{\{k\,:\; 2k+\kappa+s\le p,\; k\le s\}} (-1)^{2-n} C^{\kappa+s-k}_{s-k}\, g^{p,k}_{\kappa+s-k,\,n-3k-3},$$

$$n = 3, 5, 6, \dots, \qquad 0 \le s \le \pi(n-2)-1,$$

$$0 \le \kappa \le \pi(n-2)-s-1, \quad p \ge \kappa + s,$$

$$z^{n,s}_{0,\pi(n-2)-s} = \sum_{k=0}^{s} (-1)^{s-k} C^{\pi(n-2)-k}_{s-k}\, g^{n-3,k}_{\pi(n-2)-k,0}, \qquad s = 0, 1, \dots, \pi(n-2)-1;$$

$$z^{n,\pi(n-2)}_{0,0} = \beta(n)c_{[n/3]} + \sum_{k=0}^{[n/3]-1} (-1)^{\pi(n-2)-k} g^{n-3,k}_{\pi(n-2)-k,0}.$$

These formulas enable us to establish the following assertion:

Theorem 4.6. *For* $\varepsilon^{\lambda_2} \le \xi \le q$

$$\eta_2 = \eta_{2,0}(0) + \sum_{n=3}^{N} \mu^n \eta_{2,n}(\xi,\mu) + O(\varepsilon^{(N+1)/3-\lambda_2(N-2)}). \tag{4.82}$$

For the coordinate η_1 of system (4.4) we get in view of (4.27) and (4.82) the asymptotic series

$$\eta_1 = \sum_{n=2}^{\infty} \mu^n \gamma_n, \qquad \gamma_n = \sum_{s=0}^{\pi(n-2)} \gamma_{n,s}(\xi) \ln^s \mu. \tag{4.83}$$

Here the functions γ_n are defined by (4.75), and as $\xi \to 0$ the $\gamma_{n,s}$ can be expanded in the asymptotic series

$$\gamma_{n,s} = \frac{1}{\xi^{n-2}} \sum_{\kappa=0}^{\pi(n-2)-s-1} \sum_{p=\kappa+s}^{\infty} \varphi^{n,s}_{p,\kappa}\xi^p \ln^\kappa \xi + \varphi^{n,s}_{0,\pi(n-2)-s} \ln^{\pi(n-2)-s} \xi, \tag{4.84}$$

$$n \ge 3, \qquad \gamma_2 = \Omega_0,$$

with coefficients expressed in terms of the coefficients of (4.67) as follows:

$$\varphi_{p,\kappa}^{n,s} = \sum_{\{k\,:\,2k+\kappa+s\le p,\ k\le s\}} \beta_{\kappa+s-k,\,n-3k-3}^{p,k}(-1)^{s-k}C_{s-k}^{\kappa+s-k},$$
$$n = 3, 5, 6, \dots, \qquad s = 0, 1, \dots, \pi(n-2)-1,$$
$$\kappa = 0, 1, \dots, \pi(n-2)-s-1, \qquad p \ge \kappa + s;$$

$$\varphi_{0,\pi(n-2)-1}^{n,s} = \sum_{k=0}^{s}(-1)^{s-k}C_{s-k}^{\pi(n-2)-k}\beta_{\pi(n-2)-k,0}^{n-2,k}, \qquad s = 0, 1, \dots, \pi(n-2)-1;$$

$$\varphi_{0,0}^{n,\pi(n-2)} = \sum_{k=0}^{[n/3]-1}\beta_{\pi(n-2)-k,0}^{n-2,k}(-1)^{\pi(n-2)-k}.$$

It is interesting that the series (4.83) and (4.84) are analogous in structure to the corresponding series in the case of a relaxation system on the plane [5].

We sum up the foregoing as

Proposition 4.3. *For* $\varepsilon^{\lambda_2} \le \xi \le q$

$$\eta_1 = \sum_{n=2}^{N}\mu^n\gamma_n + O(\varepsilon^{(N+1)/3-\lambda_2(N-1)}). \tag{4.85}$$

4.7. **Conclusion.** The assertions obtained above enable us to trivialize the results of [4], where asymptotic formulas are proved for the trajectory of a relaxation system up to ε. Indeed, we divide the second and third equation in (4.4) by the first and linearize the resulting systems on the formal trajectory. The results in §§1 and 2 (see also [16]) enable us to assert that the Cauchy matrix of the system thus obtained is bounded uniformly with respect to ε. The rest of the arguments are standard: take into account the residual and the nonlinear terms, pass with the help of the Lagrange formula to an integral equation, and use the contraction mapping principle in the space of continuous functions of $\xi \in [-q, q]$.

§5. Uniqueness, asymptotics, and stability of a relaxation cycle

5.1. **Asymptotics of a relaxation cycle.** In the note [1] existence conditions were first formulated and the principal asymptotics first determined for a periodic motion of a multidimensional relaxation system (1.9). However, the question of stability and local uniquensss of a relaxation cycle was solved only much more recently (see Theorem 1.2 in [9], and [16]). Below we clarify the asymptotic properties of its initial condition and period on the basis of our assertions.

Suppose that the relay system (1.1) has a dichotomous periodic solution $y_0(t)$ of period T_0, and suppose for definiteness that $(\Phi_1(y_0(0)), y_0(0)) \in \Gamma_{1-}$. As in §1.2 we denote by Ц$_0$ the cycle of the hybrid system (1.7) obtained by "lifting" a cycle of the relay system from R^m to R^n. Then by Theorem 1.1, under Conditions 1–3 in §1 the relaxation system (1.9) has for all sufficiently small values of ε a unique cycle Ц(ε) of period $T(\varepsilon)$ in some neighborhood of Ц$_0$ independent of ε, and Ц$(\varepsilon) \to$ Ц(0) and $T(\varepsilon) \to T_0$ as $\varepsilon \to 0$. Denote by $(x(\varepsilon), y(\varepsilon))$, where $x(0) = \Phi_1(y_0(0))$ and $y(0) = y_0(0)$, its initial condition given for $t = 0$.

Theorem 5.1. *The asymptotic representations*

$$T(\varepsilon) = T_0 + \sum_{n=2}^{\infty} \varepsilon^{n/3} \sum_{\nu=0}^{\pi(n-2)} T_{n,\nu} \ln^{\nu} \frac{1}{\varepsilon}, \tag{5.1}$$

$$x(\varepsilon) = \Phi_1(y(\varepsilon)) + \sum_{n=1}^{\infty} \varepsilon^n F_n(y(\varepsilon)), \tag{5.2}$$

$$y(\varepsilon) = y_0(0) + \sum_{n=2}^{\infty} \varepsilon^{n/3} \sum_{\nu=0}^{\pi(n-2)} y_{n,\nu} \ln^{\nu} \frac{1}{\varepsilon} \tag{5.3}$$

are valid, where (5.2) *is the result of substituting* $y(\varepsilon)$ *in the series for the integral manifold of system* (1.9) *in a neighborhood of* $(\Phi_1(y_0(0)), y_0(0))$.

5.2. **Proof of Theorem 5.1.** We draw a general secant hyperplane $\mathscr{L}$ through the point $(\Phi_1(y_0(0)), y_0(0))$ belonging to the "towel" Γ_{1-}, and in a neighborhood of it we consider the return operator $\Pi(\varepsilon)$. It is conveniently represented as a composition of mappings: $\Pi_1(\varepsilon)$ acts from $\mathscr{L}$ to the hyperplane $\xi = -q$; $\Pi_2(\varepsilon)$ acts from $\xi = -q$ to $\xi = q$; $\Pi_3(\varepsilon)$ acts from $\xi = q$ to the hyperplane $\xi = -q$ on the second "towel" Γ_{2-}; and so on. By virtue of results in [8], some of these operators have expansions in asymptotic series of integer powers of ε with smooth coefficients. An asymptotic expression for the operator $\Pi_2(\varepsilon)$ (and of the others analogous to it) was found in the preceding section. Therefore, for the components x and y of the operator $\Pi(\varepsilon)$ we have asymptotic series with smooth coefficients similar to (5.2) and (5.3). This implies that the coordinates of its fixed point can be expanded in the formal series (5.2) and (5.3), and the character of the asymptotic properties of these series can be determined thanks to the following two facts. First, according to Theorem 1.1, $\Pi'(\varepsilon)_{x=x(\varepsilon), y=y(\varepsilon)} \to B$ as $\varepsilon \to 0$, where the matrix B does not have eigenvalues on the unit circle. Second, if the partial sums of (5.2) and (5.3) are substituted in the equation $\Pi(\varepsilon)(x(\varepsilon), y(\varepsilon)) = (x(\varepsilon), y(\varepsilon))$, then the residual turns out to be of a higher order of smallness. It remains to use the contraction mapping principle. Formula (5.1) is a simple consequence of the structure of the operator $\Pi(\varepsilon)$.

§6. Complex dynamics in three-dimensional relaxation systems

6.1. **Basic theorem.** We apply the theory of relaxation oscillations presented in §§1–4 to the problem of the existence of a strange attractor in the system

$$\varepsilon \dot{x} = f(x, y), \qquad \dot{y} = g(x, y), \qquad 0 < \varepsilon \ll 1, \tag{6.1}$$

where $x \in R$, $y \in R^2$, and f and g are sufficiently smooth functions. As usual, we assume that the equation $f = 0$ determines in R^3 a smooth surface Γ that splits into parts Γ_{1-}, Γ_+, and Γ_{2-} separated by curves Γ_{10} and Γ_{20}. Recall that $(x, y) \in \Gamma_{1-} + \Gamma_{2-}$ if $f'_x < 0$, $(x, y) \in \Gamma_+$, if $f'_x > 0$, and $(x, y) \in \Gamma_{10} + \Gamma_{20}$ if $f'_x = 0$. Moreover, we assume that the "towels" Γ_{1-} and Γ_{2-} are given by the respective equations $x = \Phi_1(y)$ and $x = \Phi_2(y)$, where Φ_1 and Φ_2 are smooth functions. Finally, we assume that for $(x, y) \in \Gamma_{10} + \Gamma_{20}$

$$f''_{xx} \neq 0, \qquad (\operatorname{grad}_y f, g) \neq 0. \tag{6.2}$$

It was shown in §§1 and 2 that under these conditions the relaxation system (6.1) is a special small perturbation of the relay system

$$\dot{y} = g(x, y), \qquad x = \begin{cases} \Phi_1(y), \\ \Phi_2(y), \end{cases} \tag{6.3}$$

with respect to which we assume the following:

Condition 1. A global secant l with parametrization $0 \le s \le 1$ is defined on the "towel" Γ_{1-}, and the trajectories of the relay system (6.3) with initial conditions on l intersect it in general position after a complete circuit.

Obviously, a continuously differentiable Poincaré return function is defined under Condition 1:

$$s \to F_0(s). \tag{6.4}$$

As is known [17], outside some neighborhood of the curve Γ_{10} the relaxation system (6.1) has a smooth integral manifold $x = H(y, \varepsilon)$ at a distance from Γ_{1-} not exceeding ε. We take a smooth surface Γ_c intersecting Γ_{1-} in general position along l. It intersects the integral surface $x = H(y, \varepsilon)$ along some smooth curve $l(\varepsilon)$, on which we denote the parametrization by $0 \le s \le 1$ as before. Denote by σ the coordinate along some smooth curve on Γ_c that intersects $l(\varepsilon)$ in general position. It is clear that for $0 \le s \le 1$ and $-\sigma_1 \le \sigma \le \sigma_2$, where the constants σ_1 and σ_2 are sufficiently small, the Poincaré return operator Π is defined along the trajectories of the relaxation system (6.1).

Theorem 6.1. *Under Condition* 1,

$$\Pi\colon \begin{array}{l} s \to F_1(s, \sigma, \varepsilon), \\ \sigma \to \Delta(s, \sigma, \varepsilon)\exp(-\alpha/\varepsilon), \end{array} \tag{6.5}$$

where the constant α is positive, the functions F_1 and Δ are continuously differentiable with respect to s and σ and their derivatives bounded uniformly with respect to ε, and $F_1(s, 0, \varepsilon)$ differs in the C^1-metric from the $F_0(s)$ in (6.4) *by a quantity of order $\varepsilon^{2/3}$.*

The proof follows immediately from results in §§1–4.

6.2. **The complex dynamics of system (6.1).** The basic Theorem 6.1 enables us under additional restrictions to obtain an interesting assertion about the dynamics of system (6.1).

Suppose that the mapping (6.4) has an exponentially unstable cycle

$$s_1, \dots, s_k, \tag{6.6}$$

and $F_0'(s_i) \ne 0$, $i = 1, \dots, k$. We say that it has *structurally stable homoclinic trajectory*

$$p_n, \quad -\infty < n < \infty, \tag{6.7}$$

if for $n \ge n_0$ the numbers p_n coincide with one of the numbers in (6.6); as $n \to -\infty$ the sequence p_n splits into subsequences, each of which converges to one of the numbers in (6.6); finally, $F_0'(p_n) \ne 0$.

Condition 2. The mapping (6.4) has an exponentially unstable cycle (6.6) with structurally stable homoclinic trajectory (6.7).

Theorem 6.2. *Under Conditions* 1 *and* 2 *system* (6.1) *has countably many dichotomous cycles.*

Passing to a suitable iterate, we can assume that the mapping (6.4) has an equilibrium state s_0 with structurally stable homoclinic trajectory. By (6.5), the operator Π then has a saddle equilibrium state $s_0(\varepsilon) \to s_0$, $\sigma_0(\varepsilon) \to 0$ as $\varepsilon \to 0$. Its stable and unstable manifolds are denoted by $\gamma_-(\varepsilon)$ and $\gamma_+(\varepsilon)$, respectively. On

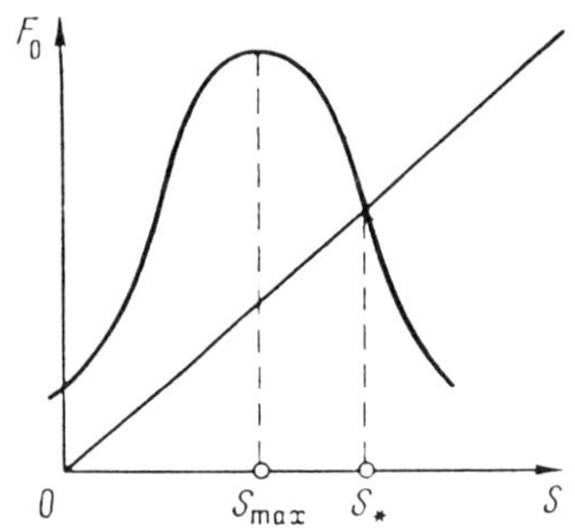

FIGURE 1

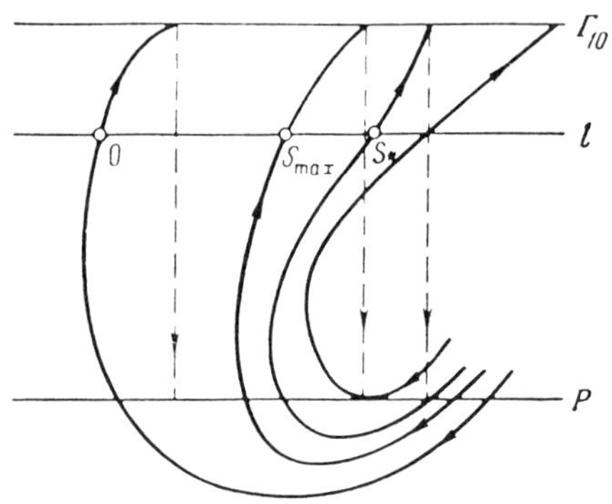

FIGURE 2

$\gamma_+(\varepsilon)$ we take the open set $\omega_+(\varepsilon)$ containing $(s_0(\varepsilon), \sigma_0(\varepsilon))$ and consider the sets $\omega_m(\varepsilon) = \Pi^m(\omega_+(\varepsilon))$, $m = 1, 2, \ldots$. It follows from Condition 2 that $\omega_{m_0}(\varepsilon)$ contains the points $\gamma_-(\varepsilon)$ for some m_0, because otherwise the equilibrium state s_0 can have only a structurally unstable homoclinic trajectory. It remains to use a theorem in [18].

Of course, the nonmonotonicity of the mapping (6.4) is a necessary prerequisite for the validity of Condition 2. The mapping will be nonmonotone (Figure 1) if, for example, the character of the trajectories on Γ_{1-} has the form shown in Figure 2, where the lower line P is a conditional representation of the drop curve, and the character of the motions on Γ_{2-} is represented by the dotted line.

Theorem 6.2 naturally does not yet imply the existence of a strange attractor for system (6.1). If, however, the mapping (6.4) has a mixing Sharkovskiĭ attractor, then, by Theorems 6.1 and 6.2 and results in [19], the behavior of the solutions of (6.1) will be sufficiently chaotic.

§7. The Pontryagin problem on tori of relaxations systems

7.1. **Statement of the problem.** In this section we consider the relaxation system

$$\varepsilon\dot{x} = f(x, y), \qquad \dot{y} = g(x, y), \qquad 0 < \varepsilon \ll 1, \tag{7.1}$$

assuming that $x \in R^n$ and $y \in R^m$, $n, m \geq 2$, and the sufficiently smooth right-hand sides f and g are defined on the whole space. We are interested in auto-oscillations of system (7.1) under the following two restrictions.

Condition 1. For each fixed y the system of fast motions

$$\dot{x} = f(x, y) \tag{7.2}$$

has an orbitally exponentially stable cycle $x_0(\tau, y)$ of period $T(y)$, where $x_0(\tau + 1, y) \equiv x_0(\tau, y)$ and $\tau = t/T(y)$.

Condition 2. The averaged system of slow motions

$$\dot{y} = \mathcal{G}_0(y), \qquad \mathcal{G}_0(y) = T(y)\int_0^1 g(x_0(\tau, y), y)\, d\tau \tag{7.3}$$

has an orbitally exponentially stable cycle $y_0(t)$ of period T_0.

Multidimensional relaxation systems were first considered under Condition 1 in [20] and [21], where it was shown that to an exponentially stable equilibrium state of system (7.3) there corresponds a periodic solution of (7.1) with the same stability. This assertion is the content of the well-known theorem of Pontryagin and Rodygin.

In the survey article [22] the possibility of generalizing the theorem to the case of a cycle in the system (7.3) is mentioned, and a corresponding conjecture is formulated. In the present article we use the apparatus of integral manifolds [17] to give a proof of this conjecture.

7.2. **An auxiliary lemma.** Making the substitution $t \to \varepsilon t$ in (7.1), we come to the system

$$\dot{x} = f(x, y), \qquad \dot{y} = \varepsilon g(x, y), \tag{7.4}$$

which for $\varepsilon = 0$ has an exponentially stable invariant integral manifold $x = x_0(\tau, y)$ with the equations

$$\dot{y} = 0, \qquad \dot{\tau} = 1/T(y) \tag{7.5}$$

on it. According to results in Chapter III of [17], this manifold is preserved also for $\varepsilon > 0$, and it can now be given by

$$x = x_0(\tau, y) + X(\tau, y, \varepsilon), \qquad X(\tau, y, 0) \equiv 0, \tag{7.6}$$

and in place of system (7.5) we have

$$\dot{y} = \varepsilon g(x_0 + X, y), \qquad \dot{\tau} = [1 + \Phi(\tau, y, \varepsilon)]/T(y). \tag{7.7}$$

Here X and Φ are functions smooth in their variables and periodic in τ with period 1, and $\Phi(\tau, y, 0) \equiv 0$.

Let us describe an algorithm for asymptotic computation of the right-hand sides of (7.6) and (7.7). We let

$$X = \sum_{k=0}^{\infty} \varepsilon^k x_k(\tau, y), \qquad \Phi = \sum_{k=1}^{\infty} \alpha_k(y)\varepsilon^k,$$

and determine the unknown coefficients in these series expansions from the identity

$$\left(\frac{\partial x_0}{\partial \tau} + \frac{\partial X}{\partial \tau}\right)(1 + \Phi)\Big/ T(y) + \varepsilon\left(\frac{\partial x_0}{\partial y} + \frac{\partial X}{\partial y}\right) g(x_0 + X, y) = f(x_0 + X, y). \tag{7.8}$$

Equating the coefficients of the same powers of ε in (7.8), we arrive at a recursive sequence of differential equations:

$$\frac{\partial x_k}{\partial \tau} + \alpha_k(y)\frac{\partial x_0}{\partial \tau} = A_0(\tau, y)x_k + F_k(\tau, y), \tag{7.9}$$

where

$$A_0(\tau, y) = T(y)\left.\frac{\partial f}{\partial x}\right|_{x = x_0(\tau, y)},$$

and the functions F_k depend only on the x_r and α_r with indices $r \le k - 1$. From the condition of solvability of (7.9) in the class of functions periodic in τ we find that

$$\alpha_k(y) = \int_0^1 (F_k(\tau, y), g_0(\tau, y))\, d\tau,$$

where g_0 is a periodic solution of the equation $dg/d\tau = -A_0^*(\tau, y)g$, $(\partial x_0/\partial \tau, g_0) \equiv 1$. After this we determine from (7.9) a unique function $x_k(\tau, y)$ periodic in τ and satisfying the condition

$$\int_0^1 (x_k(\tau, y), g_0(\tau, y))\, d\tau \equiv 0.$$

Now, the results in [17], combined with the above computations, enable us to formulate the following assertion.

Lemma 7.1. *Under Condition* 1 *system* (7.4) *has an orbitally exponentially stable invariant integral manifold* (7.6) *with the system of equations* (7.7) *on it. Moreover, for any positive integer* N *the following asymptotic representations hold*:

$$X = \sum_{k=1}^{N} \varepsilon^k x_k + \varepsilon^{N+1}\Delta_1(\tau, y, \varepsilon), \tag{7.10}$$

$$\Phi = \sum_{k=1}^{N} \varepsilon^k \alpha_k + \varepsilon^{N+1}\Delta_2(\tau, y, \varepsilon), \tag{7.11}$$

where Δ_1 *and* Δ_2 *are functions smooth in their variables and periodic in* τ.

7.3. **Main result.** We consider system (7.7), which is the result of reducing system (7.4) to the integral manifold (7.6). Dividing the first equation in (7.7) by the second, we get a nonautonomous system equivalent to it, written in standard form [17]:

$$dy/d\tau = \varepsilon \mathcal{G}(\tau, y, \varepsilon). \tag{7.12}$$

By Condition 2, the averaged equation of first approximation corresponding to (7.12) has an exponentially stable cycle. The results of [17], Chapter II, enable us to conclude that in this case system (7.12) has an exponentially stable toroidal integral manifold

$$y = y_0(T_0\sigma) + H(\tau, \sigma, \varepsilon), \qquad H(\tau, \sigma, 0) \equiv 0, \tag{7.13}$$

with equation on it

$$\frac{d\sigma}{d\tau} = \frac{\varepsilon}{T_0}[1 + \Psi(\tau, \sigma, \varepsilon)], \qquad \Psi(\tau, \sigma, 0) \equiv 0, \tag{7.14}$$

where the smooth functions H and Ψ depend periodically on τ and σ with period 1. To find the asymptotics of H and Ψ we fix an arbitrary positive integer N and make in (7.12) the substitution

$$y = \xi + \sum_{k=1}^{N} \varepsilon^k Y_k(\tau, \xi), \tag{7.15}$$

which reduces it to the form

$$\frac{d\xi}{d\tau} = \varepsilon \mathcal{G}_0(\xi) + \sum_{k=1}^{N-1} \varepsilon^{k+1} \mathcal{G}_k(\xi) + \varepsilon^{N+1}\Delta_3(\tau, \xi, \varepsilon), \tag{7.16}$$

where all the functions are smooth, and Y_k and Δ_3 are periodic in τ. An algorithm for constructing the substitution (7.15) and the right-hand side of (7.16) is contained, for example, in [23]. We next consider the truncated system

$$\frac{d\xi}{d\tau} = \varepsilon \mathcal{G}_0(\xi) + \sum_{k=1}^{N-1} \varepsilon^{k+1} \mathcal{G}_k(\xi). \tag{7.17}$$

The condition of structural stability of the periodic solution $y_0(t)$ of (7.3) implies that (7.17) has a periodic solution $\xi_0(\sigma, \varepsilon)$ with period $T_0(\varepsilon)/\varepsilon$, where $\xi_0(\sigma+1, \varepsilon) \equiv \xi_0(\sigma, \varepsilon)$ and $\sigma = \varepsilon\tau/T_0(\varepsilon)$, and that the following asymptotic expansions are valid:

$$\xi_0(\sigma, \varepsilon) = y_0(T_0\sigma) + \sum_{k=1}^{\infty} \varepsilon^k \xi_k(\sigma), \qquad T_0(\varepsilon) = T_0 + \sum_{k=1}^{\infty} \varepsilon^k T_k. \tag{7.18}$$

Since the right-hand sides of (7.16) and (7.17) differ by quantities of order ε^{N+1}, in view of the smooth dependence of the integral manifold (7.13) on the right-hand side $\mathcal{G}$ the corresponding manifolds for these equations also differ only in order ε^{N+1}. From this and (7.18) it follows that the substitution (7.15) transforms (7.13) to

$$\xi = y_0(T_0\sigma) + \sum_{k=1}^{N} \varepsilon^k \xi_k(\sigma) + \varepsilon^{N+1}\Delta_4(\tau, \sigma, \varepsilon), \tag{7.19}$$

and (7.14) takes the form

$$d\sigma/d\tau = \varepsilon\left[1\Big/\left(T_0 + \sum_{k=1}^{N}\varepsilon^k T_k\right) + \varepsilon^{N+1}\Delta_5(\tau, \sigma, \varepsilon)\right]. \tag{7.20}$$

Lemma 7.2. *Under Condition* 2 *system* (7.12) *has the exponentially stable invariant integral manifold* (7.13) *with equation* (7.14) *on it, and* (7.15), (7.19), *and* (7.20) *yield the asymptotics of the right-hand sides of* (7.13) *and* (7.14).

Returning to system (7.4), from Lemmas 7.1 and 7.2 we derive

Theorem 7.1. *Under Conditions* 1 *and* 2 *system* (7.4) *has an orbitally exponentially stable two-dimensional invariant torus* (7.6), (7.13), *and the equations on it are the second equation in* (7.7) *and equation* (7.14).

§8. On relaxation tori

8.1. **Statement of the problem and description of the result.** In this section we formulate conditions under which the relaxation system

$$\varepsilon\dot{x} = f(x, y), \qquad \dot{y} = g(x, y), \qquad x \in R^2, \quad y \in R, \tag{8.1}$$

where f and g are sufficiently smooth functions, has a relaxation torus, i.e., a torus on which the trajectory passes successively from one stable integral manifold of the form (7.6) to another in the course of time. We underscore that here the connection with the results of the preceding section is the same as the connection of the results in [24] on a cycle of a slow motion system with the results in [5] on a relaxation cycle. We note further that our restrictions, which exclude the possibility of using the Pontryagin-Rodygin theorem [21], are a natural generalization of the conditions in [5] under which it is common to study relaxation oscillations.

Condition 1. The dynamical properties of the fast motion equation

$$\dot{x} = f(x, y) \tag{8.2}$$

depend in the following way on the parameter y: for all $y < y_1$ equation (8.2) has a unique globally exponentially stable cycle, and as y passes through y_1 two more cycles are created in the domain bounded by this cycle from crowding together of trajectories—a stable cycle and an unstable cycle; for $y_1 < y < y_2$ equation (8.2) has exactly three structurally stable cycles, and for $y = y_2$ there is a confluence bifurcation of the enveloping stable cycle and the unstable cycle; for $y > y_2$ the situation is the same as for $y < y_1$.

Let $x_1(\tau, y)$ and $x_2(\tau, y)$ be stable cycles of equation (8.2) with periods $T_1(y)$ and $T_2(y)$, respectively, and let

$$x_1(\tau+1, y) \equiv x_1(\tau, y), \qquad \tau = t/T_1(y), \quad y \le y_2;$$
$$x_2(\tau+1, y) = x_2(\tau, y), \qquad \tau = t/T_2(y), \quad y \ge y_1.$$

Condition 2. Assume that

$$\mathcal{G}_1(y) = T_1(y) \int_0^1 g(x_1(\tau, y), y)\, d\tau > 0, \qquad y \le y_2,$$

$$\mathcal{G}_2(y) = T_2(y) \int_0^1 g(x_2(\tau, y), y)\, d\tau < 0, \qquad y \ge y_1.$$

For $y = y_1$ the cycle $x_2(t) = x_2(t/T_2(y_1), y_1)$ of equation (8.2) becomes multiple. Therefore, the variational equation $\dot{h} = A_0(t)h$ corresponding to it has linearly independent periodic solutions $h_1(t) = dx_2/dt$ and $h_2(t)$ such that $(h_j(t), g_k(t)) \equiv \delta_{jk}$, $j, k = 1, 2$, where g_1 and g_2 are linearly independent solutions of the conjugate equation. According to results in [25], the substitution

$$x = x_2(s) + \xi h_2(s) + \xi^2 h_3(s),$$

where $h_3(s)$ is a $T_2(y_1)$-periodic function, transforms equation (8.2) in a neighborhood of the cycle x_2 into the system

$$\dot{\xi} = \alpha\xi^2, \qquad \dot{s} = 1 + \beta\xi^2,$$

up to terms of order ξ^3.

Condition 3. The following nondegeneracy conditions hold:

$$\int_0^{T_2(y_1)} (a(t), g_2(t))\, dt \neq 0, \qquad \alpha \neq 0, \tag{8.3}$$

where

$$a(t) = \left.\frac{\partial f}{\partial y}\right|_{x = x_0(t),\ y = y_1}.$$

The inequalities analogous to (8.3) are assumed to hold also for $y = y_2$.

Condition 3 characterizes a certain generality of position, and Conditions 1 and 2 determine the global behavior of the solutions. When they hold, the trajectories of (8.1) behave as follows. Suppose, for example, that the initial conditions $x(0, \varepsilon) = x_0$ and $y(0, \varepsilon) = y_0$ are such that $y_0 < y_1$. Then the component $x(t, \varepsilon)$ first falls rapidly (in a time of order ε) into a neighborhood of the cycle $x_1(\tau, y_0)$, while the component $y(t, \varepsilon)$ gets close to the solution $y_0(t)$ of the equation $\dot{y} = \mathcal{G}_1(y)$, $y(0) = y_0$. As t increases further, the component $x(t, \varepsilon)$ performs rapid oscillations along the cycles $x_1(\tau, y)$ roughly according to the law $x_1(t/\varepsilon T_1(y_0(t)), y_0(t))$. It is clear that at some moment of time $y_0(t)$ falls in a sufficiently small neighborhood of the value y_2. Then there is a breakaway onto the cycle $x_2(\tau, y_2)$, and this leads to a motion analogous to the one already described along the cycles $x_2(\tau, y)$, with $y(t, \varepsilon)$ now decreasing. In a neighborhood of $y = y_1$ there is a breakaway onto $x_1(\tau, y_1)$, and so on. Therefore, any trajectory of system (8.1) becomes close to a two-frequency trajectory with frequencies of order 1 and ε^{-1}.

Theorem 8.1. *Under Conditions* 1–3 *system* (8.1) *has an orbitally exponentially stable two-dimensional invariant torus whose zero approximation is a closed surface* Γ_0 *consisting of the two lateral surfaces*

$$x = x_1(\tau, y), \qquad x = x_2(\tau, y), \qquad 0 \le \tau \le 1, \quad y_1 \le y \le y_2,$$

and the two bases $y = y_1$ *and* $y = y_2$, *where* x *varies in the annular domains bounded by the cycles* $x_1(\tau, y_1)$, $x_2(\tau, y_1)$, *and* $x_1(\tau, y_2)$, $x_2(\tau, y_2)$, *respectively.*

For the proof we first use the techniques developed in §§1 and 2 to investigate the variational system on an arbitrary trajectory of system (8.1), and then we employ the method of point mappings.

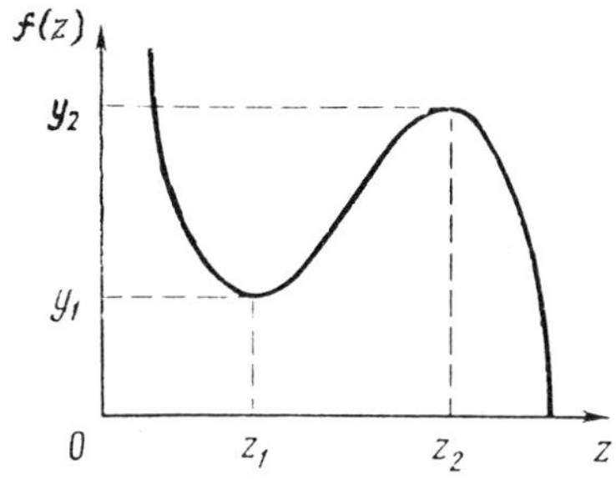

FIGURE 3

The theorem can be generalized to the multidimensional case. However, not only are Conditions 1–3 more unwieldy, but also the relaxation invariant torus can be dichotomous. It is natural that in the multidimensional case equation (8.2) can have tori of different dimensions instead of cycles.

8.2. **Example.** We describe a class of systems (8.1) for which Conditions 1–3 hold. In polar coordinates equation (8.2) is

$$dr/dt = r[f(r^2) - y], \qquad d\varphi/dt = 1,$$

where $f(z)$, $z > 0$, is some smooth function whose graph is indicated in Figure 3. We assume that the equation $f'(z) = 0$ has only two roots $z_1, z_2 > 0$, $f'(z) < 0$ for $0 < z < z_1$ and $z > z_2$, $f'(z) > 0$ for $z_1 < z < z_2$, $f''(z_1)f''(z_2) \neq 0$, $f(z_1), f(z_2) > 0$, and

$$\lim_{z\to 0} f(z) = +\infty, \qquad \lim_{z\to +\infty} f(z) = -\infty.$$

As is not hard to see, Conditions 1 and 3 hold for such a choice of the right-hand side $f(x, y)$ of system (8.1). If now we choose $g(x, y)$ to satisfy Condition 2, then we get a three-dimensional relaxation system for which Theorem 8.1 is valid.

CHAPTER 2. RELAXATION OSCILLATIONS IN A MEDIUM WITH DIFFUSION

As remarked in [26] and [27], relaxation parabolic boundary value problems of reaction-diffusion type serve as basic models for a great many biophysical problems. Therefore, the question of describing their general properties is crucial. The ideas in Chapter 1 are extended below to this class of differential equations: a parabolic relay system is introduced, and a theorem analogous to Theorem 1.1 is formulated on C^1-closeness between these objects (see [28]).

Parabolic systems of reaction-diffusion type model the process of emergence of spatially nonhomogeneous structures in homogeneous media with diffusion, which is important in many questions of biophysics, ecology, and chemistry. To describe this phenomenon it is common to invoke the Turing-Prigogine bifurcation theorem. This theorem is based on the phenomenon (discovered by Turing [29] in 1952) of diffusion instability of a spatially homogeneous equilibrium state, when for a special change in the diffusion coefficients the homogeneous equilibrium state loses stability, and in a neighborhood of it there appears a spatially nonhomogeneous equilibrium state—a so-called *dissipative structure*. However, the overwhelming majority of models of biophysical and ecological processes have a spatially homogeneous (independent of the space variables) cycle. In this connection the problem arises of extending the Turing-Prigogine bifurcation theorem to the oscillatory case, which problem was solved for parabolic relaxation systems in [30]–[32]. The explicit formulas obtained there in the problem of stability of a spatially homogeneous cycle are then used to analyze a number of mathematical models in biophysics.

§9. On a certain class of parabolic relay systems

9.1. General statement of the problem. We consider the parabolic boundary value problem

$$\varepsilon\frac{\partial x}{\partial t} = D_1\Delta x + f(x, y), \qquad \left.\frac{\partial x}{\partial \nu}\right|_{\partial\Omega} = 0, \tag{9.1}$$

$$\frac{\partial y}{\partial t} = D_2\Delta y + g(x, y), \qquad \left.\frac{\partial y}{\partial \nu}\right|_{\partial\Omega} = 0, \tag{9.2}$$

where $0 < \varepsilon \ll 1$, $x \in R$, $y \in R^m$, $D_1 > 0$, the eigenvalues of the matrix D_2 lie to the right of the imaginary axis, Δ is the Laplace operator, and ν is the direction of the outer normal to the sufficiently smooth boundary $\partial\Omega$ of a bounded domain $\Omega \subset R^N$. In what follows, it is assumed that f and g are smooth functions of their variables defined for all $(x, y) \in R^{m+1}$, $f > 0$ for $x \leq -N$, and $f < 0$ for $x \geq N$, where $N = N(y) > 0$. Assume further that f is analytic in x.

As the phase space (the space of initial conditions $(x(s), y(s))$) of the boundary value problem (9.1), (9.2) we take the space $E = C^\alpha \times C^\alpha$, where C^α consists of continuous functions satisfying a Hölder condition with exponent $0 < \alpha < 1$ in $\overline{\Omega}$. The norm in E is assumed to be introduced in the usual way.

We pose the question of which boundary value problems of the form (9.1), (9.2) it is expedient to include among relaxation systems. A corresponding concept was introduced in [5] for systems of ordinary differential equations. Below it is convenient to be guided by the approach developed in Chapter 1 to link the properties of the original system with those of the limit relay system.

9.2. Properties of parabolic relay systems. In our case an analogue of the relay system (1.1) introduced in §1 is the boundary value problem

$$\frac{\partial y}{\partial t} = D_2\Delta y + g(\Phi(y), y), \qquad \left.\frac{\partial y}{\partial \nu}\right|_{\partial\Omega} = 0, \tag{9.3}$$

where the relay operator $\Phi(y)$ is defined constructively. We fix an arbitrary function $y_0(s) \in C^\alpha$ and for $y = y_0(s)$ consider the boundary value problem

$$\frac{\partial x}{\partial t} = D_1\Delta x + f(x, y), \qquad \left.\frac{\partial x}{\partial \nu}\right|_{\partial\Omega} = 0. \tag{9.4}$$

It follows from the restrictions imposed on f and the results in [33] that the boundary value problem (9.4) has at least one asymptotically stable equilibrium state $x_0(s)$. In the general case it is exponentially stable, which is what we assume. This implies that for all $y(s)$ close to $y_0(s)$ in the C^α-norm the manifold $\Gamma_1 = \{(\Phi_1(y), y)\}$ is defined in the space E, where the operator $\Phi_1(y)$, $\Phi_1(y_0(s)) = x_0(s)$, which acts from C^α to $C^{2+\alpha}$ and has a smooth Fréchet derivative, assigns to a function $y(s)$ the exponentially stable equilibrium state $x(s)$ of the boundary value problem (9.4). We extend this manifold, including in it the points $(x(s), y(s))$, where $x(s)$ is the equilibrium state of (9.4) for $y = y(s)$, and the second coordinate $y(s)$ can be joined to $y_0(s)$ by a continuous curve in C^α in such a way that the original equilibrium state $x_0(s)$ extends continuously along this curve to $x(s)$, with preservation of structural stability. As a result we get a locally smooth manifold with boundary $\Gamma_{10} = \{(\Phi_1(y), y),\ y \in \Lambda_{10}\}$, where the set Λ_{10} is given by the equality $\lambda_0 = 0$, and $\lambda_0 = \lambda_0(y(s))$ is the largest eigenvalue of the operator

$$Lh = D_1 \Delta h + f'_x(\Phi_1(y), y)h, \qquad \left.\frac{\partial h}{\partial \nu}\right|_{\partial\Omega} = 0. \tag{9.5}$$

These constructions enable us first to define the solution of the relay system (9.3) as the solution of the mixed problem

$$\frac{\partial y}{\partial t} = D_2 \Delta y + g(\Phi_1(y), y), \qquad \left.\frac{\partial y}{\partial \nu}\right|_{\partial\Omega} = 0, \qquad y|_{t=0} = y_0(s). \tag{9.6}$$

Assume that at some first moment of time $t_0 > 0$ the point $(\Phi_1(y(t, s)), y(t, s))$ falls in Γ_{10}, the nondegeneracy condition

$$\int_\Omega f''_{xx}(\Phi_1(y(t_0, s), y(t_0, s))e_0^3(s)\, ds \neq 0 \tag{9.7}$$

holds, and the normal switching condition

$$l_0\left(\left.\frac{\partial y}{\partial t}\right|_{t=t_0-0}\right) \equiv \int_\Omega f'_y(\Phi_1(y(t_0, s)), y(t_0, s)) \left.\frac{\partial y}{\partial t}\right|_{t=t_0-0} e_0(s)\, ds \neq 0 \tag{9.8}$$

is valid. Here $e_0(s)$ is an eigenfunction corresponding to the zero eigenvalue for the operator (9.5) with $y = y(t_0, s)$.

We consider the boundary value problem (9.4) with $y = y(t_0, s)$. In a neighborhood of the multiple equilibrium state $\Phi_1(y(t_0, s))$ it has an exponentially stable one-dimensional invariant manifold, and the equation on it begins with a quadratic term in view of (9.7). We take the initial condition of the solution of (9.4) lying on the one-dimensional tendril going out from $\Phi_1(y(t_0, s))$. According to results in [33] and [34], the solution with this initial condition tends as $t \to \infty$ to an equilibrium state, which is exponentially stable in the general case. Therefore, by analogy with the foregoing, we can first introduce the manifold $\Gamma_2 = \{(\Phi_2(y), y)\}$ with boundary Γ_{20}, and then define the solution of the relay system (9.3) to be the solution of the following mixed problem for $t \geq t_0$:

$$\frac{\partial y}{\partial t} = D_2 \Delta y + g(\Phi_2(y), y), \qquad \left.\frac{\partial y}{\partial \nu}\right|_{\partial\Omega} = 0, \qquad y|_{t=t_0} = y(t_0, s). \tag{9.9}$$

Assume that at some first moment of time $t_1 > t_0$ a switching again takes place, with conditions analogous to (9.7) and (9.8). In this case we proceed as above.

Without mentioning it explicitly, we assume everywhere below that for $t = t_p$ $(p = 0, 1, \ldots)$ the conditions analogous to (9.7) and (9.8) hold. Then for $t \neq t_p$ the solutions of the relay system (9.3) depend smoothly on the initial conditions, and we have the following boundary value problem for the first variation $h(t, s)$ of the fixed solution $y(t, s)$:

$$\frac{\partial h}{\partial t} = D_2 \Delta h + A(t)h, \qquad \left.\frac{\partial h}{\partial \nu}\right|_{\partial\Omega} = 0, \tag{9.10}$$

$$\begin{aligned} h(t_p + 0, s) &= h(t_p - 0, s) \\ &+ \left(\left.\frac{\partial y}{\partial t}\right|_{t=t_p+0} - \left.\frac{\partial y}{\partial t}\right|_{t=t_p-0}\right) l_p(h(t_p - 0, s)) \Big/ l_p\left(\left.\frac{\partial y}{\partial t}\right|_{t=t_p-0}\right). \end{aligned} \tag{9.11}$$

Here the bounded operator $A(t)$ is the result of the usual linearization on the trajectory $y(t, s)$ of the nonlinear term on the right-hand side of (9.3) between successive switching times, and for $p > 0$ the linear functionals l_p are defined exactly like l_0 in (9.8). The following analogue of the Andronov-Witt theorem holds.

Lemma 9.1. *Suppose that $y(t, s)$ is a periodic solution of the relay system (9.3) of period T, and all the multipliers (except for the simple multiplier 1) of the T-periodic linear boundary value problem (9.10), (9.11) lie inside the unit disk. Then $y(t, s)$ is orbitally exponentially stable.*

The proof follows immediately from results in [15] (see also §1.1).

§10. THE THEOREM ON C^1-CLOSENESS FOR PARABOLIC RELAXATION SYSTEMS

10.1. **Description of the result.** Let $S(r_0)$ be a ball in E of sufficiently small radius r_0 about the point $(x_0(s), y_0(s)) \in \Gamma_1$. Let $(x(t, s, \varepsilon), y(t, s, \varepsilon))$ denote the solution of the parabolic system (9.1), (9.2) with initial condition $(x(s), y(s)) \in S(r_0)$ given for $t = 0$, and let $h_1(t, s, \varepsilon)$ and $h_2(t, s, \varepsilon)$ denote its first variations with initial conditions $h_1(0, s, \varepsilon) = h_1^0(s)$ and $h_2(0, s, \varepsilon) = h_2^0(s)$. In the assertion to be formulated $y(t, s)$ is the solution of the relay system (9.3) with initial condition $y(0, s) = y(s)$ and switching times $t_0, t_1, \dots$.

Theorem 10.1. *For values of t not in small fixed neighborhoods of the points $0, t_0, t_1, \dots$ the limit equalities*

$$\lim_{\varepsilon\to 0}(x(t, s, \varepsilon), y(t, s, \varepsilon)) = (\Phi(y(t, s)), y(t, s)), \tag{10.1}$$

$$\lim_{\varepsilon\to 0}(h_1(t, s, \varepsilon), h_2(t, s, \varepsilon)) = (h_{10}(t, s), h_{20}(t, s)) \tag{10.2}$$

hold uniformly with respect to $(x(s), y(s)) \in S(r_0)$ and $(h_1^0(s), h_2^0(s))$ in the unit ball about zero in the metric of E, where $h_{20}(t, s)$ is the solution of the mixed problem for (9.10), (9.11) with initial condition $h_{20}(0, s) = h_2^0(s)$, and $h_{10}(t, s) = \Phi'(y(t, s))h_{20}(t, s)$. The convergence in (10.1) and (10.2) is uniform with respect to the indicated values of t in a bounded interval.

Theorem 10.1 generalizes to the case under consideration the assertion obtained in §§1 and 2 on C^1-closeness between solutions of the limit system and the original system, and it enables us to draw conclusions about the stationary regimes of the boundary value problem (9.1), (9.2) from the structurally stable stationary regimes of system (9.3). Since the proof of Theorem 10.1 almost completely repeats (with certain natural changes connected with parabolicity) that of Theorem 1.1, in the following subsections we dwell only on its typical features.

The boundary value problem (9.1), (9.2) is called a *relaxation problem* if the relay system (9.3) has solutions with infinitely many switchings.

Suppose that for $y \in R$ we impose restrictions [5] on f and g such that the boundary value problem (9.1), (9.2) has a spatially homogeneous cycle $(x(t, \varepsilon), y(t, \varepsilon))$. In this case the boundary value problem (9.1), (9.2) is certainly a relaxation problem, and Lemma 9.1 and Theorem 10.1 enable us to describe the structure of a neighborhood, independent of ε, of the cycle $(x(t, \varepsilon), y(t, \varepsilon))$ when the latter loses stability. However, it is possible that the system of ordinary equations obtained from (9.1), (9.2) for $D_1 = 0$ and $D_2 = 0$ is not a relaxation system, while for some $D_1 \neq 0$ and $D_2 \neq 0$ the parabolic system is a relaxation system.

Obviously, the ideas developed here can be extended to more general classes of singularly perturbed parabolic systems with classical boundary conditions.

10.2. **Analysis of the system (9.1), (9.2) outside neighborhoods of the switching points.** We first consider the interval $\delta \leq t \leq t_0 - \delta$, where $\delta > 0$ is sufficiently small. For these t we find a formal integral manifold of system (9.1), (9.2) in the form of a series

(10.3)
$$x = \Phi_1(y) + \sum_{k=1}^{\infty} \varepsilon^k F_k(y).$$

Substituting (10.3) into (9.1) and (9.2) and equating coefficients of the same powers of ε, we get a recursive sequence of equations for determining the operators F_k. For example, to determine F_1 we have

(10.4)
$$LF_1 = \Phi_1'(y)(D_2\Delta y + g(\Phi_1(y), y)),$$

where L is the operator (9.5). From (10.4) we find F_1 as a Fréchet-smooth operator acting from $C^{2+\alpha}$ to $C^{4+\alpha}$. Proceeding analogously, we successively determine the Fréchet-smooth operators $F_k(y)$ acting from $C^{2k+\alpha}$ to $C^{2k+2+\alpha}$.

A rigorous meaning is assigned to the formal integral manifold (10.3) as follows. Fix the initial condition $(x(s), y(s)) \in S(r_0)$ of system (9.1), (9.2) and denote by $y_0(t, s)$ the solution of (9.6) with initial condition $y(s)$ given for $t = 0$. We then substitute (10.3) in (9.2) and look for a solution of the resulting equation in the form of a series

(10.5)
$$y(t, s, \varepsilon) = y_0(t, s) + \varepsilon y_1(t, s) + \varepsilon^2 y_2(t, s) + \cdots.$$

This procedure leads to a recursive sequence of linear nonhomogeneous boundary value problems of the form

(10.6)
$$\frac{\partial y_k}{\partial t} = D_2\Delta y_k + A(t)y_k + f_k(t, s), \qquad \left.\frac{\partial y_k}{\partial \nu}\right|_{\partial\Omega} = 0$$

with initial condition $y_k|_{t=0} = 0$ for determining the unknown coefficients $y_k(t, s)$ of (10.5). Here $A(t)$ is the Fréchet derivative of $g(\Phi_1(y), y)$ at $y = y_0(t, s)$, and f_k depends only on the y_m with indices $m \le k-1$ and is smooth with respect to s up to $t = 0$ if the initial condition satisfies $y(s) \in C^{2k+\alpha}$. According to known results in [35] and [36], this then tells us that the boundary value problem (10.6) is solvable in C^α.

Lemma 10.1. *Suppose that the initial condition $(x(s), y(s)) \in S(r_0)$ of the solution of system (9.1), (9.2) is such that $y(s) \in C^{2k+\alpha}$. Then, for $\delta \le t \le t_0 - \delta$,*

(10.7)
$$x(t, s, \varepsilon) = \Phi_1(y_0(t, s)) + \sum_{m=1}^{k-1} \varepsilon^m x_m(t, s) + O(\varepsilon^k),$$

(10.8)
$$y(t, s, \varepsilon) = y_0(t, s) + \sum_{m=1}^{k-1} \varepsilon^m y_m(t, s) + O(\varepsilon^k)$$

in the C^α-metric, where (10.8) is a partial sum of series (10.5), and (10.7) is the result of substituting (10.8) in (10.3) and then expanding again with respect to ε. Asymptotic equalities analogous to (10.7) and (10.8) are valid also for the Fréchet derivatives of the functions $x(t, s, \varepsilon)$ and $y(t, s, \varepsilon)$ with respect to the initial conditions.

The proof of the lemma is standard and reduces to an analysis of the variational system for (9.1), (9.2) on the partial sums of series (10.7) and (10.8). Without dwelling on the computations, we note only that the main difficulty here lies in getting the estimate

(10.9)
$$\|U(t, \tau, \varepsilon)\|_{C^\alpha \to C^\alpha} \le M \exp\{-\gamma(t-\tau)/\varepsilon\}, \qquad M, \gamma > 0,$$

for the solving operator of the boundary value problem

$$\varepsilon \frac{\partial h}{\partial t} = D_1 \Delta h + f_y'(\Phi_1(y_0(t, s)), y_0(t, s))h, \qquad \left.\frac{\partial h}{\partial \nu}\right|_{\partial\Omega} = 0.$$

The proof of this estimate coincides completely with the proof given, for example, in [37] of the analogous inequality for the case of ordinary differential equations.

Accordingly, Lemma 10.1 enables us to clear up the asymptotic properties of the parabolic system (9.1), (9.2) on the interval $\delta \le t \le t_0 - \delta$. Here the smoothness requirement on the component $y(s)$ of the initial condition is not a restriction, because, for example, even for $t = \delta$ the functions $x(t, s, \varepsilon)$ and $y(t, s, \varepsilon)$ have a sufficiently large (limited by the smoothness of f and g) number of derivatives with respect to s that are uniformly bounded with respect to ε (this is easily deduced from an analysis of the corresponding trajectory $(x(t, s, \varepsilon), y(t, s, \varepsilon))$ of the variational system and (10.9)). All the foregoing is valid for $t_0 + \delta \le t \le t_1 - \delta$, and so on.

10.3. **System (9.1), (9.2) in a neighborhood of a switching point.** As in §§1 and 2, the proof of Theorem 10.1 reduces in view of the results in §10.2 to an analysis of neighborhoods of switching points. Therefore, it suffices to consider a neighborhood of the point $(\Phi_1(y(t_0, s)), y(t_0, s))$.

In system (9.1), (9.2) we introduce new coordinates completely analogous to the coordinates in §1.4. With this goal we move the zero of the space E to the point $(\Phi_1(y(t_0, s)), y(t_0, s))$ and then let $x = (\xi_1, \xi_2)$ and $y = (\eta_1, \eta_2)$. Here $\xi_1 \in R$ "measures" distances along the one-dimensional integral manifold of equation (9.4) for $y = y(t_0, s)$, and the vector $\xi_2 \in C^\alpha$ is orthogonal in the sense of $L_2(\Omega)$ to the eigenfunction $e_0(s)$ of the operator (9.5); $\eta_1 \in R$ corresponds to the direction $\partial y(t_0 - 0, s)/\partial t$, and the vector $\eta_2 \in C^\alpha$ satisfies the equality $l_0(\eta_2) = 0$. After certain similarity transformations and a time normalization, the parabolic system (9.1), (9.2) takes in the new coordinates the form

$$\varepsilon \dot{\xi}_1 = F_1(\xi_1, \xi_2, \eta_1, \eta_2), \tag{10.10}$$

$$\varepsilon \frac{\partial \xi_2}{\partial t} = L\xi_2 + F_2(\xi_1, \xi_2, \eta_1, \eta_2), \qquad \left.\frac{\partial \xi_2}{\partial \nu}\right|_{\partial\Omega} = 0, \tag{10.11}$$

$$\dot{\eta}_1 = G_1(\xi_1, \xi_2, \eta_1, \eta_2), \tag{10.12}$$

$$\frac{\partial \eta_2}{\partial t} = D_2 \Delta \eta_2 + G_2(\xi_1, \xi_2, \eta_1, \eta_2), \qquad \left.\frac{\partial \eta_2}{\partial \nu}\right|_{\partial\Omega} = 0. \tag{10.13}$$

Here L is the operator (9.5) with $y = y(t_0, s)$, the functionals F_1 and G_1 and the operators F_2 and G_2 are Fréchet-smooth in a neighborhood of the origin, and

$$F_1 = \xi_1^2 + \eta_1 + \cdots, \quad F_2(\xi_1, 0, 0, 0) \equiv 0, \quad \partial \xi_2 F_2(0, 0, 0, 0) = 0, \tag{10.14}$$

$$G_1(0, 0, 0, 0) = 1, \quad G_2(0, 0, 0, 0) = 0, \tag{10.15}$$

where the dots stand for quadratic terms other than ξ_1^2 and terms of higher order of smallness.

Thus, in a neighborhood of a switching point we have obtained a system (10.10)–(10.13) analogous to system (1.14) in its properties. If we now construct a formal asymptotic expression for its trajectories, then it is not hard to verify that formulas analogous to (1.15) and (1.16) result. The sequence of subsequent actions is as follows: we linearize system (10.10)–(10.13) on the formally constructed trajectory, we study the properties of the resulting linear system (here all the computations coincide almost completely with those in §§1 and 2), and then we use the scheme described in §4.7 to prove the asymptotic formulas being used. Thus, both limit equalities in Theorem 10.1 are established at the same time.

§11. A criterion for stability of a homogeneous cycle in a medium with diffusion

11.1. **General theorem.** Suppose that $x \in R^k$ in the boundary value problem (9.1), (9.2), and the point model

$$\varepsilon \dot{x} = f(x, y), \qquad \dot{y} = g(x, y) \tag{11.1}$$

corresponding to it is such that Lemma 1.2 applies. Then by Theorem 1.2 the boundary value problem (9.1), (9.2) has a spatially homogeneous cycle $(x(t, \varepsilon), y(t, \varepsilon))$ that is exponentially stable in the framework of system (11.1). We pose the question of its stability in the framework of the distributed model (9.1), (9.2).

In the phase space E for system (9.1), (9.2) we have an analogue of the Andronov-Witt theorem reducing the question of stability of a homogeneous cycle of it to an analysis of the corresponding variational system. Representing the solutions of the latter as Fourier series in the eigenfunctions of the Laplace operator, we arrive at the equations

$$\varepsilon \dot{h}_1 = [A_{11}(t, \varepsilon) - zD_1]h_1 + A_{12}(t, \varepsilon)h_2, \tag{11.2}$$

$$\dot{h}_2 = A_{21}(t, \varepsilon)h_1 + [A_{22}(t, \varepsilon) - zD_2]h_2, \tag{11.3}$$

where $z = \lambda_k$, $k = 1, 2, \dots$, the λ_k are the eigenvalues of the operator $-\Delta$ for Neumann boundary conditions, and the matrices A_{11}, A_{12}, A_{21}, and A_{22} are the same as in (1.10). Since the domain Ω and the matrix diffusion coefficients in (9.1), (9.2) can be changed arbitrarily, it is convenient to assume that the parameter $z \geq 0$ in system (11.2), (11.3) changes continuously. The object of our further analysis is thus the linear system (11.2), (11.3) with $T(\varepsilon)$-periodic coefficients and continuous parameter $z \geq 0$.

We study the distribution of the multipliers of system (11.2), (11.3) under the following additional restriction.

Localization Condition. Assume that for all $z > 0$ the eigenvalues of the matrix $A(x, y) - zD_1$, where $A(x, y) = f'_x$, lie to the left of the imaginary axis for $(x, y) \in \Gamma_{1-} + \Gamma_{10} + \Gamma_{2-} + \Gamma_{20}$, and for $(x, y) \in \Gamma_{10} + \Gamma_{20}$ and small z an eigenvalue close to zero has order of smallness z.

We note at once that in the absence of the localization condition, when for some $z_0 > 0$ and $(x_0, y_0) \in Ц_0$ (where $Ц_0$ is a cycle of the hybrid system (1.7)) the matrix $A(x_0, y_0) - z_0 D_1$ has an eigenvalue with positive real part, and system (11.2), (11.3) has a multiplier with modulus of order $\exp(c/\varepsilon)$, $c > 0$.

Let

(11.4)

$$\varphi(\sigma) = \int_{-\infty}^{\infty} (s^2 + v_0(s))^{-2} \exp\left\{-\int_s^{\infty} (u^2 + v_0(u))^{-2}\,du - \sigma(\Omega_0 - v_0(s))\right\} ds, \qquad \sigma \geq 0,$$

where the function v_0 and the constant $\Omega_0 > 0$ are the same as in (4.14). Further, let

$$r = (D_1\alpha, \beta)(|q_1 q_2|)^{-1/3},$$

where

$$q_1 = (\dot{y}_0(t_p - 0), B^* \beta), \qquad q_2 = \frac{1}{2} \sum_{j_1, j_2, j_3 = 1}^{k} \beta_{j_1} \frac{\partial^2 f_{j_1}}{\partial x_{j_2} \partial x_{j_3}} \alpha_{j_2} \alpha_{j_3},$$

$$\alpha = (\alpha_1, \dots, \alpha_k), \qquad \beta = (\beta_1, \dots, \beta_k), \qquad f = (f_1, \dots, f_k),$$

and the derivatives are computed on the cycle Ц$_0$ at the switching times $t_p - 0$ (see §1.1 for the notation).

Theorem 11.1. *Suppose that the localization condition holds in addition to the conditions of Lemma* 1.2. *Then* k *multipliers of system* (11.2), (11.3) *tend to zero as* $\varepsilon \to 0$ *uniformly with respect to* $z \geq 0$, *and the remaining* m *multipliers are determined up to* $\varepsilon^{1/3} \ln(1/\varepsilon)$ *from the* T_0*-periodic problem*

$$\dot{h} = [D(t) - zD_2 - C(t)(A(t) - zD_1)^{-1} B(t)] h, \tag{11.5}$$

$$h(t_p + 0) = h(t_p - 0) + \varphi(rz/\varepsilon^{1/3}) a(b, h(t_p - 0)). \tag{11.6}$$

It follows from (11.6) that the most interesting case is when the quantities $(D_1\alpha, \beta)$ are small, or, more precisely, have order $\varepsilon^{1/3}$. Under this condition the multipliers of system (11.5), (11.6) change in a complicated way, in particular, passing through zero, as the diffusion coefficients change.

In large respect system (11.2), (11.3) is analogous to system (1.10), and it can be analyzed according to almost the same scheme. However, if the parameter z is small (of order $\varepsilon^{1/3}$), then a boundary layer appears, and the principal parts of the multipliers of system (11.2), (11.3) are determined from (11.5) and (11.6) up to $\varepsilon^{1/3} \ln(1/\varepsilon)$ (for finite positive z the principal parts of the multipliers of system (11.2), (11.3) are determined from (11.5) and (11.6) up to $\varepsilon^{2/3}$).

11.2. **The two-dimensional case.** For $x, y \in R$, i.e., when we consider a relaxation cycle on the plane, we can get an explicit formula for the solution to the problem of its diffusion instability (this is connected with the fact that equation (11.6) is scalar). And since it is just this case that is encountered in an overwhelming number of biophysical problems (see [26] and [27]), it makes sense to dwell separately on it.

Suppose that for $x, y \in R$ the smooth functions f and g in (11.1) are such that [5] system (11.1) has an exponentially stable relaxation cycle $(x(t, \varepsilon), y(t, \varepsilon))$. For simplicity we confine ourselves to the case when it has only two regions of slow motions (see Figure 4). Further, in place of system (11.2), (11.3) with the single parameter z it is convenient to consider here a two-parameter system

$$\begin{aligned} \varepsilon \dot{h}_1 &= -z_1 h_1 + a(t, \varepsilon) h_1 + b(t, \varepsilon) h_2, \\ \dot{h}_2 &= -z_2 h_2 + c(t, \varepsilon) h_1 + d(t, \varepsilon) h_2, \end{aligned} \tag{11.7}$$

where

$$z_1 = \lambda_k D_1, \qquad z_2 = \lambda_k D_2, \qquad k = 1, 2, \dots,$$

$$a = \partial f/\partial x, \quad b = \partial f/\partial y, \quad c = \partial g/\partial x, \quad d = \partial g/\partial y,$$

and the derivatives are taken at $x = x(t, \varepsilon)$, $y = y(t, \varepsilon)$. The object of the subsequent analysis will be system (11.7), in which the parameters z_1 and z_2 are assumed to change continuously. The study of its stability properties for all $z_1, z_2 \geq 0$ is equivalent to the study of the question of stability of a homogeneous cycle of boundary value problem (9.1), (9.2) for arbitrary diffusion coefficients D_1 and D_2 in an arbitrary domain Ω.

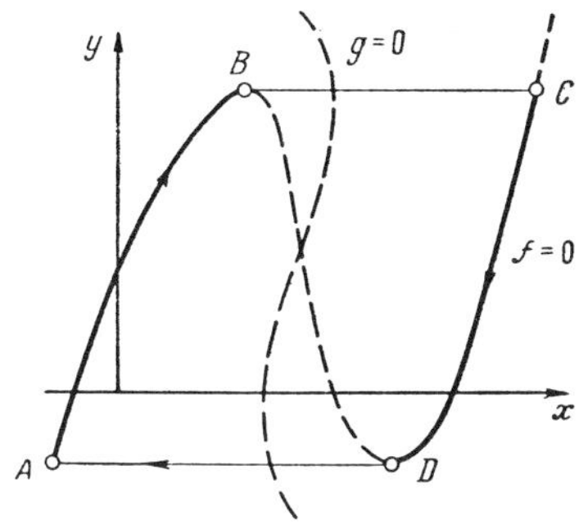

FIGURE 4

To formulate the main result we set

$$\Phi_0(\sigma) = \frac{1}{T_0}\{\ln|(1 - g(B)/g(C))\varphi(r_1\sigma) + g(B)/g(C)|$$

$$+ \ln|(1 - g(D)/g(A))\varphi(r_2\sigma) + g(D)/g(A)|\}, \qquad \sigma \geq 0, \tag{11.8}$$

$$r_1 = \left(\frac{2}{|f''_{xx}(B)f'_y(B)g(B)|}\right)^{1/3}, \qquad r_2 = \left(\frac{2}{|f''_{xx}(D)f'_y(D)g(D)|}\right)^{1/3},$$

where T_0 is the zero approximation of the period of the coefficients of (11.7), the function φ is defined by (11.4), and A, C and B, D are the drop points and breakaway points of the cycle as $\varepsilon \to 0$ (see Figure 4). In the general case the function (11.8) has two vertical asymptotes $\sigma = \sigma_1, \sigma_2$ and is monotone outside the interval $[\sigma_1, \sigma_2]$.

Let μ_0 be the multiplier of system (11.7) of greatest modulus (the modulus of the other does not exceed $N_1 \exp(-N_2/\varepsilon)$, where $N_1, N_2 > 0$, and does not depend on z_1, z_2, nor ε). The next assertion follows from Theorem 11.1.

Theorem 11.2. *For any $\delta > 0$ there exist sufficiently small positive numbers $\varepsilon_0 = \varepsilon_0(\delta)$ and $\delta_0 = \delta_0(\delta)$ such that inside the "horns"*

$$0 < \varepsilon \leq \varepsilon_0, \qquad \varepsilon^{1/3}(\sigma_j - \delta_0) \leq z_1 \leq \varepsilon^{1/3}(\sigma_j + \delta_0), \qquad j = 1, 2,$$

the inequality $|\mu_0| \leq \delta$ holds. And outside them μ_0 has order 1 as $\varepsilon \to 0$, it is positive below and above the "horns", and negative between them. Moreover, there exists a function $F(\varepsilon, z_1)$ ($F(\varepsilon, 0) \equiv 0$) such that for $z_2 - F(\varepsilon, z_1) > 0$ (< 0) this multiplier is less (greater) in modulus than 1; the function has the representation (which is asymptotically best possible with respect to ε)

$$F(\varepsilon, z_1) = F_0(z_1) + \Phi_0(z_1/\varepsilon^{1/3}) - F_0(+0) + \varepsilon^{1/3}\ln\frac{1}{\varepsilon}\Phi(\varepsilon, z_1), \tag{11.9}$$

where

$$F_0(z_1) = \frac{1}{T_0}\int_0^{T_0}(d_0(t) - b_0(t)c_0(t)[a_0(t) - z_1]^{-1})\,dt \tag{11.10}$$

(a_0, b_0, c_0, and d_0 are the zero approximations of a, b, c, and d), and the function Φ is bounded uniformly with respect to its variables.

We describe some properties of the function (11.10) necessary for what follows. First, it is obvious that

$$\lim_{z_1\to\infty} F_0(z_1) = \frac{1}{T_0}\int_0^{T_0}\frac{\partial g}{\partial y}\,dt. \tag{11.11}$$

Second, after certain transformations we see that

$$\lim_{z_1 \to +0} F_0(z_1) = \frac{1}{T_0} \ln \frac{g(B)g(D)}{g(A)g(C)}, \tag{11.12}$$

where the points A, B, C, and D are shown in Figure 4. We note further that the integral

$$F_0'(z_1) = -\frac{1}{T_0} \int_0^{T_0} \frac{\partial f}{\partial y} \frac{\partial g}{\partial x} \left(\frac{\partial f}{\partial x} - z_1 \right)^{-2} dt \tag{11.13}$$

diverges as $z_1 \to +0$ in the general case.

Theorems 10.1 and 11.2 imply

Theorem 11.3. *Suppose that $\lambda_k D_2 > F_0(\lambda_k D_1)$ for all $k = 1, 2, \dots$ (here the λ_k are the positive eigenvalues, in increasing order, of the operator Δ for Neumann boundary conditions). Then there exists an $\varepsilon_0 > 0$ such that for $0 < \varepsilon \le \varepsilon_0$ a homogeneous relaxation cycle of the boundary value problem (9.1), (9.2) is orbitally exponentially stable, and as $t \to \infty$ all the solutions of (9.1), (9.2) with initial conditions in some (independent of ε) neighborhood of it in the phase space E tend to it. But if $\lambda_{k_0} D_2 < F_0(\lambda_{k_0} D_1)$ for some k_0, then this cycle is unstable.*

We dwell separately on the case (which is important from the applied point of view [26]) when $D_2 = 0$ and a boundary condition for y becomes unnecessary.

Theorem 11.4. *Suppose that $D_2 = 0$ and the right-hand side of (11.12) is positive. Then there exist a positive number ε_0 and a positive integer k_0 independent of ε_0 such that for $0 < \varepsilon \le \varepsilon_0$ and for each $k \ge k_0$ the system (11.7) has an exponentially increasing solution.*

This assertion also follows from Theorem 11.2. We underscore that under the conditions of Theorem 11.4 the characteristic exponents on the high modes $e_k(s)$ ($e_k(s)$ is an eigenfunction of $-\Delta$ corresponding to the eigenvalue λ_k) have positive real parts of order 1, and this circumstance allows us to refer to explosive diffusion instability of the homogeneous cycle. It is interesting that an analogous situation can hold also in the case of harmonic auto-oscillations (see [38] and [39]).

In conclusion we mention that under the conditions of Theorem 11.4 the parabolic system (9.1), (9.2) with $D_2 = 0$ has in a neighborhood of a homogeneous cycle an exponentially stable infinite-dimensional invariant integral manifold corresponding to the characteristic exponents of the varational system with nonnegative real parts. The proof is based on knowledge of the spectral properties of the monodromy operator generated by the variational system, and on the standard technique for proving the existence of integral manifolds [40], which was extended to parabolic boundary value problems in [41].

11.3. **Complete asymptotics of the function $F(\varepsilon, z_1)$ for $z_1 \ge z_{10} > 0$.** Here we use the results in §5 together with a certain special device to clarify in full scope the asymptotic properties of the function (11.9) for $z_1 \ge z_{10}$, where $z_{10} > 0$ is an arbitrarily small fixed number.

Let us consider the auxiliary four-dimensional system

$$\begin{aligned} \varepsilon \dot{x}_1 &= z_1(x_2 - x_1)/2 + f(x_1, y_1), \\ \varepsilon \dot{x}_2 &= z_1(x_1 - x_2)/2 + f(x_2, y_2), \\ \dot{y}_1 &= z_2(y_2 - y_1)/2 + g(x_1, y_1), \\ \dot{y}_2 &= z_2(y_1 - y_2)/2 + g(x_2, y_2), \end{aligned} \tag{11.14}$$

which is called the *biolocal model* for the boundary value problem (9.1), (9.2). Obviously, system (11.14) has a relaxation cycle with components $x_1 \equiv x_2 = x(t, \varepsilon)$ and $y_1 \equiv y_2 = y(t, \varepsilon)$, and we can see after an uncomplicated check that all the conditions of §1 hold here for $z_1 \geq z_{10} > 0$. It is important that the linear system (11.7) is obtained upon linearization of system (11.14) on the indicated cycle and projection on the eigenvectors of the Laplace difference operator. Therefore, the computation of the function $F(\varepsilon, z_1)$ reduces to an analysis of the Poincaré return operator $\Pi(\varepsilon, z_1, z_2)$ introduced in §5.

We noted in §5 that for the components $x = (x_1, x_2)$ and $y = (y_1, y_2)$ of the operator $\Pi(\varepsilon, z_1, z_2)$ there are formulas analogous to the asymptotic formulas (5.2) and (5.3). Therefore, $F(\varepsilon, z_1)$, which is determined from the equation

$$\det(\Pi'|_{x=x(\varepsilon),\ y=y(\varepsilon)} - I) = 0,$$

where $x(\varepsilon)$ and $y(\varepsilon)$ are the initial conditions of the relaxation cycle and I the identity matrix, admits the asymptotic expansion

$$F(\varepsilon, z_1) = F_0(z_1) + \sum_{n=2}^{\infty} \varepsilon^{n/3} \sum_{\nu=0}^{\pi(n-2)} F_{n,\nu}(z_1) \ln^{\nu} \frac{1}{\varepsilon}, \tag{11.15}$$

valid uniformly with respect to $z_1 \geq z_{10}$. Moreover, dividing the first equation in (11.7) by z_1 and letting z_1 go to ∞, we conclude that

$$\lim_{z_1 \to \infty} F(\varepsilon, z_1) = \frac{1}{T(\varepsilon)} \int_0^{T(\varepsilon)} \frac{\partial g}{\partial y}(x(t, \varepsilon), y(t, \varepsilon))\, dt. \tag{11.16}$$

If we now compare (11.15) and (11.16), we see that all the coefficients $F_{n,\nu}(z_1)$ in (11.15) have as $z_1 \to \infty$ finite limits that are equal to the corresponding terms of the expansion of the right-hand side of (11.16).

Theorem 11.5. *The function* $F(\varepsilon, z_1)$ *has the asymptotic expansion* (11.5), *which is uniform with respect to* $z_1 \geq z_{10} > 0$, *where* $z_{10} > 0$ *is any fixed number.*

We point out that Theorem 11.2 cannot be obtained with the help of the constructions presented above, because for $z_1 = 0$ the matrix $A(x, y) = f'_x$ of system (11.14) at a breakaway point of the relaxation cycle has zero eigenvalue of multiplicity two, i.e., one of the conditions of §1 is violated.

§12. Some mathematical models in biophysics

12.1. **Analysis of the mathematical models.** The results in §11 make it simple for us to analyze a number of nominal mathematical models in biophysics [26]. Suppose that in (9.1), (9.2)

$$f = y + x - \tfrac{1}{3}x^3, \qquad g = -x - by + a, \qquad a, b > 0, \tag{12.1}$$

i.e., consider the Bonhoeffer-van der Pol model. It can be verified without trouble that in the case (12.1) the right-hand sides of (11.11) and (11.12) are negative, while the right-hand side of (11.13) is positive. Therefore, $F_0(z_1)$ increases monotonically and tends to a negative limit as $z_1 \to \infty$. Therefore, $F_0(z_1) < 0$ for all $z_1 > 0$, and in view of Theorem 11.3 the relaxation cycle is stable for any diffusion coefficients. The analogous fact is valid also for the boundary value problems (9.1), (9.2) with

(12.2) $$f = x - x^3 - y, \qquad g = x - B,$$

(12.3) $$f = x - x^3 - y, \qquad g = B - y + \gamma x,$$

(12.4) $$f = A - x(B+1) + x^2(y - x), \qquad g = A - x,$$

where all the parameters are positive. We remark that in the case (12.2) the corresponding boundary value problem is called the Fitzhugh-Nagumo model, in the case (12.3) it is called the Nagumo model, and in the case (12.4) the "Brusselator" model.

From the other boundary value problems (9.1), (9.2) encountered in the collection [26] of papers we single out the article of Yakhno, who assumed that $f = x - x^3 - y$ and $g = x + 1.5y - 2.7y^2$ or $g = x + 3y - 2.7y^2$, with $D_2 = 0$. By Theorem 11.4, explosive diffusion instability of a homogeneous relaxation cycle is observed for these f and g. This circumstance enabled Yakhno to numerically construct relatively complex spatially nonhomogeneous steady-state regimes.

We remark further that in the case of the models (12.1)–(12.4) the conclusion of stability for a homogeneous cycle is confirmed also by Theorem 11.2; however, the analysis here is fairly complicated and was carried out with the use of numerical methods.

12.2. **Some conclusions.** In [42]–[44] parabolic systems with auto-oscillatory ordinary part were divided according to the character of their dynamics into simple physical systems, complex physical systems, and biological systems. In particular, the simple physical systems included the parabolic systems with a homogeneous cycle stable under any change of diffusion coefficients. There cannot be complex spatially nonhomogeneous steady-state regimes for this class of parabolic systems, because appearance of the latter is connected with desynchronization of auto-oscillations, which, of course, does not exclude such systems having well-ordered regimes of the "traveling wave" type on the circle or "reverberators" in the disk and close domains. In complex physical and biological parabolic systems the spatially nonhomogeneous steady-state regimes are qualitatively more complex in their structure, and it turns out [45] that their appearance is connected not only with a loss of stability of the homogeneous cycle under a suitable decrease of the diffusion coefficients, but also with the phenomenon of self-organization, when regimes organized in a very complex and ordered manner are created "from thin air".

As already mentioned in [26] and [27], the boundary value problem (9.1), (9.2) is a basic model for a large number of biophysical problems. The situation is in stark contrast to the one presented above. This is connected mainly with the fact that the authors of [26] and [27] took modifications of known models of generators of discontinuous oscillations for modeling complex phenomena [6], i.e., they actually realized a transition to simple physical objects. At the same time, it is a simple matter to avoid trivial models. By Theorem 11.3, it is necessary only that $F_0(z_1)$ be positive for relatively small z_1 and negative for all sufficiently large z_1—it is only for $F_0(z_1)$ constructed in this way that the homogeneous cycle is stable for all sufficiently large values of D_1, while it loses stability when D_1 decreases.

Chapter 3. The structure of a neighborhood of the relaxation cycle

The assertions of Theorems 1.2 and 5.1 on the existence, uniqueness, and asymptotics of the relaxation cycle $x(t, \varepsilon)$, $y(t, \varepsilon)$ with fixed zero approximation $(x_0(t), y_0(t))$ of period T are valid also in the case when some of the multipliers of the periodic problem (1.4), (1.5) (which are different from 1) lie on the unit circle. However, the questions of the structure of neighborhoods of the cycles $x(t, \varepsilon)$, $y(t, \varepsilon)$, and $y_0(t)$ of the respective systems (1.9) and (1.1) remain open here.

Recall that for general systems of ordinary differential equations and for parabolic boundary value problems of reaction-diffusion type the question posed can be answered, for example, with the help of the algorithm developed in [25] and [46] for constructing the normal form in a neighborhood of a cycle. But this way of proceeding does not take into account the singularity of the problem for relaxation systems. Below we show that in this case the Poincaré operator of return along trajectories of the system in a neighborhood of the cycle serves as a regular object of investigation: as $\varepsilon \to 0$ it is not only C^1-convergent to the corresponding object of the relay system, but also C^k-convergent for any $k \geq 1$.

Accordingly, the problem arises of extending the normalization algorithm in [25] and [46] to mappings, and this problem is solved in the present chapter. Here we give some applications to relay and relaxation systems.

§13. The normal form of a mapping

13.1. **Statement of the problem.** In a real Banach space E we consider the mapping

$$u \mapsto \Pi(u, \varepsilon), \qquad 0 < \varepsilon \ll 1, \tag{13.1}$$

where Π is an operator smooth with respect to its variables, and $\Pi(0, \varepsilon) \equiv 0$. We introduce the linear operator $V_0 = D_u\Pi|_{u=0, \varepsilon=0}$ (D is the Fréchet derivative) and assume that it has $2m$ eigenvalues, equal to $\exp(\pm i\omega_k)$, $0 < \omega_k < \pi$, $k = 1, \dots, m$, while the remainder of its spectrum lies in a disk of radius less than 1. Further, corresponding to them are exactly $2m$ linearly independent eigenvectors $a_k = a_k^1 + ia_k^2$, $\overline{a}_k = a_k^1 - ia_k^2$, with

$$l_k(a_n) = \delta_{kn}, \qquad l_k(\overline{a}_n) = 0, \qquad k, n = 1, \dots, m,$$

where $V_0^* l_k = \exp(-i\omega_k) l_k$, $k = 1, \dots, m$.

Under these assumptions the mapping (13.1) has in a neighborhood of the zero fixed point a $2m$-dimensional exponentially stable local invariant manifold (see [40] and [41]), and this allows us to reduce the problem to a finite-dimensional one. Recall that in [25] and [26] the familiar Poincaré-Lyapunov method was used to construct an integral manifold in a neighborhood of a cycle along with a system of ordinary differential equations on it, with the latter constructed directly in Poincaré-Dulac normal form. Below, this device is extended to the case when we are interested in the structure of a neighborhood of a fixed point of the mapping.

We look for the equation of the $2m$-dimensional invariant manifold of the mapping (13.1) in the form

$$\begin{aligned} u = {} & \sum_{j=1}^{m} \xi_j[a_j \exp(i\omega_j\tau_j) + \overline{a}_j \exp(-i\omega_j\tau_j) + \varepsilon w_j] \\ & + \sum_{k,j=1}^{m} w_{kj}\xi_k\xi_j + \sum_{k,j,p=1}^{m} w_{kjp}\xi_k\xi_j\xi_p + U(\omega\tau, \xi, \varepsilon), \end{aligned} \tag{13.2}$$

and we try to determine the mapping on it directly in normal form

$$\xi_k \mapsto \xi_k + \varphi_k, \qquad \tau_k \mapsto \tau_k + 1 + \psi_k, \qquad k = 1, \dots, m, \tag{13.3}$$

where

$$\varphi_k = \varepsilon\varphi'_k\xi_k + \varepsilon \sum_{\omega_j\in\Omega_k^1} a_{kj}\cos(\omega_j\tau_j - \omega_k\tau_k + \gamma_{kj})\xi_j$$
$$+ \sum_{\omega_j,\,\omega_l\in\Omega_k^2} a_{kjl}\cos(k_j\omega_j\tau_j + k_l\omega_l\tau_l - \omega_k\tau_k + \gamma_{kjl})\xi_j^{|k_j|}\xi_l^{|k_l|}$$
$$+ \sum_{\omega_j,\,\omega_l,\,\omega_p\in\Omega_k^3} a_{kjlp}\cos(k_j\omega_j\tau_j + k_l\omega_l\tau_l + k_p\omega_p\tau_p - \omega_k\tau_k + \gamma_{kjlp})\xi_j^{|k_j|}\xi_j^{|k_l|}\xi_p^{|k_p|}$$
$$+ \xi_k\sum_{j=1}^{m}\varphi_{kj}^3\xi_j^2 + \Phi_k(\omega\tau,\,\xi,\,\varepsilon)$$

and

$$\psi_k = \varepsilon\psi_k^1 + \varepsilon \sum_{\omega_j\in\Omega_k^1} \frac{a_{kj}}{\omega_k}\sin(\omega_j\tau_j - \omega_k\tau_k + \gamma_{kj})\frac{\xi_j}{\xi_k}$$
$$+ \sum_{\omega_j,\,\omega_l\in\Omega_k^2} \frac{a_{kjl}}{\omega_k}\sin(k_j\omega_j\tau_j + k_l\omega_l\tau_l - \omega_k\tau_k + \gamma_{kjl})\frac{\xi_j^{|k_j|}\xi_l^{|k_l|}}{\xi_k}$$
$$+ \sum_{\omega_j,\,\omega_l,\,\omega_p\in\Omega_k^3} \frac{a_{kjlp}}{\omega_k}\sin(k_j\omega_j\tau_j + k_l\omega_l\tau_l + k_p\omega_p\tau_p - \omega_k\tau_k + \gamma_{kjlp})\frac{\xi_j^{|k_j|}\xi_l^{|k_l|}\xi_p^{|k_p|}}{\xi_k}$$
$$+ \sum_{j=1}^{m}\psi_{kj}^2\xi_j^2 + \frac{1}{\xi_k}\Psi_k(\omega\tau,\,\xi,\,\varepsilon).$$

Here $\Omega_k^1 = \{\omega_j\colon \omega_k = \omega_j,\ j \neq k\}$, $\Omega_k^2 = \{\omega_j,\,\omega_l\colon \omega_k = k_j\,\omega_j + k_l\omega_l \pmod{2\pi},\ |k_j|+|k_l| = 2\}$, the set Ω_k^3 is defined similarly (identity resonances $\omega_k = \omega_k+\omega_j-\omega_j$ are excluded, of course), and the smooth functions Φ_k and Ψ_k and the vector-valued function U (with values in E) are periodic with period 2π in the vector argument $\omega\tau = (\omega_1\tau_1,\,\dots,\,\omega_m\tau_m)$ and satisfy the inequalities

$$\|U\|_E,\ |\Phi_k|,\ |\Psi_k| \leq M(\varepsilon^2\|\xi\| + \varepsilon\|\xi\|^2 + \|\xi\|^4), \tag{13.4}$$

where M is a constant, $\|*\|_E$ and $\|*\|$ are the norm in E and the Euclidean norm, and $\xi = (\xi_1,\,\dots,\,\xi_m)$.

13.2. **A description of the algorithm and the main result.** The principal terms of (13.2) and (13.3) are found in parallel upon realization of the following algorithm. We form the expression

$$\tilde{u} = F(u,\,\varepsilon), \tag{13.5}$$

where u is the right-hand side of (13.2), and $\tilde{u}$ is the right-hand side of (13.2) with ε_k and τ_k replaced by $\xi_k+\varphi_k$ and $\tau_k+1+\psi_k$. Equating successively the coefficients of the powers $\varepsilon\xi_1,\,\dots,\,\varepsilon\xi_m,\,\xi_1^2,\,\dots,\,\xi_m^3$ in (13.5), we arrive at equations of the form

$$v(\omega(\tau+1)) = V_0 v(\omega\tau) + f(\omega\tau), \tag{13.6}$$

where $\omega(\tau+1) = (\omega_1(\tau_1+1),\,\dots,\,\omega_m(\tau_m+1))$, and f is a trigonometric polynomial in $\omega\tau$ with coefficients in E. We look for a solution of (13.6) with the same dependence on $\omega\tau$ as f. As a result the problem is reduced to the solution of several equations of the form

$$\exp(i\kappa)z = V_0 z + g, \tag{13.7}$$

where κ is a linear combination of the numbers ω_k, with integer coefficients. The constants on the right-hand side of (13.3), on which the vector g depends linearly at each step, are determined from the solvability conditions (13.7).

The main result is the following assertion.

Theorem 13.1. *In a neighborhood of the zero fixed point, formula* (13.2) *determines a* $2m$*-dimensional exponentially stable local invariant manifold of the mapping* (13.1), *and* (13.3) *is the mapping on it.*

The proof is based on a method in [47] that combines the Bogolyubov-Mitropol'skiĭ reduction principle [17] with the normalizing substitution technique. We describe its main steps.

In (13.1) let $u = \eta + v$, where $\eta = P(\varepsilon)u$, $v = u - P(\varepsilon)u$, and $P(\varepsilon)$ is the root projection of the operator $V(\varepsilon) = D_u\Pi(0, \varepsilon)$, corresponding to the eigenvalues of it that become $\exp(\pm i\omega_k)$, $k = 1, \ldots, m$, when $\varepsilon = 0$. Further, we identify the finite-dimensional space $P(\varepsilon)E$ with R^{2m} and assume that $\eta \in R^{2m}$. In the new coordinates the mapping (13.1) becomes

$$\eta \mapsto B(\varepsilon)\eta + G_1(\eta, v, \varepsilon), \tag{13.8}$$

$$v \mapsto L(\varepsilon)v + G_2(\eta, v, \varepsilon), \tag{13.9}$$

where $L(\varepsilon) = (I - P(\varepsilon))V(\varepsilon)(I - P(\varepsilon))$, the matrix $B(0)$ has eigenvalues $\exp(\pm i\omega_k)$, $k = 1, \ldots, m$, with a complete collection of eigenvectors e_k and $\overline{e}_k$, respectively, and the mappings

$$G_1(*, *, \varepsilon)\colon R^{2m} \oplus (I - P(\varepsilon))E \to R^{2m},$$

$$G_2(*, *, \varepsilon)\colon R^{2m} \oplus (I - P(\varepsilon))E \to (I - P(\varepsilon))E,$$

which are smooth with respect to their variables, are defined in a sufficiently small neighborhood of the zero in the space $R^{2m} \oplus (I - P(\varepsilon))E$ and vanish at $\eta = 0$, $v = 0$ together with their first Fréchet derivatives with respect to η and v.

It follows from results in [40] and [41] that the $2m$-dimensional exponentially stable invariant manifold of the mapping (13.1) is given by

$$v = H(\eta, \varepsilon). \tag{13.10}$$

Here the vector-valued function H, which is smooth in its variables and takes values in $(I - P(\varepsilon))E$, is such that

$$H(0, \varepsilon) \equiv 0, \qquad D_\eta H(0, \varepsilon) \equiv 0.$$

Substituting (13.10) in (13.8), we get a finite-dimensional mapping on the integral manifold of the mapping (13.1):

$$\eta \mapsto B(\varepsilon)\eta + G(\eta, \varepsilon), \tag{13.11}$$

where the vector-valued function G has order of smallness greater than 1 at zero.

To conclude the proof we should make in (13.11) the change of variables

$$\eta = \sum_{j=1}^{m} \xi_j[e_k \exp(i\omega_k\tau_k) + \overline{e}_k \exp(-i\omega_k\tau_k) + \varepsilon\eta_j] + \sum_{k,j=1}^{m} \eta_{kj}\xi_k\xi_j + \sum_{k,j,p=1}^{m} \eta_{kjp}\xi_k\xi_j\xi_p,$$

which is constructed with the help of the algorithm described at the beginning of the section and reduces it to the form (13.3). Returning back to the mapping (13.1), we get the theorem.

In conclusion we note that Theorem 13.1 can be generalized without trouble to the case when the operator V_0 has eigenvalues equal to -1 and when there are Jordan blocks. Moreover, the parameter ε in (13.1) can be a vector parameter.

13.3. **On the invariants of a normal form.** We note that all the terms written in (13.3) are invariants up to similarity transformations. Here, however, we consider the question of the number of invariants of the normal form of a mapping in a neighborhood of the zero fixed point, which loses stability through the eigenvalue -1, or through the pair $\exp(\pm i\omega)$, where $2\pi/\omega$ is irrational.

Suppose that the operator V_0 has the simple eigenvalue -1, corresponding to the eigenvector h_1, and the remainder of its spectrum lies in a disk of radius less than 1. Then by Theorem 13.1 the "truncated" normal form of the mapping (13.1) has for $\varepsilon = 0$ the form

$$\xi \mapsto -\xi + d_0\xi^3.$$

The main restriction imposed on the mapping (13.1) and characterizing a certain generality of position is the condition that $d_0 \neq 0$. In this case on the formally analytic integral manifold

$$u = \xi h_1 + \xi^2 h_2 + \xi^3 h_3 + \cdots \tag{13.12}$$

the mapping (13.1) can be transformed for $\varepsilon = 0$ to the form

$$\xi \mapsto -\xi + d_0\xi^3 + d_1\xi^5, \tag{13.13}$$

which contains only invariants.

We explain a method for constructing the indicated manifold. The terms of (13.12) are determined successively from the series of linear nonhomogeneous equations

$$(-1)^k h_k = V_0 h_k + f_k, \qquad k = 2, 3, \dots, \tag{13.14}$$

which are obtained by comparing coefficients of like power of ξ in the expression analogous to (13.5). We remark that for even k equation (13.14) is always solvable, while a solvability condition is necessary at an odd step. Therefore, we choose solutions h_{2m+1}, $m \geq 1$, containing the terms $c_m h_1$, where the c_m are arbitrary constants. These constants are determined from the solvability conditions for equations (13.14).

We note that in the fifth step the solvability conditions cannot be attained by suitably choosing the constant c_1. In this connection a term of fifth power appears in (13.14). For $k > 5$ this situation no longer arises, and the necessary solvability of (13.14) is ensured by the constants c_2, c_3, etc.

These facts remain true when $V_0 a = \exp(i\omega)a$, $V_0\bar{a} = \exp(-i\omega)\bar{a}$, $0 < \omega < \pi$, the ratio $2\pi/\omega$ is irrational, and $d_0 \neq 0$. In this case, on the formally analytic invariant manifold (its construction is similar to that of (13.12))

$$u = \xi[a\exp(i\omega\tau) + \bar{a}\exp(-i\omega\tau)] + \xi^2 h_2(\omega\tau) + \cdots$$

the mapping (13.1) becomes for $\varepsilon = 0$

$$\xi \mapsto \xi + d_0\xi^3 + d_1\xi^5, \qquad \tau \mapsto \tau + 1 + c_0\xi^2,$$

and this is a generalization of the well-known result on the number of invariants of a normal form in a neighborhood of a compound focus.

§14. Internal resonance on loss of stability of a homogeneous cycle of a relay system

14.1. **The case of a closed interval.** Here we apply the results of §13 to a relay system (9.3) on the interval $[0, 1]$ whose diffusion matrix depends smoothly on a small positive parameter μ.

Accordingly, we assume that the relay system

$$\partial y/\partial t = D(\mu)\frac{\partial^2 y}{\partial s^2} + g(\Phi(y), y), \tag{14.1}$$

$$\left.\frac{\partial y}{\partial s}\right|_{s=0} = \left.\frac{\partial y}{\partial s}\right|_{s=1} = 0 \tag{14.2}$$

has a simple homogeneous cycle $y_0(t)$, $(\Phi_1(y_0(0), y_0(0)) \in \Gamma_1$ (see §9). Suppose below that for $\mu = 0$ all the multipliers of the periodic boundary value problem (9.10), (9.11) lie inside a disk of radius 1, except for 1 and the pair $\exp(\pm i\omega)$, $0 < \omega < \pi$, to which the Lyapunov-Floquet solutions $a(t)\exp(i\omega t)\cos \pi s$ and $\overline{a}(t)\exp(-i\omega t)\cos \pi s$ correspond. Under these restrictions we draw through the point $y_0(0)$ in the phase space $E = C^\alpha$ of the boundary value problem (14.1), (14.2) a secant hyperplane to the cycle $y_0(t)$ in general position, and we then take $y_0(0)$ to the zero of E and consider the Poincaré operator of return along trajectories of system (14.1), (14.2). The conditions of Theorem 13.1 hold for Π in view of the restrictions imposed on the boundary value problem (14.1), (14.2). This implies that in the absence of strong resonances the normal form of Π has the following form in a neighborhood of the zero fixed point:

$$\xi \mapsto \xi + \mu\tau_0'\xi + d_0\xi^2 + \Phi(\omega\tau, \xi, \mu), \tag{14.3}$$

$$\tau \mapsto \tau + 1 + \mu\omega_0' + c_0\xi^2 + \frac{1}{\xi}\Psi(\omega\tau, \xi, \mu), \tag{14.4}$$

where the functions Φ and Ψ have all the properties described in §13.

Theorem 14.1. *Suppose that $\tau_0' > 0$ and $d_0 < 0$ in (14.3). Then there is a $\mu_0 > 0$ such that for $0 < \mu \le \mu_0$ the mapping (14.3), (14.4) has an exponentially stable invariant manifold*

$$\xi = \sqrt{-\mu\tau_0'/d_0}[1 + \sqrt{\mu}\varphi(\omega\tau, \mu)], \tag{14.5}$$

where the function φ, which is smooth in its variables and periodic in $\omega\tau$ with period 2π, is uniformly bounded. The mapping on it is obtained from (14.4) by using (14.5).

The proof follows directly from results in [48].

Returning to system (14.1), (14.2), we conclude from Theorem 14.1 that in a neighborhood of a homogeneous cycle it has an orbitally exponentially stable two-dimensional invariant torus. The same is true in the case of a soft loss of stability of the homogeneous cycle when the resonance has order three, i.e., when $\omega = 2\pi/3$. This has to do with the fact that the normal form here remains as before (the resonance terms in it vanish in view of the equality $\int_0^1 \cos^3 \pi s\, ds = 0$).

For $\omega = \pi/2$ the character of the stationary regimes of the boundary value problem (14.1), (14.2) located at a distance of order $\sqrt{\mu}$ from the cycle $y_0(t)$ is determined from an analysis of the truncated mapping:

$$\xi \mapsto \xi + \mu d_1\xi + d_2\xi^3 + d_3\xi^3\cos 2\pi(\tau + \gamma),$$

$$\tau \mapsto \tau + 1 + \mu\alpha_0 + \alpha_1\xi^2 - \frac{2}{\pi}d_3\xi^2 \sin 2\pi(\tau + \gamma),$$

where all the constants are nonzero in the general case. The results of this investigation are contained in [49].

The above facts are based in a very essential way on the algorithm described in §13 for constructing a normal form; this algorithm enables us to take into account the specifics of the problem. For example, Theorem 13.1 leads to a normal form

$$(14.6)\qquad \begin{aligned} &\xi \mapsto \xi + \mu\alpha_1\xi + \alpha_2\xi^2\cos 2\pi(\tau+\gamma) + \alpha_3\xi^3 + \Phi(\tau, \xi, \mu), \\ &\tau \mapsto \tau + 1 + \varepsilon\beta_1 - \frac{3}{2\pi}\alpha_2\xi\sin 2\pi(\tau+\gamma) + \beta_2\xi^2 + \frac{1}{\xi}\Psi(\tau, \xi, \mu) \end{aligned}$$

with fairly complex dynamics (see [49] and [50]) for a third-order resonance in the case of an arbitrary mapping (13.1), and in the boundary value problem (14.1), (14.2) this resonance only affects the character of behavior of the solutions on the invariant torus.

In conclusion we note that everything said about third-order resonance remains in force also for a multidimensional domain when the smallest positive eigenvalue λ_1 of the operator $-\Delta$ is simple, the eigenfunction $e_1(s)$ corresponding to it satisfies the equality

$$(14.7)\qquad \int_\Omega e_1^3(s)\,ds = 0,$$

and the Lyapunov-Floquet solutions $a(t)\exp(i\omega t)e_1(s)$ and $\bar{a}(t)\exp(-i\omega t)e_1(s)$ correspond to the multipliers $\exp(\pm i\omega)$ of the boundary value problem (9.10), (9.11) for $\mu = 0$. The equality (14.7) characterizes a certain symmetry of the domain Ω. If this equality fails, then the normal form for $\omega = 2\pi/3$ takes the general form (14.6).

14.2. **The case of a disk.** The results of §14.1 relate to the simplest situation, when the eigenvalue λ_1 of $-\Delta$ is simple. The absence of this requirement leads to new dynamical effects. All their typical features are very clear in the example of the boundary value problem

$$(14.8)\qquad \frac{\partial y}{\partial t} = D(\mu)\left[\frac{1}{r}\frac{\partial}{\partial r}\left(r\frac{\partial y}{\partial r}\right) + \frac{1}{r^2}\frac{\partial^2 y}{\partial \varphi^2}\right] + g(\Phi(y), y),$$

$$(14.9)\qquad \left.\frac{\partial y}{\partial r}\right|_{r=1} = 0,$$

where r, φ are polar coordinates, and $\Omega = \{(r, \varphi): r \le 1,\ 0 \le \varphi \le 2\pi\}$. As above, we assume that for $\mu = 0$ the boundary value problem (9.10), (9.11) has on the circle of radius 1 the multiplier 1 and the pair $\exp(\pm i\omega)$, to which there correspond the linearly independent Lyapunov-Floquet solutions

$$(14.10)\qquad \begin{aligned} &a(t)\exp(i\omega t)J_1(x_{11}r)\cos\varphi, \qquad &a(t)\exp(i\omega t)J_1(x_{11}r)\sin\varphi, \\ &\bar{a}(t)\exp(-i\omega t)J_1(x_{11}r)\cos\varphi, \qquad &\bar{a}(t)\exp(-i\omega t)J_1(x_{11}r)\sin\varphi, \end{aligned}$$

while all the remaining multipliers of it are interior to the unit disk. Here J_1 is the first-order Bessel function of the first kind, and x_{11} is the first positive zero of J_1'.

In the absence of strong resonances the truncated mapping takes the following form in this situation (the equations for the phase variables are not written):

$$\xi_1 \mapsto \xi_1 + \varepsilon\alpha_1\xi_1 + \alpha_2(\xi_1^2 + \xi_2^2)\xi_1, \qquad \xi_2 \mapsto \xi_2 + \varepsilon\alpha_1\xi_2 + \alpha_2(\xi_1^2 + \xi_2^2)\xi_2.$$

Thus, for $\alpha_1 > 0$ and $\alpha_2 < 0$ in the complete mapping there exists an exponentially stable two-dimensional invariant torus, given by the equality $\xi_1^2 + \xi_2^2 = -\varepsilon\alpha_1/\alpha_2$ in first approximation. To clear up the behavior of the iterates of the mapping on it we use the following device.

Let us consider (14.8) in the half-disk, assuming that

$$\left.\frac{\partial y}{\partial r}\right|_{r=1} = 0, \qquad \left.\frac{\partial y}{\partial \varphi}\right|_{\varphi=0} = \left.\frac{\partial y}{\partial \varphi}\right|_{\varphi=\pi} = 0. \tag{14.11}$$

The truncated normal form for the amplitude variable under the boundary conditions (14.1) can be obtained in a simple way from (14.10): it is necessary to set $\xi_2 = 0$. We extend to $[-\pi, 0]$ the solutions lying on a two-dimensional exponentially stable torus of the boundary value problem (14.8), (14.11) (which exists in view of the inequalities $\alpha_1 > 0$ and $\alpha_2 < 0$) in an even way with respect to φ, and then extend them to the whole axis by periodicity. In the end they also satisfy the boundary value problem (14.8), (14.9). From this we conclude that the orbitally exponentially stable three-dimensional torus of the boundary value problem (14.8), (14.9), whose existence has already been established, is entirely filled by two-dimensional tori that are carried one into another under translations with respect to the variable φ.

We underscore that these properties of the three-dimensional torus are entirely connected with circular symmetry, i.e., are specific for parabolic boundary value problems. Closely related effects were noted earlier in [38] and [46] in somewhat different situations.

§15. Some applications to relaxation systems

15.1. **Doubling bifurcation of a relaxation cycle.** Assume that the system (1.9) has a structurally stable relaxation cycle $x(t, \varepsilon)$, $y(t, \varepsilon)$ and that all the conditions in §1 hold. Suppose further that $\Pi(\varepsilon)$ is the Poincaré operator of return along trajectories of (1.9) introduced in the proof of Theorem 5.1, and let Π_0 be the corresponding operator of return along trajectories of the hybrid system (1.7).

Lemma 15.1. *Suppose that the right-hand sides f and g of the system (1.9) are infinitely differentiable. Then the limit equality*

$$\lim_{\varepsilon\to 0} \Pi(\varepsilon) = \Pi_0 \tag{15.1}$$

holds in the C^k-metric for any fixed $k \geq 1$.

We represent the operator $\Pi(\varepsilon)$ as a composition of mappings just as in the proof of Theorem 5.1. As a result the problem is reduced to a verification of the equality analogous to (15.1) for the correspondence operator from the plane $\xi = -q$ to the plane $\xi = q$ in a neighborhood of a breakaway point (for the remaining mappings, (15.1) follows from results in [8]). But the latter holds in view of the fact that all the asymptotic formulas in §4 for the trajectory of the relaxation system in a neighborhood of a breakaway point can be differentiated any number of times with respect to the initial conditions. Indeed, differentiating the trajectory of system (4.22) with respect to the initial conditions, we arrive at linear nonhomogeneous systems that can be integrated in a way analogous to that for (4.22) with the help of the method developed in §4.

Lemma 15.1 shows that the operator of return along trajectories of the singularly perturbed system (1.9) depends regularly on ε. Theorem 13.1 is thereby applicable to it.

We assume in addition that the right-hand sides f and g in (1.9) depend smoothly on the parameter $0 < \mu \ll 1$, i.e., we have

$$\varepsilon \dot{x} = f(x, y, \mu), \qquad \dot{y} = g(x, y, \mu). \tag{15.2}$$

Moreover, we assume that for $\mu = 0$ the periodic problem (1.4), (1.5) has simple multiplier -1, while its remaining multipliers (except for the simple multiplier 1) are less than 1 in modulus. From the results of §4 and Lemma 15.1 it then follows that the linear system (1.10) has a multiplier

$$\lambda(\mu, \varepsilon) = \lambda_0(\mu) + \sum_{n=2}^{\infty} \varepsilon^{n/3} \sum_{k=0}^{\pi(n-2)} \lambda_{n,k}(\mu) \ln^k \frac{1}{\varepsilon} \tag{15.3}$$

smoothly dependent on μ, where $\lambda_0(0) = -1$. Further, suppose that $\lambda_0'(0) \neq 0$. Then, by the implicit function theorem, we find from the equality $\lambda(\mu, \varepsilon) = -1$ a critical value $\mu_{\rm cr}(\varepsilon)$ of μ, for which the analogous asymptotic representation is valid in view of (15.3).

Let $\mu = \mu_{\rm cr}(\varepsilon) + \nu$, $0 < \nu \ll 1$, in (15.2). The result of using Theorem 13.1 here is the mapping

$$\xi \mapsto -\xi + \alpha_1(\varepsilon)\nu\xi + \alpha_2(\varepsilon)\xi^3 + \Phi(\xi, \varepsilon, \nu), \tag{15.4}$$

where the function Φ satisfies

$$|\Phi| \leq M[\xi^4 + \nu^2|\xi| + \nu\xi^2] \tag{15.5}$$

with a constant M independent of ε, and the coefficients α_1 and α_2 become for $\varepsilon = 0$ the corresponding coefficients of the normal form of the mapping along trajectories of the relay system (1.1).

Theorem 15.1. *Suppose that* $\alpha_1(0) < 0$ *and* $\alpha_2(0) > 0$ *in* (15.4). *Then there are positive constants* ε_0 *and* ν_0 *such that for* $0 < \varepsilon \leq \varepsilon_0$ *and* $0 < \nu \leq \nu_0$ *the mapping* (15.4) *has an exponentially stable cycle of period* 2 *in a neighborhood of the unstable zero fixed point.*

Theorem 15.1 is proved by direct verification with the use of inequality (15.5).

15.2. **Bifurcation of spatially nonhomogeneous relaxation cycles.** We consider the two-component parabolic system

$$\varepsilon\frac{\partial x}{\partial t} = D_1\frac{\partial^2 x}{\partial s^2} + f(x, y), \qquad \left.\frac{\partial x}{\partial s}\right|_{s=0} = \left.\frac{\partial x}{\partial s}\right|_{s=1} = 0, \tag{15.6}$$

$$\frac{\partial y}{\partial t} = (D_2(\varepsilon) - \nu)\frac{\partial^2 y}{\partial s^2} + g(x, y), \qquad \left.\frac{\partial y}{\partial s}\right|_{s=0} = \left.\frac{\partial y}{\partial s}\right|_{s=1} = 0, \tag{15.7}$$

where the right-hand sides f and g are the same as in §11.2, $0 < \nu \ll 1$, and the parameters D_1, $D_2 > 0$ are chosen so that (see §11.2)

$$\pi^2 D_2(\varepsilon) = F(\varepsilon, \pi^2 D_1), \qquad F_0(\pi^2 D_1) > 0,$$
$$k^2 F_0(\pi^2 D_1) > F_0(k^2\pi^2 D_1), \qquad k = 2, 3, \dots.$$

Considering the operator of return along trajectories of the relaxation system (15.6), (15.7) (for which the analogue of Lemma 15.1 is also valid) in a neighborhood of a homogeneous cycle of it, and using Theorem 13.1, we get the mapping

$$\xi \mapsto \xi + \alpha_1(\varepsilon)\nu\xi + \alpha_2(\varepsilon)\xi^3 + \Phi(\xi, \varepsilon, \nu), \tag{15.8}$$

where the function Φ satisfies (15.5), and $\alpha_1(0) = \pi^2$ and $\alpha_2(0)$ are the coefficients of the mapping analogous to (15.8) for the relay system (9.3). Here the following assertion is a direct consequence of Theorem 13.1.

Theorem 15.2. *Suppose that $\alpha_2(0) < 0$ in (15.8). Then there exist positive constants ε_0 and ν_0 such that for $0 < \varepsilon \leq \varepsilon_0$ and $0 < \nu \leq \nu_0$ the boundary value problem (15.6), (15.7) has in some (independent of ε) neighborhood of an unstable homogeneous cycle two stable spatially nonhomogeneous cycles (the system (15.6), (15.7) does not have other attractors in this neighborhood).*

CHAPTER 4. DUCK TRAJECTORIES OF RELAXATION SYSTEMS

Under restrictions of general position character a theorem on C^1-closeness of the solutions of a multidimensional relaxation system and the corresponding relay system was proved in Chapter 1, i.e., a basic case was analyzed. These results enable us to draw conclusions about the dynamics of a relaxation system from properties of the limit relay system. In the present chapter we consider the situation when one of the important restrictions in §1 is absent—the normal switching condition. This case is also basic, since the indicated feature can arise in relaxation systems in an unremovable way. It turns out that so-called duck trajectories appear. Here the well-known van der Pol system with a harmonic external action on the slow variable serves as a typical representative.

We mention another general theoretical question that has not yet been analyzed—the mechanism for the creation of an ordinary relaxation cycle from an equilibrium state. This problem has been investigated only for systems on the plane, in which, as shown in several papers using techniques of nonstandard analysis (see [7] and [51]), duck cycles preceding the relaxation cycle appear as the equilibrium state passes through a breakaway point. The geometric justification of this phenomenon is connected with the phenomenon, discovered by Pontryagin and demonstrated by a concrete example in [22], of delay of the loss of stability in relaxation systems with one slow variable. Therefore, the mechanism for creation of relaxation oscillations discovered for two-dimensional systems must operate also in the case of arbitrary relaxation systems with a single slow variable, as shown below. The analysis here is based on a technique developed in [5], general methods of asymptotic integration of singularly perturbed differential equations [8], and the Bogolyubov-Mitropol'skiĭ reduction principle [17]. Moreover, the problem of bifurcation of cycles is solved by combining the reduction principle with the classical theorem of Pontryagin on cycles of weakly perturbed Hamiltonian systems [52], and this leads to a determination of a more complete idea of the dynamics of a relaxation system when an equilibrium state of it passes through a breakaway point.

It turns out that in the case of multidimensional systems with an arbitrary number of slow variables the emergence of periodic relaxation oscillations takes place in a more complex way, and the duck cycles appearing here are unstable in the general case.

§16. THE EMERGENCE OF DUCK TRAJECTORIES IN THE ABSENCE OF THE NORMAL SWITCHING CONDITION

16.1. **Statement of the problem and description of the result.** For simplicity we confine ourselves to the three-dimensional case, in which our assertions have content, i.e., for $x \in R$ and $y \in R^2$ we consider the relaxation system

$$(16.1) \qquad \varepsilon\dot{x} = f(x, y), \qquad \dot{y} = g(x, y), \qquad 0 < \varepsilon \ll 1,$$

in a neighborhood of the zero breakaway point. The generality of position usually assumed means (see §1) that the smooth right-hand sides f and g satisfy the restrictions

$$(16.2)\qquad f(0,0)=0,\qquad f'_x(0,0)=0,\qquad f''_{xx}(0,0)\neq 0,$$

$$(16.3)\qquad l(g(0,0))\neq 0,\qquad l(*)=(\operatorname{grad}_y f(0,0),*).$$

We assume, however, that inequality (16.3), which is the normal switching condition, fails in a general way, i.e.,

$$(16.4)\qquad l(g(0,0))=0,\qquad g(0,0)\neq 0,\qquad \frac{d}{ds}l(g(0,sg(0,0)))|_{s=0}\neq 0.$$

Under the conditions (16.2) and (16.4) we transform system (16.1) to normal form in a neighborhood of the breakaway point. With this goal we first make a linear change of variables in (16.1): instead of x we introduce a variable ξ, changing the direction of the x-axis according to the motion of the phase point in the equation $\dot{x}=f(x,0)$, and instead of y we introduce the variables η_1 and η_2, directing the η_1-axis along a vector e_1 such that $l(e_1)=1$ and the η_2-axis along the vector $e_2=g(0,0)$. Then making in (16.1) the substitutions described in [5] that carry the surface $f=0$ into $\eta_1=-\xi^2$, we arrive at the normal form

$$(16.5)\qquad \varepsilon\dot{\xi}=\xi^2+\eta_1,\qquad \dot{\eta}_1=g_1(\xi,\eta_1,\eta_2),\qquad \dot{\eta}_2=g_2(\xi,\eta_1,\eta_2),$$

where, by (16.4) and the choice of the coordinates η_1 and η_2,

$$(16.6)\qquad g_1=\pm\eta_2+a\xi+b\eta_1+\cdots,\qquad g_2(0,0,\eta_2)\equiv 1,$$

and the dots stand for terms of a higher order of smallness. Below we assume that in the first equality (16.6) the minus sign is taken for η_2, which corresponds to the third quantity in (16.4) being negative. The other case is investigated similarly and will be discussed separately.

We consider the degenerate system corresponding to (16.5). With this goal let $\varepsilon=0$ in (16.5) and take ξ as the new time. As a result we get

$$(16.7)\qquad \eta_{10}(\varepsilon)=-\xi^2,\qquad d\eta_{20}/d\varphi=-[2\xi g_2(\xi,-\xi^2,\eta_{20})]/g_1(\xi,-\xi^2,\eta_{20}).$$

We point out that in view of the first equality in (16.6) the equation for determining η_{20} is singular for $\xi=\eta_{20}=0$. We shall be interested in solutions of it that are smooth in a neighborhood of $\xi=0$ and satisfy $\eta_{20}(0)=0$. Substituting $\eta_{20}(\xi)=\kappa\xi+\cdots$ in (16.7) and taking into account the properties (16.6) of the functions g_1 and g_2, we get the branching equation

$$(16.8)\qquad \kappa(\kappa-a)=2$$

for determining κ. It turns out that solutions $\eta^1_{20}(\xi)$ and $\eta^2_{20}(\xi)$ of the differential equation for η_{20} that are smooth in a neighborhood of the point $\xi=0$ correspond to the roots of (16.8). For a proof we consider the auxiliary system

$$(16.9)\qquad \dot{\xi}=g_1(\xi,-\xi^2,\eta_{20}),\qquad \dot{\eta}_{20}=-2\xi g_2(\xi,-\xi^2,\eta_{20}).$$

By (16.6) and (16.8), its zero equilibrium state is a saddle state. Therefore, there exist smooth functions $\eta_{20}=\eta^1_{20}(\xi)$ and $\eta_{20}=\eta^2_{20}(\xi)$ that determine its stable and unstable manifolds.

We note that in the degenerate system the phase point goes in a finite amount of time from the unstable slow motion surface $\eta_1=-\xi^2$, $\xi\geq 0$, along the trajectory $\eta_{10}=-\xi^2$, $\eta_{20}=\eta^2_{20}(\xi)$ to the stable surface $\eta_1=-\xi^2$, $\xi\leq 0$, and back along the trajectory $\eta_{10}=-\xi^2$, $\eta_{20}=\eta^1_{20}(\xi)$, i.e., we have duck trajectories. It is natural to expect that system (16.5) has similar trajectories.

System (16.5) will be considered for $|\xi|, |\eta_1|, |\eta_2| \leq q$, where q is sufficiently small. According to results in [17], it has invariant integral manifolds $\eta_1 = H_1(\xi, \eta_2, \varepsilon)$ and $\eta_1 = H_2(\xi, \eta_2, \varepsilon)$ in sufficiently small neighborhoods of the points $\xi = -q$ and $\xi = q$, respectively, where the functions H_1 and H_2 are smooth in their variables and satisfy

$$H_1(\xi, \eta_2, 0) \equiv H_2(\xi, \eta_2, 0) \equiv -\xi^2.$$

Here the manifold $\eta_1 = H_1$ is locally exponentially stable with exponent of order ε^{-1}, but the second manifold is unstable with the same exponent. It turns out that the duck trajectories—the trajectories with zero approximations $(-\xi^2, \eta_{20}^1(\xi))$ or $(-\xi^2, \eta_{20}^2(\xi))$—of system (16.5) are closely connected with these manifolds. In particular, any solution of the following nonlinear boundary value problem is a duck trajectory:

$$(16.10)\qquad \frac{d\eta_1}{d\xi} = \frac{\varepsilon g_1(\xi, \eta_1, \eta_2)}{\xi^2 + \eta_1}, \qquad \frac{d\eta_2}{d\xi} = \frac{\varepsilon g_2(\xi, \eta_1, \eta_2)}{\xi^2 + \eta_1};$$

$$(16.11)\qquad \eta_1|_{\xi=-q} = H_1|_{\xi=-q}, \qquad \eta_1|_{\xi=q} = H_2|_{\xi=q}.$$

Moreover, in view of the nonuniqueness of the choice of the manifolds H_1 and H_2, all the duck trajectories of system (16.5) can be constructed by means of the boundary value problem (16.10), (16.11). Indeed, if we have a duck trajectory, then one of its endpoints must lie in some manifold $\eta_1 = H_1$ (since there is a "fall" onto one of them). The other end lies on some manifold $\eta_1 = H_2$ (otherwise, there would be a "break" onto one of the stable manifolds).

Theorem 16.1. *The boundary value problem* (16.10), (16.11) *has exactly two solutions smoothly dependent on* ε *and with zero approximations* $\eta_{10}(\xi) = -\xi^2$, $\eta_{20}(\xi) = \eta_{20}^1(\xi)$ *and* $\eta_{10}(\xi) = -\xi^2$, $\eta_{20}(\xi) = \eta_{20}^2(\xi)$, *respectively.*

From the proof given below for the theorem it will follow that the duck trajectories are preserved under perturbations of g_1 and g_2 that are small in smooth metrics. For comparison we recall [7] that in a relaxation system on the plane the duck trajectories arising due to coalescence of integral manifolds analogous to H_1 and H_2 into a single manifold are destroyed under arbitrarily smooth perturbations of order $\exp(-c/\varepsilon)$, $c > 0$.

16.2. **Formal asymptotics of duck trajectories.** In this subsection we present an algorithm for asymptotic computation of solutions of the boundary value problem (16.10), (16.11). For definiteness we always take $\eta_{20}^1(\xi)$ as η_{20} below, since all the arguments are completely the same for $\eta_{20}^2(\xi)$.

Setting

$$(16.12)\qquad \eta_1 = -\xi^2 + \sum_{k=1}^{\infty} \varepsilon^k a_k(\xi), \qquad \eta_2 = \sum_{k=0}^{\infty} \varepsilon^k b_k(\xi), \qquad b_0 = \eta_{20}(\xi)$$

in (16.10) and equating successively the coefficients of like powers of ε, we arrive at the equalities

$$(16.13)\qquad \frac{g_1(\xi, -\xi^2, \eta_{20})}{a_1^2} a_{k+1} = \frac{g'_{1\eta_2}(\xi, -\xi^2, \eta_{20})}{a_1} b_k + \Phi_k(\xi),$$

$$(16.14)\qquad \frac{db_k}{d\xi} = \frac{g'_2\eta_2(\xi, -\xi^2, \eta_{20})}{a_1} b_k - \frac{g_2(\xi, -\xi^2, \eta_{20})}{a_1} a_{k+1} + \Psi_k(\xi),$$

where $a_1 = -g_1(\xi, -\xi^2, \eta_{20})/(2\xi)$, and the functions Φ_k and Ψ_k depend only on the a_r and b_m with indices $r \le k$ and $m \le k-1$. Suppose that the a_r and b_m with $r \le k$ and $m \le k-1$ have already been determined and are smooth in a neighborhood of $\xi = 0$. Then the functions Φ_k and Ψ_k are also smooth. Expressing the a_{k+1} from (16.13) and substituting in (16.14), we get for the determination of b_k the differential equation

$$db_k/d\xi = w(\xi)b_k + \varphi_k(\xi), \tag{16.15}$$

where, by (16.6),

$$w(\xi) \sim -\kappa^2/2\xi, \qquad \varphi_k(\xi) = O(1/\xi), \qquad \xi \to 0. \tag{16.16}$$

It follows from the asymptotic representations (16.16) that

$$\exp\left\{(\mathrm{H})\int_0^\xi w(s)\,ds\right\} = \gamma(\xi)/|\xi|^{\kappa^2/2}, \qquad \gamma(0) = 1,$$

where $(\mathrm{H})\int$ is the Hadamard regularization [5], and γ is some smooth function. Therefore, the formula

$$b_k = \int_0^\xi \exp\left\{(\mathrm{H})\int_\sigma^\xi w(s)\,ds\right\}\varphi_k(\sigma)\,d\sigma$$

supplies a unique solution of (16.15) smooth at $\xi = 0$, and a smooth function a_{k+1} is determined from (16.14), and so on.

16.3. **General scheme of proof.** The central point of the proof is the justification of the estimate

$$\sup_{-q \le \xi, s \le q} \|G(\xi, s, \varepsilon)\| \le M \tag{16.17}$$

for the matrix Green function of the linear boundary value problem

$$\begin{aligned} \varepsilon dh_1/d\xi &= a(\xi, \varepsilon)h_1 + b(\xi, \varepsilon)h_2, \\ dh_2/d\xi &= c(\xi, \varepsilon)h_1 + d(\xi, \varepsilon)h_2, \end{aligned} \tag{16.18}$$

$$h_1(-q, \varepsilon) = \delta_1(\varepsilon)h_2(-q, \varepsilon), \qquad h_1(q, \varepsilon) = \delta_2(\varepsilon)h_2(q, \varepsilon), \tag{16.19}$$

obtained upon linearization of the boundary value problem (16.10), (16.11) on the $(n+1)$st and nth $(n \ge 3)$ partial sums of the series in (16.12) and upon the substitution $h_1/\varepsilon \to h_1$. In (16.17), $\| * \|$ is the matrix norm in R^2, and the same letter M is used here and below to denote positive constants independent of ε and with exact values not important to us.

Suppose that (16.17) has already been established. Then, setting

$$\eta_1 = -\xi^2 + \sum_{k=1}^{n+1} \varepsilon^k a_k(\xi) + \varepsilon h_1, \qquad \eta_2 = \eta_{20}(\xi) + \sum_{k=1}^{n} \varepsilon^k b_k(\xi) + h_2$$

in (16.10) and (16.11), we arrive at the nonlinear boundary value problem

$$\begin{aligned} \varepsilon\, dh_1/d\xi &\equiv a(\xi, \varepsilon)h_1 + b(\xi, \varepsilon)h_2 + \Delta_1(\xi, \varepsilon, h_1, h_2) + \Delta_2(\xi, \varepsilon), \\ dh_2/d\xi &= c(\xi, \varepsilon)h_1 + d(\xi, \varepsilon)h_2 + \Delta_3(\xi, \varepsilon, h_1, h_2) + \Delta_4(\xi, \varepsilon), \end{aligned} \tag{16.20}$$

$$\begin{aligned} h_1(-q, \varepsilon) &= \delta_1(\varepsilon)h_2(-q, \varepsilon) + \Omega_1(h_2(-q, \varepsilon), \varepsilon) + \Omega_2(\varepsilon), \\ h_1(q, \varepsilon) &= \delta_2(\varepsilon)h_2(q, \varepsilon) + \Omega_3(h_2(q, \varepsilon), \varepsilon) + \Omega_4(\varepsilon), \end{aligned} \tag{16.21}$$

where the functions Δ_j and Ω_k, $j, k = 1, 2, 3, 4$, are smooth in their variables and such that

$$\max_{-\varepsilon\le\xi\le q} |\Delta_2(\xi, \varepsilon)|, \ \max_{-q\le\xi\le q} |\Delta_4(\xi, \varepsilon)|, \ |\Omega_2|, \ |\Omega_4| \le M\varepsilon^n, \tag{16.22}$$

and Δ_1, Δ_3, Ω_1, and Ω_3 vanish for $h_1 = h_2 = 0$ together with their first derivatives with respect to h_1 and h_2. Next, using the substitution

$$\begin{aligned} h_2(\xi, \varepsilon) &+ \frac{1}{\delta_1(\varepsilon)}(\Omega_1(h_2(\xi, \varepsilon), \varepsilon) + \Omega_2(\varepsilon))\frac{q-\xi}{2q} \\ &+ \frac{1}{\delta_2(\varepsilon)}(\Omega_3(h_2(\xi, \varepsilon), \varepsilon)) + \Omega_4(\varepsilon))\frac{q+\xi}{2q} \to h_2 \end{aligned}$$

to move the nonhomogeneity in the boundary conditions (16.21) to the right-hand sides of (16.20), we pass with the help of the Green function of the boundary value problem (16.18), (16.19) from a system of differential equations to an integral equation. The estimate

$$\max_{-q\le\xi\le q} \frac{1}{\varepsilon}\int_{-q}^{q} \|G(\xi, s, \varepsilon)\|\, ds \le M/\varepsilon$$

(which follows from (16.17)) and the properties (16.22) enable us to conclude that the right-hand side of the integral equation obtained transforms into itself some ball with radius of order ε^{n-1} in the space $C[-q, q)$ and is a contraction mapping.

Thus, the proof of the theorem reduces to a proof of (16.17). In this connection we dwell in greater detail on the properties of the boundary value problem (16.18), (16.19). Obviously, its coefficients are smooth with respect to their variables, and in view of the properties (16.6) of the functions g_1 and g_2 and the asymptotic formulas

$$\begin{aligned} \partial H_1/\partial\eta_2 &= -\varepsilon g'_{1\eta_2}(\xi, -\xi^2, \eta_2)/(2\xi) + O(\varepsilon^2), \\ \partial H_2/\partial\eta_2 &= -\varepsilon g'_{1\eta_2}(\xi, -\xi^2, \eta_2)/(2\xi) + O(\varepsilon^2) \end{aligned}$$

we have the equalities

$$a(\xi, 0) = 2\kappa\xi + \cdots, \qquad b(0, 0) = -\kappa, \qquad c(0, 0) = -\kappa^2, \tag{16.23}$$

$$d(0, 0) = 0, \qquad \delta_1(0) = g'_{1\eta_2}(-q, -q^2, \eta_{20}(-q))/(2q), \tag{16.24}$$

$$\delta_2(0) = -g'_{1\eta_2}(q, -q^2, \eta_{20}(q))/(2q). \tag{16.25}$$

Further, let $h_{11}(\xi, \varepsilon)$, $h_{21}(\xi, \varepsilon)$ and $h_{12}(\xi, \varepsilon)$, $h_{22}(\xi, \varepsilon)$ be two solutions of (16.18) with the initial conditions

$$\begin{aligned} h_{11}(-q, \varepsilon) &= \delta_1(\varepsilon), \qquad h_{21}(-q, \varepsilon) = 1, \\ h_{12}(q, \varepsilon) &= \delta_2(\varepsilon), \qquad h_{22}(q, \varepsilon) = 1. \end{aligned} \tag{16.26}$$

As it turns out, they will be shown to be linearly independent below. Therefore, for the Green function of the boundary value problem (16.18), (16.19) we have

$$G(\xi, s, \varepsilon) = \begin{cases} U(\xi, \varepsilon)P_1C(s, \varepsilon), & -q \le \xi \le s, \\ -U(\xi, \varepsilon)P_2C(s, \varepsilon), & s < \xi \le q, \end{cases} \tag{16.27}$$

where the indicated solutions of (16.18) serve as the columns of the matrix U, $C(s, \varepsilon) = U^{-1}(s, \varepsilon)$, and

$$P_1 = \begin{pmatrix} 1 & 0 \\ 0 & 0 \end{pmatrix}, \qquad P_2 = \begin{pmatrix} 0 & 0 \\ 0 & 1 \end{pmatrix}.$$

It follows from (16.27) that to justify (16.17) we must analyze the asymptotic properties of the solutions of (16.18) with the initial conditions (16.26). Below is a construction of the required asymptotics on the regions

$$-q \le \xi \le -\varepsilon^\lambda, \qquad -\varepsilon^\lambda \le \xi \le \varepsilon^\lambda, \qquad \varepsilon^\lambda \le \xi \le q,$$

where $0 < \lambda < 1/2$, with a subsequent matching at the points $\xi = -\varepsilon^\lambda$ and $\xi = \varepsilon^\lambda$. We note at once that it suffices to construct the asymptotics of the first of these solutions, because in view of (16.23)–(16.25) the asymptotics of the second solution is obtained by replacing ξ by $-\xi$.

16.4. **Asymptotic integration of the variational system on a duck trajectory.** It follows from the results of [17] that for $-q \le \xi \le -q_1 < 0$ system (16.18) has an exponentially stable (with exponent of order ε^{-1}) invariant integral manifold

$$h_1 = \omega(\xi, \varepsilon)h_2. \tag{16.28}$$

Here ω is a function smooth in its variables. Further, since system (16.18) is the linearization of (16.7),

$$\omega = \varepsilon^{-1}\partial H_1/\partial \eta_2 + O(\varepsilon^n), \tag{16.29}$$

where the derivative is taken at $\eta_2 = \eta_{20}(\xi) + \sum_{k=1}^n \varepsilon^k b_k(\xi)$. It follows from (16.29) that

$$\omega(-q, \varepsilon) - \delta_1(\varepsilon) = O(\varepsilon^n). \tag{16.30}$$

It follows from (16.26) that with the accuracy indicated in (16.30) the initial condition of the solution $h_{11}(\xi, \varepsilon)$, $h_{21}(\xi, \varepsilon)$ lies on the integral manifold (16.28). This circumstance enables us to reduce the problem to integration of the scalar linear equation obtained from (16.18) with (16.28) taken into account. Therefore, our immediate problem is to construct the asymptotics of the function ω. In passing we show that the manifold (16.28) can be extended to the point $\xi = q$.

Using (16.28) in (16.18), we get the equation

$$\varepsilon\, d\omega/d\xi - a(\xi, \varepsilon)\omega = -\varepsilon c(\xi, \varepsilon)\omega^2 - \varepsilon d(\xi, \varepsilon)\omega + b(\xi, \varepsilon) \tag{16.31}$$

for the function ω. Setting

$$\omega = \sum_{k=0}^{\infty} \varepsilon^k \omega_k(\xi) \tag{16.32}$$

in (16.31), where

$$\omega_0(\xi) = -b(\xi, 0)/a(\xi, 0) = 1/(2\xi) + \cdots, \tag{16.33}$$

and equating the coefficients of like powers of ε, we have a recursive sequence of algebraic equations of the form

$$-a(\xi, 0)\omega_k = R_k(\xi) \tag{16.34}$$

for determining the unknown functions ω_k, where R_k depends only on the ω_m with indices $m \le k-1$. A more detailed analysis of the structure of the right-hand side of (16.34) together with the induction method lead to the asymptotic expansion

$$\omega_k(\xi) = \frac{1}{\xi^{2k+1}} \sum_{m=0}^{\infty} \varphi_{km}\xi^m, \qquad \xi \to -0. \tag{16.35}$$

The uncomplicated computations are omitted here.

By the general properties of the manifold (16.28), the asymptotic representation (16.32) automatically holds for $-q \le \xi \le -q_1 < 0$. For a proof of its validity up to the point $\xi = -\varepsilon^{\lambda}$, $0 < \lambda < 1/2$, we set

$$\omega = \sum_{k=0}^{N} \varepsilon^k \omega_k + z, \qquad N \le n-1,$$

in (16.31) and pass to the integral equation

(16.36)
$$z = \omega(-q, \varepsilon) - \sum_{k=0}^{N} \varepsilon^k \omega_k(-q) + \frac{1}{\varepsilon}\int_{-q}^{\xi} \exp\left\{\frac{1}{\varepsilon}\int_{s}^{\xi} a(\sigma, \varepsilon)\, d\sigma\right\} (\varepsilon F_1(s, \varepsilon) z + \varepsilon F_2(s, \varepsilon) z^2 + F_3(s, \varepsilon))\, ds.$$

Here the functions F_1 and F_2 are bounded uniformly with respect to their variables, and in view of (16.35) F_3 has the estimate

$$\max_{-q \le \xi \le -\varepsilon^{\lambda}} |F_3(\xi, \varepsilon)| \le M\varepsilon^{(1-2\lambda)(N+1)}.$$

This inequality, in combination with the directly verifiable estimate

$$\max_{-q \le \xi \le -\varepsilon^{\lambda}} \frac{1}{\varepsilon}\int_{-q}^{\xi} \exp\left\{\frac{1}{\varepsilon}\int_{s}^{\xi} a(\sigma, \varepsilon)\, d\sigma\right\} ds \le M\varepsilon^{-\lambda}$$

enables us to conclude that the integral operator generated by the right-hand side of (16.36) transforms into itself some ball in $C[-q, -\varepsilon^{\lambda}]$ with radius of order $\varepsilon^{(1-2\lambda)(N+1)-\lambda}$ and is a contraction there. It remains to use the contraction mapping principle, leading to the following assertion.

Lemma 16.1. *Uniformly with respect to* $-q \le \xi \le -\varepsilon^{\lambda}$, $0 < \lambda < 1/2$,

(16.37)
$$\omega(\xi, \varepsilon) = \sum_{k=0}^{N} \varepsilon^k \omega_k(\xi) + O(\varepsilon^{(1-2\lambda)(N+1)-\lambda}).$$

We show that the integral manifold (16.28) can be extended to the point $\xi = \varepsilon^{\lambda}$.

For $-\varepsilon^{\lambda} \le \xi \le \varepsilon^{\lambda}$ we make in (16.31) the substitutions $\xi = \mu u$, $\omega = \mu^{-1} v$, and $\mu = \varepsilon^{1/2}$, which transform it to

(16.38)
$$dv/du = \mu^{-1} a(\mu u, \mu^2) v - c(\mu u, \mu^2) v - \mu d(\mu u, \mu^2) v + b(\mu u, \mu^2).$$

Then we let

(16.39)
$$v = \sum_{k=0}^{\infty} \mu^k \nu_k$$

in (16.38) and equate the coefficients of like powers of μ. As a result we get a recursive sequence of differential equations

(16.40)
$$dv_0/du = \kappa^2 v_0^2 + 2\kappa u v_0 - \kappa,$$

(16.41)
$$dv_k/du = 2\kappa[\kappa v_0 + u] v_k + V_k(u), \qquad k \ge 1,$$

where the functions V_k depend only on the v_m with indices $m \le k-1$. In (16.32) we let $\xi = \mu u$ and $\varepsilon = \mu^2$, take into account the expansions (16.35) of the functions ω_k, and group together the terms with the same powers of μ. It is clear that the coefficients obtained for μ^{k-1} must serve as the asymptotic expansions of the functions v_k as $u \to -\infty$. Therefore, v_0 should be taken to be the unique solution of (16.40) tending to zero as $u \to -\infty$, and v_k should be determined from

$$(16.42) \qquad v_k = \int_{-\infty}^{u} \exp\left\{\int_s^u 2\kappa(\kappa v_0(\sigma) + \sigma)\, d\sigma\right\} V_k(s)\, ds.$$

We note that, as in §4, where an asymptotic expression was constructed for the trajectories of a relaxation system in a neighborhood of a breakaway point, the choice of the particular solution (16.42) of (16.41) has the same meaning here: the solutions are taken that deviate least from bounded solutions as $u \to -\infty$.

We deduce from (16.40) that

$$(16.43) \qquad v_0 = u^{-1} \sum_{m=0}^{\infty} \frac{\alpha_{m0}}{u^{2m}}, \quad u \to -\infty, \qquad v_0 = u \sum_{m=0}^{\infty} \frac{\beta_{m0}}{u^{2m}}, \quad u \to +\infty,$$

where $\alpha_{00} = 1/2$ and $\beta_{00} = -2/\kappa$. It can then be established by induction that for $k \geq 1$

$$(16.44) \qquad v_k = u^{k-1} \sum_{m=0}^{\infty} \frac{\alpha_{mk}}{u^{2m}}, \qquad u \to -\infty,$$

$$(16.45) \qquad v_k = u^{k+1} \sum_{m=0}^{\infty} \frac{\beta_{mk}}{u^{2m}}, \qquad u \to +\infty.$$

We confine ourselves to a proof of the second of these equalities, since the proof of the first is analogous. The same letter P will be used to denote expressions of the form

$$\sum_{m=0}^{\infty} \frac{P_m}{u^{2m}}.$$

Then (16.45) can be written as $v_k = u^{k+1} P$. Moreover, we have the easily verified relations

$$(16.46) \qquad P^2 = P, \qquad \int s^m P\, ds = u^{m+1} P,$$

from which we conclude, integrating by parts, that

$$(16.47) \qquad \int_{-\infty}^{u} \exp\left\{\int_s^u 2\kappa(\kappa v_0(\sigma) + \sigma)\, d\sigma\right\} s^m P\, ds = u^{m-1} P.$$

Suppose that (16.45) is valid for all $m \leq k - 1$. Then, taking into account the expansions

$$a, b, c, d = \sum_{r=0}^{\infty} \mu^r u^r P$$

in (16.38), we get the representation $V_k = u^{k+2} P$ for the nonhomogeneity in (16.41). It remains to use (16.47).

A precise meaning is given to the asymptotic expansion (16.39) just as in the case of (16.32): in (16.38) we let

$$v = \sum_{k=0}^{N} \mu^k v_k + z,$$

we pass from the differential equation to an integral equation, inverting the corresponding linear differential operator with the help of a formula analogous to (16.42), and then we use the contraction mapping principle. In this way we arrive at the following assertion.

Lemma 16.2. *Uniformly with respect to* $-\varepsilon^{\lambda-1/2} \le u \le \varepsilon^{\lambda-1/2}$, $0 < \lambda < 1/2$,

$$v(u, \mu) = \sum_{k=0}^{N} \mu^k v_k(u) + O(\mu^{2\lambda(N+2)-1}), \tag{16.48}$$

and for $\xi = -\varepsilon^\lambda$ *the functions* $\omega(\xi, \varepsilon)$ *and* $\mu^{-1}v(\xi/\mu, \mu)$ *coincide up to* ε^α, *where* $\alpha > 0$ *is arbitrary.*

We confine ourselves what is new in comparison to the preceding lemma—matching the asymptotic formulas (16.32) and (16.39) at $\xi = -\varepsilon^\lambda$. Consider the interval $-2\varepsilon^\lambda \le \xi \le -\varepsilon^\lambda$, on which both $\omega(\xi, \varepsilon)$ and $\mu^{-1}v(\xi/\mu, \mu)$ are defined. Suppose that $\omega(-\varepsilon^\lambda, \varepsilon) - \mu^{-1}v(-\varepsilon^{\lambda-1/2}, \mu) \sim \varepsilon^\beta$, $\beta > 0$. Then we take $\mu^{-1}v(-\varepsilon^{\lambda-1/2}, \mu)$ as the initial condition of (16.31) with $\xi = -\varepsilon^\lambda$, and we integrate it in the direction of decreasing ξ. Since the deviation of this solution from $\omega = \omega(\xi, \varepsilon)$ computed from the linear approximation grows like $\exp(\kappa\xi^2/\varepsilon)$, it becomes of order $\varepsilon^\beta \exp(4\kappa\varepsilon^{2\lambda-1})$ for $\xi = -2\varepsilon^\lambda$, i.e., is exponentially large. On the other hand, by (16.32) and (16.39) and the properties (16.35) and (16.44) of the functions ω_k and v_k, this deviation remains a quantity of order ε^{β_1} for some β_1 on the whole interval $-2\varepsilon^\lambda \le \xi \le -\varepsilon^\lambda$. We have come to a contradiction.

Therefore the integral manifold (16.28) can be extended to the point $\xi = \varepsilon^\lambda$, $0 < \lambda < 1/2$. For $\varepsilon^\lambda \le \xi \le q$ the substitution $\omega = \varepsilon^{-1}w$ can be used to pass from (16.31) to the equation

$$\varepsilon\, dw/d\xi = a(\xi, \varepsilon)w - c(\xi, \varepsilon)w^2 - \varepsilon d(\xi, \varepsilon)w + \varepsilon b(\xi, \varepsilon), \tag{16.49}$$

the solution of which is sought in the form

$$w = \sum_{k=0}^{\infty} \varepsilon^k w_k(\xi), \tag{16.50}$$

where

$$w_0 = a(\xi, 0)/c(\xi, 0). \tag{16.51}$$

Substituting (16.50) in (16.49) and equating the coefficients of like powers of ε, we get a recursive sequence of algebraic equations from which we find the unknown functions w_k. The algorithmic part and the proof of (16.50) here are completely analogous to the case $-q \le \xi \le -\varepsilon^\lambda$, and the matching of (16.50) and (16.39) at $\xi = \varepsilon^\lambda$ is carried out just as in Lemma 16.2. Therefore, we present only the final result.

Lemma 16.3. *Uniformly with respect to* $\varepsilon^\lambda \le \xi \le q$, $0 < \lambda < 1/2$,

$$w = \sum_{k=0}^{N} \varepsilon^k w_k(\xi) + O(\varepsilon^{(1-2\lambda)N+1-\lambda}), \tag{16.52}$$

where as $\xi \to +0$ *the functions* w_k *have the asymptotic expansions*

$$w_k(\xi) = \frac{1}{\xi^{2k-1}} \sum_{m=0}^{\infty} \psi_{km}\xi^m.$$

Lemmas 16.1–16.3 enable us without trouble to construct the asymptotics of the solution $h_{11}(\xi, \varepsilon)$, $h_{21}(\xi, \varepsilon)$ of (16.18) with any degree of accuracy. However, for our purposes only the principal asymptotics is important, and it has the following form: for $-q \le \xi \le -\varepsilon^{\lambda}$, $0 < \lambda < 1/3$,

$$h_{11} = [\omega_0(\xi) + O(\varepsilon^{1-3\lambda})]h_{21}, \tag{16.53}$$

$$h_{21} = |\xi|^{-\kappa^2/2}[h_{21,0}(\xi) + O(\varepsilon^{1-3\lambda})], \tag{16.54}$$

where

$$h_{21,0}(\xi) = q^{\kappa^2/2} \exp\left\{\int_{-q}^{\xi} \left[c(s, 0)\omega_0(s) + \frac{\kappa^2}{2s} + d(s, 0)\right] ds\right\};$$

for $-\varepsilon^{\lambda} \le \xi \le \varepsilon^{\lambda}$, $1/4 < \lambda < 1/2$,

$$h_{11} = \mu^{-1}[v_0(u) + O(\mu^{4\lambda-1})]h_{21}, \tag{16.55}$$

$$h_{21} = c_0 \exp\left\{-(H)\int_{-\infty}^{u} \kappa^2 v_0(s)\, ds\right\}(1 + O(\mu^{4\lambda-1})), \tag{16.56}$$

where $c_0 = \mu^{-\kappa^2/2}h_{21,0}(0)$ and $u = \xi/\mu$; and for $\varepsilon^{\lambda} \le \xi \le q$, $0 < \lambda < 1/2$,

$$h_{11} = \varepsilon^{-1}[a(\xi, 0)/c(\xi, 0) + O(\varepsilon^{1-\lambda})]h_{21}, \tag{16.57}$$

$$h_{21} = \exp\left\{\frac{1}{\varepsilon}\int_0^{\xi} a(s, 0)\, ds(1 + O(\varepsilon^{1-\lambda}))\right\}. \tag{16.58}$$

As mentioned above, the structure of the asymptotics for the solution $h_{12}(\xi, \varepsilon)$, $h_{22}(\xi, \varepsilon)$ of (16.18) can be obtained from (16.53)–(16.58) by replacing ξ by $-\xi$. This means that in the given case (16.53) and (16.58) must be considered for $\varepsilon^{\lambda} \le \xi \le q$, and the function

$$h_{22,0}(\xi) = q^{\kappa^2/2} \exp\left\{-\int_{\xi}^{q} \left[c(s, 0)\omega_0(s) + \frac{\kappa^2}{2s} + d(s, 0)\right] ds\right\} \tag{16.59}$$

should be taken instead of $h_{21,0}(\xi)$. Further, for $-\varepsilon^{\lambda} \le \xi \le \varepsilon^{\lambda}$ we have

$$h_{12} = \mu^{-1}[-v_0(-u) + O(\mu^{4\lambda-1})]h_{22}, \tag{16.60}$$

$$h_{22} = c_0 \exp\left\{-(H)\int_{u}^{+\infty} \kappa^2 v_0(-s)\, ds\right\}(1 + O(\mu^{4\lambda-1})), \tag{16.61}$$

where $c_0 = \mu^{-\kappa^2/2}h_{22,0}(0)$, and for $-q \le \xi \le -\varepsilon^{\lambda}$ we have

$$h_{12} = \varepsilon^{-1}[a(\xi, 0)/c(\xi, 0) + O(\varepsilon^{1-\lambda})]h_{22}, \tag{16.62}$$

$$h_{22} = \exp\left\{-\frac{1}{\varepsilon}\int_{\xi}^{0} a(s, 0)\, ds(1 + O(\varepsilon^{1-\lambda}))\right\}. \tag{16.63}$$

Using (16.53)–(16.63) in (16.27), we see that (16.17) holds. The proof of Theorem 16.1 is complete.

16.5. **The case** $(d/ds)l(g(0, sg(0, 0)))|_{s=0} > 0$. We now consider the case when the plus sign is taken for η_2 in the first equality in (16.6). Then instead of the branching equation (16.8) we have

$$\kappa(\kappa + a) = -2. \tag{16.64}$$

For $|a| < 2\sqrt{2}$ the roots of (16.64) are complex, and hence for these a there are no duck trajectories for system (16.5).

Suppose that $|a| > 2\sqrt{2}$. Then (16.6) has two real roots κ_1 and κ_2 of the same sign, and $0 < |\kappa_1| < \sqrt{2}$, $|\kappa_2| > \sqrt{2}$, and $\kappa_1\kappa_2 = 2$. According to results in [40], the smooth solution $\eta_{20} = \eta_{20}^1(\xi)$ always corresponds to the root κ_1 : the equality $\eta_{20} = \eta_{20}^1(\xi)$ determines the integral manifold of (16.9) in a neighborhood of the origin corresponding to the eigenvalue $\lambda_1 = -\kappa_2$ of the zero equilibrium state. And in the case of the inequalities

(16.65) $$\kappa_2/\kappa_1 \neq n, \qquad n = 2, 3, \dots$$

(i.e., when there are no resonances) the unique smooth integral manifold $\eta_{20} = \eta_{20}^2(\xi)$ of (16.9) corresponds to the eigenvalue $\lambda_2 = -\kappa_1$ [40]. Thus, using the techniques developed in the preceding part of this section, we obtain the following assertion.

Theorem 16.2. *Suppose that* $|a| > 2\sqrt{2}$, *and the inequalities* (16.65) *hold. Then the boundary value problem* (16.10), (16.11) *has exactly two solutions. These solutions depend smoothly on* ε, *and their zero approximations have the form* $\eta_{10}(\xi) = -\xi^2$, $\eta_{20}(\xi) = \eta_{20}^1(\xi)$ *and* $\eta_{10}(\xi) = -\xi^2$, $\eta_{20}(\xi) = \eta_{20}^2(\xi)$, *where* η_{20}^1 *and* η_{20}^2 *are the solutions of* (16.7) *corresponding to the roots* κ_1 *and* κ_2 *of the branching equation* (16.64).

We dwell on the topological meaning of the resonances (16.65). Suppose that $n + 1 < \kappa_2/\kappa_1 < n + 2$ under the conditions of Theorem 16.2. On the curve $\eta_1 = H_1(-q, \eta_2, \varepsilon)$ in the plane $\xi = -q$ we consider the segment $|\eta_2 - \eta_2(\varepsilon)| \leq \delta$, where $\eta_2(\varepsilon)$ is the initial condition of the solution of the boundary value problem (16.10), (16.11) with zero approximation $(-\xi^2, \eta_{20}^2(\xi))$. For $\delta = \exp(-c/\varepsilon)$ and sufficiently large $c > 0$ the solutions of (16.5) with initial conditions in this segment can be extended to the point $\xi = q$, and the image of the segment in the plane $\xi = q$ intersects the curve $\eta_1 = H_2(q, \eta_2, \varepsilon)$ versally. Moreover, the band formed by these trajectories is twisted like the Möbius band n times in a neighborhood of the point $\xi = 0$. The latter follows from the fact that in a neighborhood of $\xi = 0$ the variational system on the duck trajectory with initial condition $\eta_2(\varepsilon)$ is transformed, after the normalizations described in §16.4 and after terms with higher order of smallness are discarded, to the so-called Weber equation

(16.66) $$d^2v/du^2 - u\,dv/du + (\kappa_2^2/2 - 1)v = 0.$$

Each nontrivial solution of it has exactly n zeros. Therefore, the deviations from the duck trajectory under consideration computed from the linear approximation will have n alternating zeros each with respect to the coordinates η_1 and η_2, which corresponds to the twisting described above for the band.

We remark that for $\kappa_2^2/2 = \kappa_2/\kappa_1 = n + 1$ equation (16.66) has a solution $v = H_n(u)$, where $H_n(u)$ is the corresponding Chebyshev-Hermite polynomial. Therefore, upon passage through a resonance the number of zeros of the solutions of (16.66) changes by 1, and this leads to the next twisting of the band. This implies, in particular, that for $\kappa_2/\kappa_1 = n$ the versality of the intersection of the band with the manifold $\eta_1 = H_2$ is violated for $\xi = q$. Thus, system (16.5) does not have duck trajectories corresponding to the root κ_2 of the branching equation (16.64) in the general case when $\kappa_2/\kappa_1 = n$ (the duck trajectories with zero approximation $(-\xi^2, \eta_{20}^1(\xi))$ are preserved). In addition we note that, first, for $\kappa_2/\kappa_1 = n$ and for a certain generality of position all the solutions of (16.7) satisfying the conditions $\eta_{20}(0) = 0$ and $\eta_{20}^{1\prime}(0) = \kappa_2$ have only finite smoothness [40], and, second, if $\kappa_2/\kappa_1 \to n$ due to a change in the parameters of the system, then the coefficient of ξ^n in the Taylor expansion of the function $\eta_{20}^2(\xi)$ at $\xi = 0$ tends to infinity. The latter leads to the condition that the curve $(-\xi^2, \eta_{20}^2(\xi))$ "flows into" the curve $(-\xi^2, \eta_{20}^1(\xi))$.

§17. On the mechanism for destruction of the invariant torus of a van der Pol relaxation system with harmonic external force

17.1. Existence and stability of the invariant torus. In this section the object of our analysis is the van der Pol relaxation system

$$\varepsilon\dot{x} = y - \tfrac{1}{3}x^3 + x, \qquad \dot{y} = -x + A\cos\varphi, \qquad \dot{\varphi} = \omega, \tag{17.1}$$

which it is natural to consider on the cylinder $(x, y, \varphi) \in R^2 \times S^1$, where S^1 is the unit circle. Here the parameters A and ω are greater than 0, and $0 < \varepsilon \ll 1$. As shown in Chapter 1, the properties of system (17.1) are closely connected with the relay system

$$\dot{y} = -x + A\cos\varphi, \qquad \dot{\varphi} = \omega, \qquad x = \begin{cases} \Phi_1(y), \\ \Phi_2(y), \end{cases} \tag{17.2}$$

where $\Phi_1(y)$ $(-\infty < y < 2/3)$ and $\Phi_2(y)$ $(-2/3 < y < \infty)$ are roots of the equation $y = \frac{1}{3}x^3 - x$ for $|x| > 1$. We recall how the solutions of (17.2) are defined. For example, for $y(0) = y_0 < 2/3$ and $\varphi(0) = \varphi_0$ the solution of (17.2) is first taken to be the solution of the Cauchy problem

$$\dot{y} = -\Phi_1(y) + A\cos\varphi, \qquad \dot{\varphi} = \omega, \qquad y|_{t=0} = y_0, \qquad \varphi|_{t=0} = \varphi_0.$$

Since $\Phi_1(y) < -1$, there is a first moment of time t_0 such that $y(t_0) = 2/3$. At this moment there is a switching, i.e., thereafter $y(t)$, $\varphi(t)$ is the solution of the Cauchy problem

$$\dot{y} = -\Phi_2(y) + A\cos\varphi, \qquad \dot{\varphi} = \omega, \qquad y|_{t=t_0} = 2/3, \qquad \varphi|_{t=t_0} = \varphi(t_0).$$

From analogous considerations there is a first moment of time $t_1 > t_0$ for which $y(t_1) = -2/3$. Then there is another switching, and so on.

Assume that the normal switching conditions

$$1 + A\cos\varphi(t_{2s}) \neq 0, \qquad -1 + A\cos\varphi(t_{2s+1}) \neq 0, \qquad s = 0, 1, \dots, \tag{17.3}$$

hold at the switching times t_p, $p = 0, 1, \dots$. Then by Theorem 1.1 the solution $x(t, \varepsilon)$, $y(t, \varepsilon)$, $\varphi(t, \varepsilon)$ of (17.1) with initial conditions $x(0, \varepsilon) = x_0 \leq -1$, $y(0, \varepsilon) = y_0$, and $\varphi(0, \varepsilon) = \varphi_0$ converges as $\varepsilon \to 0$ to the solution $x(t)$, $y(t)$, $\varphi(t)$ of the degenerate system in the C^1-metric, with certain small neighborhoods of the times $0, t_0, t_1, \dots$ excluded.

Theorem 17.1. *Suppose that $A < 1$. Then there exists an $\varepsilon_0 > 0$ such that for all $0 < \varepsilon \leq \varepsilon_0$ system (17.1) has a globally exponentially stable two-dimensional invariant torus.*

For a proof in the plane $y = 0$ we consider the Poincaré operator of return along trajectories of the relay system (17.2). By the conditions (17.3) and the inequality $\dot{y}|_{y=0} = -\Phi_1(0) + A\cos\varphi > 0$, it induces a smooth mapping $\varphi \mapsto f_0(\varphi)$, $f_0(\varphi + 2\pi) = f_0(\varphi) + 2\pi$, of the circle into itself with the property that

$$f_0'(\varphi) > 0. \tag{17.4}$$

Indeed, it follows from the symmetry of system (17.2) that the mapping $\varphi \to f_0(\varphi)$ is the square of the mapping $\varphi \to f_1(\varphi)$, $f_1 = f_3 \circ f_2$, where f_2 is the correspondence mapping along trajectories of the system from the plane $y = 0$ to the plane $y = 2/3$, and f_3 is the correspondence mapping from the plane $y = 2/3$ to the plane $y = 0$ after the switching. Differentiating the equality $y(f_2(\varphi), \varphi) = 2/3$, where $y(s, \varphi)$ is the solution of the Cauchy problem

$$\frac{dy}{ds} = \frac{1}{\omega}(-\Phi_1(y) + A\cos s), \qquad y|_{s=\varphi} = 0,$$

we arrive at the formula

$$f_2'(\varphi) = -\omega h(f_2(\varphi))/[1 + A\cos(f_2(\varphi))] > 0,$$

where $h = h(s)$ is the solution of the Cauchy problem

$$\frac{dh}{ds} = -\frac{1}{\omega}\Phi_1'(y(s, \varphi))h, \qquad h|_{s=\varphi} = -\frac{1}{\omega}(-\Phi_1(0) + A\cos\varphi) < 0.$$

The monotonicity of f_3 is verified analogously.

In the same plane $y = 0$ we consider the operator $\Pi(\varepsilon)$ of return along trajectories of the original relaxation system (17.1). By the foregoing, the following limit equality holds in the C^1-metric:

$$(17.5) \qquad \lim_{\varepsilon\to 0}\Pi(\varepsilon) = \Pi_0, \qquad \Pi_0 = \begin{cases} x \to \Phi_1(0), \\ \varphi \to f_0(\varphi). \end{cases}$$

The limit equality (17.5) together with condition (17.4) ensures [23] the existence for $\Pi(\varepsilon)$ of an exponentially stable invariant curve on which it has the form $\varphi \to f(\varphi, \varepsilon)$, where the function $f(\varphi, \varepsilon)$ converges to $f_0(\varphi)$ in the C^1-metric as $\varepsilon \to 0$. The theorem is proved.

Below we shall be interested in the question of the mechanism for destruction of the invariant torus of system (17.1) as the parameter A passes through 1.

17.2. **Duck trajectories of the van der Pol system.** It was shown in §16 that duck trajectories appear for a relaxation system when the normal switching condition is violated. We apply the technique developed there to system (17.1) for $A > 1$, confining ourselves in view of symmetry to the consideration of singular points with coordinate φ determined from the equality

$$(17.6) \qquad 1 + A\cos\varphi = 0.$$

We have from (17.6) that $\varphi_1 = \pi - \arccos(1/A)$ and $\varphi_2 = \pi + \arccos(1/A)$. Let us first consider the singular point $x = -1$, $y = 2/3$, $\varphi = \varphi_1$. In a neighborhood of it we transform system (17.1) to normal form, making the substitutions $x = -1 + \psi(\xi)$, $y = 2/3 + \eta_1$, and $\varphi = \varphi_1 + \eta_2$, where the function $\psi(\xi)$, $\psi(0) = 0$, is determined from the equation $\psi\sqrt{1 - \frac{1}{3}\psi} = \xi$, $\alpha\eta_2 \to \eta_2$, where $\alpha = (\sqrt{A^2 - 1}/\omega)^{1/2}$, $\beta t \to t$, where $\beta = (\omega\sqrt{A^2 - 1})^{1/2}$. Multiplying the right-hand sides by $\psi'(\xi)$, we get the system

$$(17.7) \qquad \varepsilon\dot{\xi} = \xi^2 + \eta_1, \qquad \dot{\eta}_1 = g_1(\xi, \eta_1, \eta_2), \qquad \dot{\eta}_2 = g_2(\xi, \eta_1, \eta_2),$$

where the functions g_1 and g_2 are analytic in their variables and such that

$$(17.8) \qquad g_1(\xi, \eta_1, \eta_2) = -\eta_2 + a\xi + b\eta_1 + \cdots, \qquad g_2(0, 0, 0) = 1.$$

Here $a = -(\omega\sqrt{A^2 - 1})^{-1/2}$, and the dots stand for terms of higher order of smallness.

The properties (17.8) of the functions g_1 and g_2 show that Theorem 16.1 is applicable, i.e., in a neighborhood of the singular point of (17.7) under consideration there exist two types of duck trajectories: for one the motion takes place from stable slow motion manifolds to unstable manifolds, and for the other the opposite. It is natural to call this kind of singular point a *saddle point*.

We proceed to a consideration of the second singular point $(-1, 2/3, \varphi_2)$ of system (17.1). In a neighborhood of it the normal form has the previous form (17.7), but the first equality in (17.8) must be replaced by

$$(17.9) \qquad g_1(\xi, \eta_1, \eta_2) = \eta_2 + a\xi + b\eta_1 + \cdots.$$

It follows from (17.9) that for $|a| < 2\sqrt{2}$, i.e., for $A > \sqrt{1 + 1/(64\omega^2)}$, the roots of the branching equation (16.64) are complex, and hence system (17.1) does not have duck trajectories in a neighborhood of the singular point under consideration.

Let $A < \sqrt{1 + 1/(64\omega^2)}$. Then equation (16.64) has two positive roots $0 < \kappa_1 < \sqrt{2}$ and $\kappa_2 > \sqrt{2}$, and the statement of Theorem 16.2 is valid in the absence of the resonances (16.65) (by analogy with the above, this singular point will be called a *node*). We remark further that to the equality $\kappa_2/\kappa_1 = n$ there corresponds

$$A_n = \sqrt{1 + 4n/(64\omega^2(n+1)^2)}, \qquad n = 2, \dots, \tag{17.10}$$

i.e., as $A \to 1 + 0$ we successively pass through all resonances.

17.3. **Properties of an unstable cycle.** Denote by $\Phi_+(y)$ $(-2/3 < y < 2/3)$ the root of the equation $y = x^3/3 - x$ with modulus less than 1. Then on the unstable slow motion manifold system (17.1) can be written for $\varepsilon = 0$ in the form

$$\frac{dy}{d\varphi} = \frac{1}{\omega}[-\Phi_+(y) + A\cos\varphi]. \tag{17.11}$$

An uncomplicated check shows that for equation (17.11) with $A > 0$ an unstable cycle $y_0(\varphi, A)$ with period 2π is created from the zero equilibrium state, and it is preserved for all $0 < A < 1$. Indeed, it cannot vanish, because the multiplier of it given by

$$\mu_0(A) = \exp\left\{\int_0^{2\pi} \frac{d\varphi}{\omega[1 - x_0^2(\varphi, A)]}\right\}, \tag{17.12}$$

where $x_0(\varphi, A) = \Phi_+(y_0(\varphi, A))$, is greater than 1 for all the A under consideration. In the coordinates (φ, x) the phase portrait of the slow motion system

$$\dot{x} = -x + A\cos\varphi/(x^2 - 1), \qquad \dot{\varphi} = \omega \tag{17.13}$$

is shown for $A < 1$ in Figure 5. We point out that for all $0 < A < 1$ the cycle $x_0(\varphi, A)$ stays in the domain $|x| < 1$. However, its amplitude grows as A increases, and for $A = 1$ we have the phase portrait shown in Figure 6.

We return to system (17.1). According to results in [24], for $0 < A < 1$ it has an unstable cycle with zero approximation $(x_0(\varphi, A), y_0(\varphi, A), \varphi)$ for which one multiplier $\mu_0(A, \varepsilon)$ tends to the finite limit (17.12) as $\varepsilon \to 0$, while for the other we get the asymptotic representation

$$\mu_1(A, \varepsilon) = \mu_0^{-1}(A)\exp\left\{\frac{1}{\varepsilon}\int_0^{2\pi} \frac{1}{\omega}(1 - x_0^2(\varphi, A))\,d\varphi\right\}(1 + o(1)) \tag{17.14}$$

from the Liouville formula.

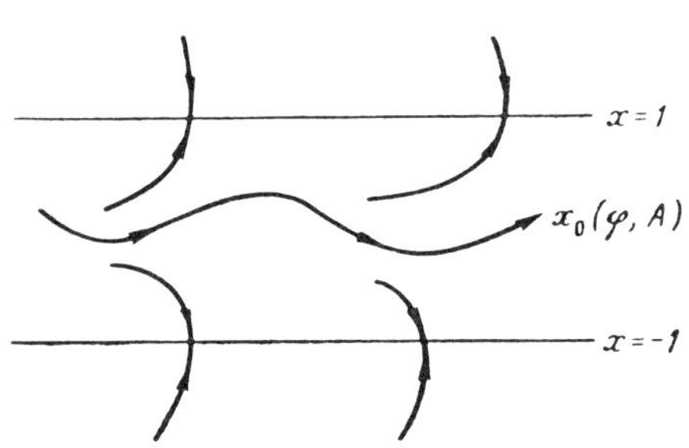

FIGURE 5

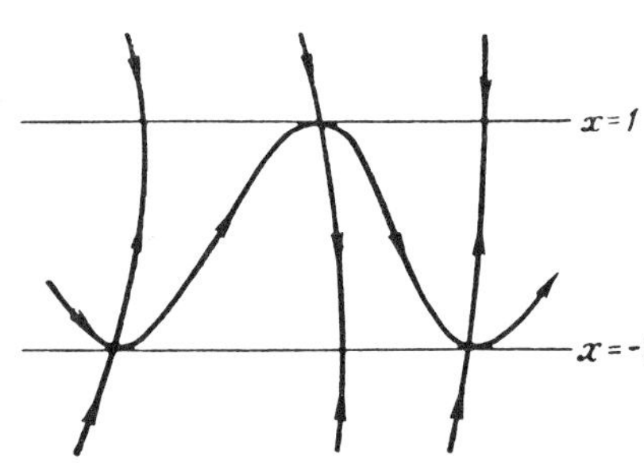

FIGURE 6

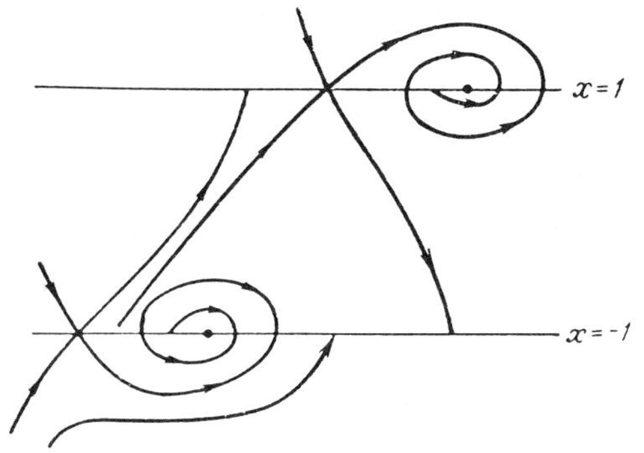

FIGURE 7

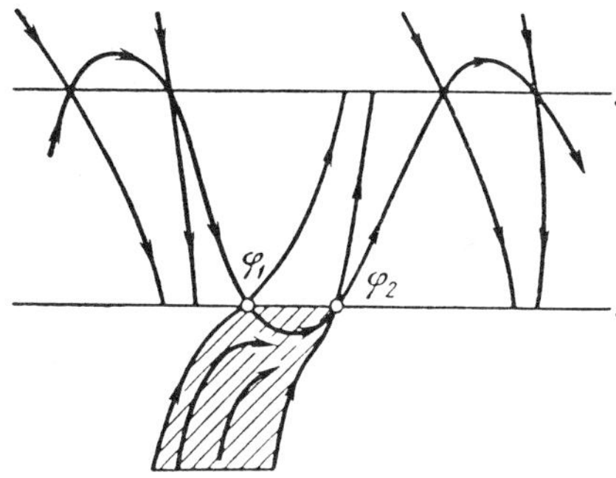

FIGURE 8

We summarize all the foregoing as a theorem.

Theorem 17.2. *Let $A < 1$. Then there is an $\varepsilon_0 > 0$ such that for all $0 < \varepsilon \leq \varepsilon_0$ system (17.1) has an unstable cycle with zero approximation $(x_0(\varphi, A), y_0(\varphi, A), \varphi)$. One of the multipliers of this cycle tends to the limit (17.12) as $\varepsilon \to 0$, while the asymptotic formula (17.14) holds for the other.*

Let $1 < A < \sqrt{1 + 1/(64\omega^2)}$. For these A the phase portrait of system (17.13) is shown in Figure 7. However, now a cycle of the original system (17.1) no longer corresponds to a closed trajectory of the degenerate system passing through the points $\varphi = \varphi_1$ and $\varphi = \varphi_2$ in the general case. Indeed, for the A under consideration two points where the normal switching condition fails appear on each of the breakaway lines $x = -1$ and $x = 1$, one a saddle point (denoted by φ_1 in Figure 8), and the other a node. At the nodal singular point $\varphi = \varphi_2$ the conditions of Theorem 16.2 hold for $A \neq A_n$ (see (17.10)), i.e., we have two duck trajectories, while for $A = A_n$ system (17.1) does not have duck trajectories corresponding to the leading direction of the zero equilibrium state of system (16.9) in the general case. As the parameter A increases from A_{n+1} to A_n the duck trajectory with zero approximation $(-\xi^2, \eta_{20}^2(\xi))$ (see §16.5) wanders in the shaded sector in Figure 8. If a cycle of system (17.1) corresponds to a cycle of the degenerate system, then it must be a duck cycle including the duck trajectory with zero approximation $(-\xi^2, \eta_{20}^2(\xi))$. But the latter is impossible in view of the dynamics described above for this duck trajectory.

We remark that the dissipative property of system (17.1) and the Bol-Brouwer fixed point theorem give us that for all A and ω the system has a cycle of period $2\pi/\omega$. In combination with the foregoing we conclude that for $A > 1$ this cycle is a duck cycle with zero approximation of fairly complex construction.

According to results in §17.2, for $A > \sqrt{1 + 1/(64\omega^2)}$ system (17.1) does not have duck trajectories at the singular point $\varphi = \varphi_2$, and the phase portrait of the degenerate system (17.13) corresponding to it is shown in Figure 8.

17.4. **Destruction of the invariant torus.** We show that for $A > 1$ system (17.1) no longer has the invariant torus mentioned in Theorem 17.1. Assuming the opposite, we consider a neighborhood of the singular point $(-1, 2/3, \varphi_1)$ and take as the manifold $\eta_1 = H_1$ in Theorem 16.1 a piece of the surface of the torus, and as $\eta_1 = H_2$ the manifold containing the duck trajectory going from the chosen stable manifold through the whole unstable manifold (the zero approximation of this duck trajectory is shown in Figure 6). But then the operator $\Pi(\varepsilon)$ of return along trajectories of (17.1), which is defined in the plane $\xi = -q$, will have on the curve $\eta_1 = H_1(-q, \eta_2, \varepsilon)$ a point of discontinuity $\eta_2(\varepsilon)$, where $\eta_2(\varepsilon)$ differs from the initial condition of the duck trajectory described above by a quantity of order $\exp(-\delta/\varepsilon)$ $\delta > 0$. This has to do with the fact that all three of the following possibilities are realized for trajectories of system (17.1) with initial conditions taken on the curve $\eta_1 = H_1(-q, \eta_2, \varepsilon)$ and exponentially close to $\eta_2(\varepsilon)$: the trajectory, proceeding some distance in a neighborhood of the indicated duck trajectory, can break away onto one of the two stable manifolds or can continue moving along the duck trajectory.

Accordingly, the mapping $\Pi(\varepsilon)$ has a point of discontinuity on the assumed invariant curve. We have come to a contradiction, which proves the following assertion.

Theorem 17.3. *Let* $A > 1$. *Then there is an* $\varepsilon_0 > 0$ *such that for all* $0 < \varepsilon \leq \varepsilon_0$ *system* (17.1) *does not have the invariant torus mentioned in Theorem* 17.1.

This theorem, in combination with the results of §§17.2 and 17.3, enables us to give the following intuitive mechanism for destruction of the torus: as $A \to 1 - 0$ the unstable cycle approaches the invariant torus, the motion along the former being perpendicular to the motions on the torus; because of this the torus loses smoothness and tears, and the discontinuities must appear on the duck trajectory. The question of a more detailed mechanism for destruction of the torus and of the accompanying dynamical effects remains open here. Open also is the question of the character of the attractor arising in its place. In particular, the following problem is of interest: is it possible for system (17.1) to have a strange attractor containing duck trajectories?

§18. The connection between the Pontryagin delay phenomenon and the stability of duck cycles of multidimensional relaxation systems with one slow variable

18.1. **Statement of the problem.** For $x \in R^n$ and $y \in R$ we consider a singularly perturbed system of equations

$$\varepsilon\dot{x} = f(x, y), \qquad \dot{y} = g(x, y, \mu), \qquad 0 < \varepsilon \ll 1, \tag{18.1}$$

depending on an auxiliary small parameter μ, with the right-hand sides assumed to be smooth in their variables. Suppose that the equation $f = 0$ determines in R^{n+1} a smooth curve Γ that splits into disjoint parts $\Gamma_{1-}, \Gamma_{10}, \Gamma_+, \Gamma_{20}, \Gamma_{2-}$. By definition, $(x, y) \in \Gamma_{1-} + \Gamma_{2-}$ if all the eigenvalues of the matrix f'_x lie to the left of the imaginary axis; $(x, y) \in \Gamma_+$ if exactly one simple eigenvalue of f'_x is positive and all the others are in the left complex half-plane as before; and at the points Γ_{10} and Γ_{20} the matrix f'_x has a simple zero eigenvalue, and the real parts of all the remaining eigenvalues are negative. Assume that the parts Γ_{1-}, Γ_+, and Γ_{2-}, of the curve Γ are given by the equalities $x = \Phi_{1-}(y)$, $x = \Phi_+(y)$, and $x = \Phi_{2-}(y)$, respectively, where the smooth functions Φ_{1-}, Φ_+, and Φ_{2-} are defined for $(-\infty, 0)$, $(y_0, 0)$, and $(y_0, +\infty)$ (Figure 9), the surface $g(x, y, 0) = 0$ has the unique point Γ_{10} of

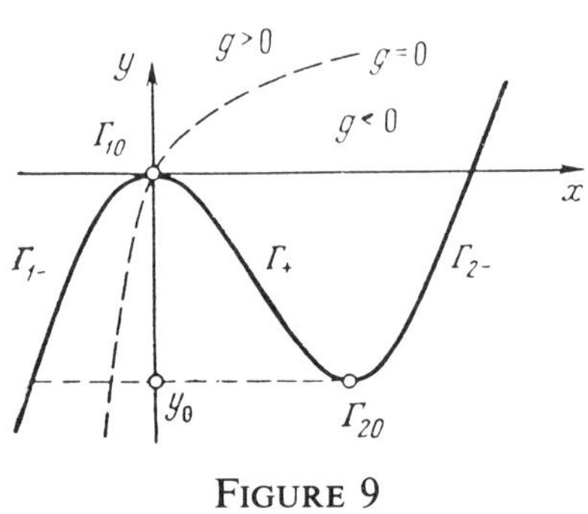

FIGURE 9

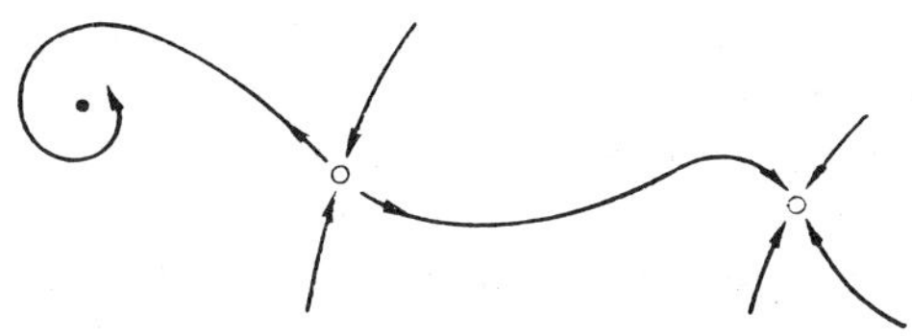

FIGURE 10

intersection with Γ, $g > 0$ on Γ_{1-}, and $g < 0$ at the remaining points of Γ.

The rest of the restrictions are mainly connected with the fast motion differential equation

$$\dot{x} = f(x, y). \tag{18.2}$$

In view of results in [17] this equation has in a neighborhood of the multiple zero equilibrium state an exponentially stable one-dimensional invariant integral manifold on which the motion is described with the help of a scalar equation.

Condition 1. The right-hand side of the scalar equation mentioned above begins with a quadratic term, and the unstable tendril going out from the point $x = 0$ tends as $t \to \infty$ to the exponentially stable equilibrium state $x = \Phi_{2-}(0)$. What happens at the point Γ_{20} is analogous.

Condition 2. At the points of Γ_+ in a neighborhood of the saddle equilibrium state $x = \Phi_+(y)$ of system (18.2) it tends to the equilibrium state $\Phi_{1-}(y)$ as $t \to \infty$ in one unstable direction, and to the equilibrium state $x = \Phi_{2-}(y)$ in the other direction (for $n = 2$ the situation is shown in Figure 10 when Φ_{1-} is a focus and Φ_{2-} a node).

Denote by e and h, $(e, h) = 1$, eigenvectors corresponding to the zero eigenvalue of the matrix $A = f'_x(0, 0)$ and A^*, respectively, assuming that the direction of e coincides with that of the indicated unstable tendril of system (18.2) for $y = 0$.

Condition 3. Assume that $(f'_y(0, 0), h) \neq 0$, and also the analogous inequality at the point Γ_{20}.

Condition 4. Assume that $g'_x(0, 0, 0)e \neq 0$ (here the row vector is applied to a column vector).

Condition 5. Assume that $g'_\mu(0, 0, 0) \neq 0$.

The first three restrictions are typical in this topic (see §1). Condition 4 ensures that the surface $g(x, y, 0) = 0$ intersects the curve Γ in general position, and Condition 5 ensures that the equilibrium state of system (18.1) passes through the point Γ_{10} with nonzero velocity as μ changes.

18.2. **A lemma on the existence of a trajectory joining a stable integral manifold and an unstable integral manifold.** According to results in [17], outside small neighborhoods of the points Γ_{10} and Γ_{20} shown in Figure 9 system (18.1) has an exponentially stable invariant manifold $x = M_{1-}(y, \varepsilon)$ and a dichotomous (with one unstable direction) manifold $x = M_+(y, \varepsilon)$, where the smooth functions M_{1-} and M_+ are such that $M_{1-}(y, 0) = \Phi_{1-}(y)$ and $M_+(y, 0) = \Phi_+(y)$. We remark that these manifolds are not uniquely determined. This is not essential for what follows. We only assume that M_{1-} is defined for values of y somewhat less than y_0, and we must extend the manifold M_+ below to an asymptotically small neighborhood of Γ_{20}. Here, however, we show that there is a value $\mu_{\mathrm{cr}}(\varepsilon)$ of the parameter μ such that the manifolds M_{1-} and M_+ coalesce in a neighborhood of the origin. To prove this fact we use a technique developed in [5].

Let us first transform system (18.1) to normal form in some small neighborhood of the origin. To do this we introduce the variables (ξ_1, ξ_2), $\xi_1 \in R$, $\xi_2 \in R^{n-1}$, in place of the variable x, where the coordinate ξ_1 corresponds to the direction of the vector e, and the group of coordinates ξ_2 corresponds to the directions orthogonal to the vector h. It follows from [17] that system (18.1) has an exponentially stable (with exponent of order ε^{-1}) invariant integral manifold

$$\xi_2 = V(\xi_1, y, \varepsilon, \mu), \qquad |\xi_1|, |y| \le q, \tag{18.3}$$

where the function V depends smoothly on all the arguments, and $q > 0$ is sufficiently small (indeed, after the subsitution $t \to \varepsilon t$ we come to the thoroughly studied question of the center integral manifold). Using (18.3) and (18.1), we get a system of differential equations on the integral manifold

$$\varepsilon\dot{\xi} = f(\xi, \eta, \varepsilon, \mu), \qquad \dot{\eta} = g(\xi, \eta, \varepsilon, \mu), \qquad \xi = \xi_1, \quad \eta = y.$$

With the help of substitutions described in [5] the last system reduces to the normal form (cf. (4.4))

$$\varepsilon\dot{\xi} = \xi^2 + \eta, \qquad \dot{\eta} = \gamma(\xi, \eta, \varepsilon, \mu), \tag{18.4}$$

where

$$\gamma(\xi, \eta, 0, 0) = -\xi + b\eta + \cdots, \qquad \gamma'_\mu(0, 0, 0, 0) \ne 0, \tag{18.5}$$

and the dots stand for terms of higher order of smallness.

Lemma 18.1. *There exists a unique value $\mu_{\text{cr}}(\varepsilon)$ of the parameter μ for which the integral manifolds $x = M_{1-}(y, \varepsilon)$ and $x = M_+(y, \varepsilon)$ of system (18.1) coalesce in a neighborhood of the point Γ_{10}, and it has the asymptotic expansion*

$$\mu_{\text{cr}}(\varepsilon) = \sum_{k=1}^{\infty} \mu_k \varepsilon^k. \tag{18.6}$$

Obviously, the proof reduces to the case of the assertion for the normal form (18.4). We begin by describing a formal scheme for finding the coefficients of (18.6). We pass from (18.4) to the equation

$$d\eta/d\xi = \varepsilon\gamma(\xi, \eta, \varepsilon, \mu)/(\xi^2 + \eta). \tag{18.7}$$

Then we use the expansion (18.6) and find the solutions of (18.7) in the form

$$\eta = -\xi^2 + \sum_{k=1}^{\infty} \varepsilon^k a_k(\xi). \tag{18.8}$$

Finally, equating the coefficients of like powers of ε, we get a recursive sequence of algebraic equations for the functions $a_k(\xi)$, where the constants μ_k are determined from the condition that the functions $a_{k+1}(\xi)$ be bounded as $\xi \to 0$. Note that the function

$$a_1(\xi) = -\gamma(\xi, -\xi^2, 0, 0)/(2\xi)$$

is bounded as $\xi \to 0$ by virtue of the properties of the function γ. For the coefficient $a_2(\xi)$ we have

$$a_2(\xi) = -\frac{\gamma^2(\xi, -\xi^2, 0, 0) - \xi\gamma'_\xi(\xi, -\xi^2, 0, 0)\gamma(\xi, -\xi^2, 0, 0)}{8\xi^4} - \frac{\gamma'_\varepsilon(\xi, -\xi^2, 0, 0) + \gamma'_\mu(\xi, -\xi^2, 0, 0)\mu_1}{2\xi},$$

which gives us that

$$\mu_1 = -\frac{1}{\gamma'_\mu(0,0,0,0)}(\gamma'_\varepsilon(0,0,0,0) + \tfrac{1}{4}\gamma'_\eta(0,0,0,0) + \tfrac{1}{8}\gamma''_{\xi\xi}(0,0,0,0)).$$

The remaining coefficients of (18.6) and (18.8) are determined similarly.

In (18.7) let $\mu = \mu_{k_0}(\varepsilon) + \varepsilon^{k_0-1}\delta$, where

$$\mu_{k_0}(\varepsilon) = \sum_{k=1}^{k_0-2} \mu_k \varepsilon^k, \qquad k_0 \geq 3,$$

and denote by $\eta = F_1(\xi, \varepsilon, \delta)$ and $\eta = F_2(\xi, \varepsilon, \delta)$ solutions of it with initial conditions belonging to the stable manifold and the unstable manifold of system (18.4) and given for $\xi = -q$ and $\xi = q$, respectively. We show that these solutions can be extended to the point $\xi = 0$, and the function $\delta = \delta(\varepsilon)$ is uniquely determined from the equation

$$\varphi(\delta, \varepsilon) \equiv F_1(0, \varepsilon, \delta) - F_2(0, \varepsilon, \delta) = 0. \tag{18.9}$$

Let us make the substitution

$$\eta = -\xi^2 + \sum_{k=1}^{k_0-1} \varepsilon^k a_k(\xi) + \varepsilon z$$

in (18.7), transforming it to the form

$$\varepsilon\, dz/d\xi = b(\xi, \varepsilon, \delta)z + \Phi(z, \xi, \varepsilon, \delta) + \varepsilon^{k_0-1}\psi(\xi, \varepsilon, \delta), \tag{18.10}$$

where the functions b, Φ, and ψ are smooth in their variables and satisfy the conditions

$$\begin{gathered} b(\xi, \varepsilon, \delta) = b_0(\xi) + O(\varepsilon), \qquad \Phi(0, \xi, \varepsilon, \delta) \equiv \partial\Phi(0, \xi, \varepsilon, \delta)/\partial z \equiv 0, \\ b_0(\xi) = 4\xi + O(\xi^2). \end{gathered} \tag{18.11}$$

From (18.10) we pass to the integral equation

$$\begin{aligned} z(\xi, \varepsilon, \delta) = z(-q, \varepsilon, \delta) \exp\left\{\frac{1}{\varepsilon}\int_{-q}^{\xi} b(s, \varepsilon, \delta)\, ds\right\} \\ + \frac{1}{\varepsilon}\int_{-q}^{\xi} \exp\left\{\frac{1}{\varepsilon}\int_{\tau}^{\xi} b(s, \varepsilon, \delta)\, ds\right\} \\ \times [\Phi(z, \tau, \varepsilon, \delta) + \varepsilon^{k_0-1}\psi(\tau, \varepsilon, \delta)]\, d\tau, \end{aligned} \tag{18.12}$$

where

$$z(-q, \varepsilon, \delta) = \varepsilon^{-1}\left(F_1(-q, \varepsilon, \delta) + q^2 - \sum_{k=1}^{k_0-1} \varepsilon^k a_k(-q)\right) = O(\varepsilon^{k_0-1}). \tag{18.13}$$

Using properties (18.11) and (18.13) and the inequality

$$\max_{-q \leq \xi \leq 0} \frac{1}{\varepsilon}\int_{-q}^{\xi} \exp\left\{\frac{1}{\varepsilon}\int_{\tau}^{\xi} b(s, \varepsilon, \delta)\, ds\right\} d\tau \leq M/\sqrt{\varepsilon}, \qquad M > 0, \tag{18.14}$$

we establish in an elementary way that the operator generated by the right-hand side of (18.12) transforms into itself a ball with radius of order $\varepsilon^{k_0-3/2}$ about zero in $C[-q, 0]$ and is a contraction. This uniquely determines a function $z_1(\xi, \varepsilon, \delta)$,

$-q \le \xi \le 0$, smooth in its variables. The function $z_2(\xi, \varepsilon, \delta)$, $0 \le \xi \le q$, is constructed similarly.

Thus the solutions $\eta = F_1(\xi, \varepsilon, \delta)$ and $\eta = F_2(\xi, \varepsilon, \delta)$ of (18.7) have been extended to the point $\xi = 0$. To investigate (18.9) we make more detailed use in (18.12) of the structure of the nonhomogeneity ψ:

$$\psi = \gamma'_\mu(\xi, -\xi^2, 0, 0)\delta/a_1(\xi) + \Psi_0(\xi) + O(\varepsilon), \tag{18.15}$$

where Ψ_0 is a smooth function whose explicit expression is not of interest to us. After this we perform the substitutions $s/\sqrt{\varepsilon} \to s$ and $\tau/\sqrt{\varepsilon} \to \tau$ in the integrals on the right-hand side of (18.12) and use properties (18.11) and (18.15). As a result we arrive at the asymptotic equalities

$$\varphi(\delta, \varepsilon) = \varepsilon^{k_0 - 1/2}(\kappa_1\delta + \kappa_2) + O(\varepsilon^{k_0}), \tag{18.16}$$

$$\partial\varphi/\partial\delta = \varepsilon^{k_0 - 1/2}\kappa_1 + O(\varepsilon_{k_0}), \tag{18.17}$$

where the value of the constant κ_2 is not essential, and

$$\kappa_1 = 2\gamma'_\mu(0, 0, 0, 0)\int_{-\infty}^{\infty} \exp(-2s^2)\, ds \ne 0. \tag{18.18}$$

The lemma is proved.

The proof of an assertion analogous to our lemma is sketched roughly in the monograph [51]. We cite it in connection with the fact that the constructions used will be needed in the next part.

18.3. **Conditions for the existence of stable duck cycles.** Let $\lambda_{1-}(y)$ and $\lambda_{2-}(y)$ be the real parts of those eigenvalues of the matrix f'_x at the branch points Γ_{1-} and Γ_{2-}, respectively, that are closest to the imaginary axis, and let $\lambda_+(y)$ be a simple positive eigenvalue of f'_x with $(x, y) \in \Gamma_+$. We introduce the balance functions

$$H_1(y) = \int_y^0 \frac{\lambda_{1-}(s)}{g(\Phi_{1-}(s), s, 0)}\, ds + \int_0^y \frac{\lambda_+(s)}{g(\Phi_+(s), s, 0)}\, ds, \tag{18.19}$$

$$H_2(y) = \int_{y_0}^0 \frac{\lambda_{1-}(s)}{g(\Phi_{1-}(s), s, 0)}\, ds + \int_0^y \frac{\lambda_+(s)}{g(\Phi_+(s), s, 0)}\, ds + \int_y^{y_0} \frac{\lambda_{2-}(s)}{g(\Phi_{2-}(s), s, 0)}\, ds, \tag{18.20}$$

where $y \in [y_0, 0]$, which play a fundamental role in the question of the existence and stability of the cycles of system (18.1) whose zero approximations are shown in Figures 11 and 12. Here the cycles pictured in Figures 11 and 12a will be called *ducks without a head of height* y_0 *and* y_*, respectvely, while the cycle in Figure 12b will be called a *duck cycle with a head of height* y_*.

In system (18.1) let $\mu = \mu_{\text{cr}}(\varepsilon) + \Delta\mu$, where $\Delta\mu = \pm\exp(-a/\varepsilon)$, $a > 0$. In this case the manifolds M_{1-} and M_+ of system (18.1) break apart, but since $\Delta\mu$ is small the phenomenon (discovered by Pontryagin) of delay of loss of stability is preserved. Denote by $x(t, \varepsilon)$, $y(t, \varepsilon)$ the trajectory of system (18.1) corresponding to the single integral manifold for $\mu = \mu_{\text{cr}}(\varepsilon)$, assuming that $\xi = 0$ corresponds to the time $t = 0$ on it (see §18.2). Then we consider the Cauchy problem with zero initial conditions for the variational system

$$\begin{aligned} \varepsilon\dot{h}_1 &= A(t, \varepsilon)h_1 + B(t, \varepsilon)h_2, \\ \dot{h}_2 &= C(t, \varepsilon)h_1 + D(t, \varepsilon)h_2 + K(t, \varepsilon)\Delta\mu, \end{aligned} \tag{18.21}$$

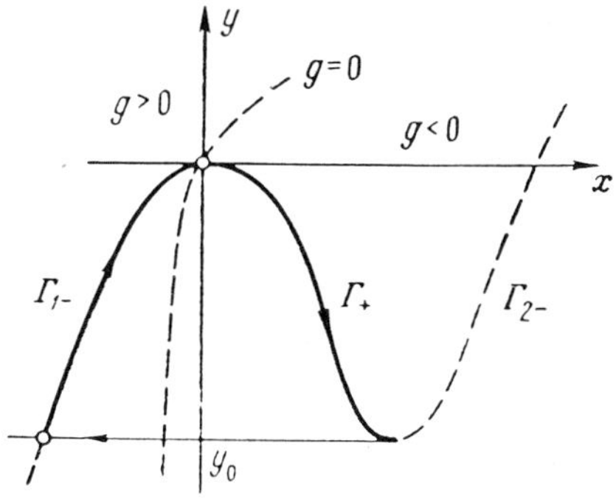

FIGURE 11

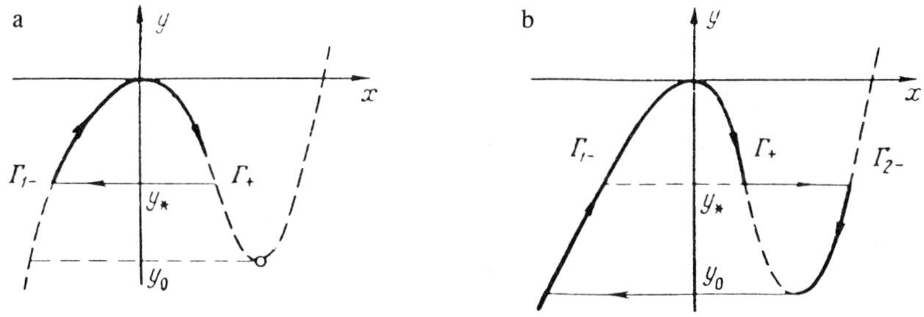

FIGURE 12

where $A = f'_x$, $B = f'_y$, $C = g'_x$, $D = g'_y$, and $K = g'_\mu$, and the derivatives are taken at $x = x(t, \varepsilon)$, $y = y(t, \varepsilon)$, and $\mu = \mu_{\rm cr}(\varepsilon)$. It follows from the restrictions imposed on system (18.1) that for $t \geq 0$ such that the point $(x(t, \varepsilon), y(t, \varepsilon))$ moves along the unstable branch Γ_+ of the curve Γ the matrix $A(t, \varepsilon)$ has a simple nonnegative eigenvalue $\lambda(t, \varepsilon)$ $(\lambda(0, 0) = 0)$ with eigenvector $e(t, \varepsilon)$ $(e(0, 0) = e)$, while the rest of its eigenvalues lie in the left complex half-plane. This circumstance enables us to make in (18.21) a linear substitution with respect to h_1 that splits the matrix $A(t, \varepsilon)$, and to get for the variable $\theta(t, \varepsilon)$ corresponding to the direction $e(t, \varepsilon)$ the asymptotic formula

$$\theta(t, \varepsilon) = (1 + o(1))\theta_0(\varepsilon) \exp\left\{\frac{1}{\varepsilon}\int_0^t \lambda(\tau, 0)\, d\tau\right\}, \tag{18.22}$$

where $\lambda(\tau, 0) = \lambda_+(y(\tau, 0))$, and

$$\theta_0(\varepsilon) = g'_\mu(0, 0, 0)\Delta\mu. \tag{18.23}$$

Formulas (18.22) and (18.23) obviously imply the existence of a duck trajectory of height y_* (perhaps not closed), where $y_* \in [y_0, 0)$ is a root (if it exists) of the equation

$$\int_0^y \frac{\lambda_+(s)}{g(\Phi_+(s), s, 0)}\, ds = a, \tag{18.24}$$

and this is a duck trajectory without head if $\theta_0(\varepsilon) < 0$, and a duck trajectory with head if $\theta_0(\varepsilon) > 0$. The answer to the natural question of whether there are closed duck trajectories is given by the following assertions.

Theorem 18.1. *Suppose that* $\mu = \mu_{\rm cr}(\varepsilon) + \Delta\mu$ *in system* (18.1), *where* $\Delta\mu = \pm\exp(-a/\varepsilon)$, $a > 0$, *and let* y_* *be a root of equation* (18.24). *Then system* (18.1) *has a stable duck cycle without head of height* y_* *for*

$$\theta_0(\varepsilon) < 0, \qquad H_1(y_*) < 0,$$

and a stable duck cycle with head of height y_* *for*

$$\theta_0(\varepsilon) > 0, \qquad H_2(y_*) < 0.$$

Theorem 18.2. *Suppose that in system* (18.1)

$$|\mu - \mu_{\text{cr}}(\varepsilon)| \leq \exp\left\{\frac{1}{\varepsilon}\int_0^{y_0} \frac{\lambda_+(s)}{g(\Phi_+(s), s, 0)}\, ds\right\}.$$

Then for $H_1(y) < 0$ *a duck cycle of maximal height with type shown in Figure* 11 *exists and is stable.*

Let us begin by proving Theorem 18.1. We introduce a hyperplane L intersecting Γ_{1-} in general position at a point with coordinate $\tilde{y} \in (y_*, 0)$ such that

$$\int_{\tilde{y}}^0 \frac{\lambda_{1-}(s)}{g(\Phi_{1-}(s), s, 0)}\, ds + \int_0^{y_*} \frac{\lambda_+(s)}{g(\Phi_+(s), s, 0)}\, ds < 0 \tag{18.25}$$

in the case of a duck without head, while $\tilde{y} \in (y_0, 0)$ is chosen so that (18.26)

$$\int_{\tilde{y}}^0 \frac{\lambda_{1-}(s)}{g(\Phi_{1-}(s), s, 0)}\, ds + \int_0^{y_*} \frac{\lambda_+(s)}{g(\Phi_+(s), s, 0)}\, ds + \int_{y_*}^{y_0} \frac{\lambda_{2-}(s)}{g(\Phi_{2-}(s), s, 0)}\, ds < 0$$

in the case of a duck with head. We remark that in view of the conditions in Theorem 18.1 inequality (18.25) or (18.26) can always be satisfied by choosing $\tilde{y}$ sufficiently close to y_* in the case of a duck without head, and sufficiently close to y_0 in the case of a duck with head. The advantage of such a choice of secant hyperplane is that in this case the operator Π of return along trajectories of system (18.1) transforms into itself some neighborhood (independent of ε) of a point of intersection of L with the duck trajectory (otherwise this property is absent). This follows from the fact that under the inequalities (18.25) (or (18.26)) any trajectory with initial condition in a sufficiently small neighborhood of an intersection of L with the duck trajectory has the same asymptotics. Indeed, first there is a fall onto the stable manifold M_{1-}, which can be described by means of a technique in [8]. We then extend the trajectory up to the point $\xi = 0$ by using the construction in §18.2. After this the deviation of the trajectory from M_+ is estimated from the linear approximation, and by the choice of $\tilde{y}$ and inequalities (18.25) and (18.26) the initial condition for the variational system (18.21) is much less than $\Delta\mu$ in norm, and this allows us to take it to be zero.

Accordingly, the existence of a duck cycle has been established. To prove its uniqueness and stability we note that the variational system on it has the same property as the original nonlinear system: any solution of it with initial conditions independent of ε has the same asymptotics, coinciding with the asymptotics of $\dot{x}(t, \varepsilon)$ and $\dot{y}(t, \varepsilon)$, where $x(t, \varepsilon)$ and $y(t, \varepsilon)$ are the components of the cycle. Therefore, the eigenvalues of the monodromy matrix different from 1 tend to zero as $\varepsilon \to 0$.

The proof of Theorem 18.2 is completely analogous. The only new point here is the possibility (which follows from results in [5]) of extending the integral manifold M_+ to an asymptotically small neighborhood of Γ_{20}.

18.4. **The bifurcation theorem.** Since the equilibrium state in system (18.4) loses stability in an oscillatory way, the well-known Andronov bifurcation theorem is applicable to it. However, it is a priori clear that the limits of its applicability are not great. In this connection we carry out an analysis taking into account the specifics of singularly perturbed equations.

Making the normalizing substitutions

$$\xi = \nu u, \quad \eta = \nu^2 v, \quad \varepsilon = \nu^2, \quad \mu = \alpha\nu^2, \quad t \to \nu t$$

in (18.4), we get (up to ν^2) that

$$\dot{u} = u^2 + v, \qquad \dot{v} = -u + \nu(b_1 v + b_2 u^2 + b_3 + \alpha b_4), \tag{18.27}$$

where $b_1 = \gamma'_\eta$, $b_2 = \frac{1}{2}\gamma''_{\xi\xi}$, $b_3 = \gamma'_\varepsilon$, and $b_4 = \gamma'_\mu$, and the derivatives are taken at the zero arguments. For $\nu = 0$ we multiply the right-hand sides of the system (18.27) by $2(\exp 2v)$, after first replacing u by $-u$. Finally, it becomes a Hamiltonian system with Hamiltonian function

$$H(u, v) = 1/2 + (u^2 + v - 1/2)\exp 2v,$$

for which the level curves $H(u, v) = \delta$ are closed for $0 < \delta < 1/2$. For the indicated δ we need the functions

$$\begin{gathered} I_1(\delta) = \iint_{\Omega_\delta} \exp(2v)\,du\,dv, \qquad I_2(\delta) = -\iint_{\Omega_\delta} v\exp(2v)\,du\,dv, \\ I_3(\delta) = \iint_{\Omega_\delta} u^2\exp(2v)\,du\,dv, \end{gathered} \tag{18.28}$$

where the domain Ω_δ is bounded by the curve $H(u, v) = \delta$.

Theorem 18.3. *System* (18.27) *has a stable* (*unstable*) *cycle with zero approximation* $H(u, v) = \delta_0$ *if*

$$I(\delta_0) = 0, \qquad I'(\delta_0) < 0\ (> 0),$$

where

$$I(\delta) = (b_1 + 2b_3 + 2b_4\alpha)I_1 - 2b_1 I_2 + 2b_2 I_3. \tag{18.29}$$

The proof follows immediately from a well-known theorem of Pontryagin [52].

It is useful to keep in mind that

$$I_1(\delta) \sim \pi\delta, \qquad I_2(\delta) \sim \pi\delta^{3/2}/3, \qquad I_3(\delta) \sim \pi\delta^{3/2}/3, \qquad \delta \to +0, \tag{18.30}$$

$$\begin{gathered} I_1'(\delta) \sim \sqrt{2}\left(\ln\frac{1}{1/2-\delta}\right)^{1/2}, \qquad I_2'(\delta) \sim \frac{1}{3\sqrt{2}}\left(\ln\frac{1}{1/2-\delta}\right)^{3/2}, \\ I_3'(\delta) \sim \frac{1}{3\sqrt{2}}\left(\ln\frac{1}{1/2-\delta}\right)^{3/2}, \qquad \delta \to 1/2 - 0. \end{gathered} \tag{18.31}$$

Formulas (18.28)–(18.31) and Theorems 18.1–18.3 imply the possibility of the relaxation system (18.1) having a stable ordinary relaxation cycle and simultaneously the stable cycle supplied by Theorem 18.3.

Thus, the use of standard methods in the theory of relaxation oscillations enables us to give a sufficiently thorough treatment of the question of stable duck cycles of relaxation systems with a single slow variable.

18.5. **Diffusion instability of duck cycles.** For $x \in R$ we take into account the diffusion terms in system (18.1) and pose Neumann boundary conditions. As a result we get the parabolic relaxation system ($D_1, D_2 > 0$)

$$\varepsilon\frac{\partial x}{\partial t} = D_1\Delta x + f(x, y), \qquad \left.\frac{\partial x}{\partial \nu}\right|_{\partial\Omega} = 0, \tag{18.32}$$

$$\frac{\partial y}{\partial t} = D_2\Delta y + g(x, y, \mu), \qquad \left.\frac{\partial y}{\partial \nu}\right|_{\partial\Omega} = 0. \tag{18.33}$$

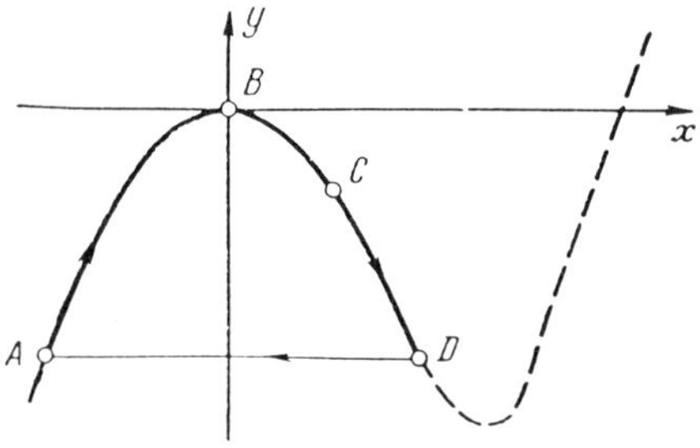

FIGURE 13

Suppose that in (18.32) and (18.33) the functions f and g are such that Theorem 18.1 is applicable, and the parameter $\mu = \mu(\varepsilon)$ is chosen so that the system of ordinary differential equations corresponding to the boundary value problem (18.32), (18.33) has the orbitally exponentially stable duck cycle without head shown in Figure 13. We pose the problem of whether it is stable in the framework of the boundary value problem (18.32), (18.33). Recall (see §11.2) that this problem is equivalent to studying the stability properties of the following linear $T(\varepsilon)$-periodic system ($T(\varepsilon)$ is the period of the duck cycle $x(t, \varepsilon), y(t, \varepsilon)$):

$$\text{(18.34)} \qquad \begin{aligned} \varepsilon\dot{h}_1 &= -z_1h_1 + a(t, \varepsilon)h_1 + b(t, \varepsilon)h_2, \\ \dot{h}_2 &= -z_2h_2 + c(t, \varepsilon)h_1 + d(t, \varepsilon)h_2 \end{aligned}$$

with arbitrary parameters $z_1, z_2 \geq 0$, which is derived just as system (11.7) was derived. Therefore, system (18.34) will be the object of our analysis in what follows.

We assume that the time on the zero approximation $(x_0(t), y_0(t))$ of the duck cycle is chosen so that the point A corresponds to time zero (see Figure 13), the time $t_0 > 0$ corresponds to the point B, and the time $t = T_0$ corresponds to the point D, where T_0 is the zero approximation of $T(\varepsilon)$. Further, as in 11.2, the zero approximations of the coefficients of (18.34) are denoted by $a_0(t)$, $b_0(t)$, $c_0(t)$, and $d_0(t)$.

We note that on the part AB of the duck cycle the coefficient $a_0(t)$ is negative, on the part BD it is positive, and at the point t_0 it has a simple zero. Therefore, if the parameter z_1 in (18.34) is so large that

$$\text{(18.35)} \qquad z_1 > \max_{t_0 \leq t \leq T_0} a_0(t),$$

then the standard technique of asymptotic integration in [8] is applicable to system (18.34), and this leads to the following assertion.

Theorem 18.4. *Suppose that the parameters z_1 and z_2 in* (18.34) *are chosen so that* (18.35) *holds and*

$$z_2 - F_0(z_1) > 0 \ (< 0),$$

where the function F_0 is defined by (11.10). *Then both its multipliers are less than (greater than)* 1 *in modulus.*

It follows from Theorem 18.4 that if the diffusion coefficient D_1 in (18.32) is sufficiently large, then the answer to the question of stability of a homogeneous duck cycle can be formulated just as in the case of an ordinary relaxation cycle (see Theorem 11.3).

Suppose now that inequality (18.35) is violated. For simplicity we assume that on the interval (t_0, T_0) the function $a_0(t) - z_1$ has a unique simple zero t_1, and the point C in Figure 13 corresponds to the time t_1. Then a homogeneous duck cycle of the boundary value problem (18.32), (18.33) is always unstable when there is a certain generality of position.

Theorem 18.5. *If the conditions in the preceding paragraph hold, and*

(18.36) $$c_0(T_0-0)\neq 0, \qquad na_0'(t_1)+b_0(t_1)c_0(t_1)\neq 0, \qquad n=0,1,\dots,$$

then system (18.34) *has a multiplier of order* $\exp(c/\varepsilon)$, $c>0$.

Theorem 18.5 is proved with the help of asymptotic integration of the linear system (18.34). This is carried out in a way completely analogous to the integration of system (16.18): the asymptotic integration technique in [8] is generalized for $0\le t\le t_1-\varepsilon^\lambda$, $0<\lambda<1/2$; for $t_1-\varepsilon^\lambda\le t\le t_1+\varepsilon^\lambda$ we get after the normalizations described in §16 the Weber equation

$$\frac{d^2v}{du^2}-\frac{u\,dv}{du}-\left[1+\frac{b_0(t_1)c_0(t_1)}{a_0'(t_1)}\right]v=0,$$

whose solutions grow like $\exp(u^2/2)$ as $|u|\to\infty$ when the inequalities (18.36) hold. But the latter means that at the time $t_1+\delta$, $\delta>0$, the functions h_1 and h_2 take values of order $\exp(c/\varepsilon)$, $c>0$. For $t_1+\delta\le t\le T(\varepsilon)$ the technique in [8] again works, and together with the inequality $c_0(T_0-0)\neq 0$ leads to the statement of the theorem.

In conclusion we note that the results given here are not of interest from an applied point of view, since the duck cycles are observed in a very narrow region of the parameters. However, Theorem 18.5 together with the content of the next section show that stable duck cycles are not typical for relaxation systems with many slow variables.

§19. Duck trajectories appearing as an equilibrium state of a relaxation system passes through a breakaway manifold. Instability of duck cycles

19.1. **Statement of the problem.** Below we confine ourselves for simplicity to the case of a single fast variable, i.e., for $x\in R$ and $y\in R^m$, $m\ge 2$, we consider the relaxation system

(19.1) $$\varepsilon\dot x=f(x,y), \qquad \dot y=g(x,y,\mu), \qquad 0<\varepsilon\ll 1, \quad \mu\in R^m,$$

with smooth right-hand sides. We assume that $(0,0)$ is a breakaway point, i.e., [5]

(19.2) $$f(0,0)=0, \qquad f_x'(0,0)=0, \qquad f_{xx}''(0,0)\neq 0.$$

In addition we assume that

(19.3) $$g(0,0,0)=0, \qquad \det g_\mu'(0,0,0)\neq 0,$$

(19.4) $$(\operatorname{grad}_y f(0,0),\, g_x'(0,0,0))\operatorname{sgn} f_{xx}''(0,0)<0.$$

The conditions (19.3) mean that system (19.1) has an equilibrium state that passes through the breakaway manifold in general manner as the parameter μ changes. Inequality (19.4), which is analogous to Condition 4 of §18, ensures that the degenerate system corresponding to (19.1) has a duck trajectory [7], i.e., a trajectory along which the phase point falls in finite time from the stable slow motion surface $f=0$, $f_x'<0$, to the unstable surface $f=0$, $f_x'>0$. For a proof we use the substitutions described in §4.1 to transform system (19.1) to the normal form

(19.5) $$\varepsilon\dot\xi=\xi^2+\eta_1, \qquad \dot\eta_1=g_1(\xi,\eta_1,\eta_2,\varepsilon,\mu), \qquad \dot\eta_2=g_2(\xi,\eta_1,\eta_2,\varepsilon,\mu).$$

Here $\eta_1 \in R$, $\eta_2 \in R^{m-1}$, and

(19.6) $g_1|_{\varepsilon=0,\mu=0} = -\xi + a_{11}\eta_1 + a_{12}\eta_2 + \cdots$, $\qquad g_2|_{\varepsilon=0,\mu=0} = a_{21}\eta_1 + A_{22}\eta_2 + \cdots$,

where the dots stand for terms of higher order of smallness. Setting $\varepsilon = 0$ and $\mu = 0$ in (19.5) and taking ξ as the new time, we get the inequalities

$$\eta_{10}(\xi) = -\xi^2, \qquad \frac{d\eta_{20}}{d\xi} = -\frac{2\xi g_2(\xi, -\xi^2, \eta_{20}, 0, 0)}{g_1(\xi, -\xi^2, \eta_{20}, 0, 0)}. \tag{19.7}$$

The differential equation for η_{20} has in a neighborhood of $\xi = 0$ a unique smooth solution $\eta_{20}(\xi) = -\frac{2}{3}a_{21}\xi^3 + \cdots$ with $\dot{\xi} = -g_1(\xi, -\xi^2, \eta_{20}(\xi), 0, 0)/(2\xi) \sim 1$ as $\xi \to 0$. This follows from the fact that in a neighborhood of the zero equilibrium state of the system

$$\dot{\xi} = g_1(\xi, -\xi^2, \eta_{20}, 0, 0), \qquad \eta_{20} = -2\xi g_2(\xi, -\xi^2, \eta_{20}, 0, 0),$$

there is a unique smooth invariant curve $\eta_{20} = \eta_{20}(\xi)$ along which the solutions fade exponentially.

For purely technical reasons it is convenient in what follows to work with a system (19.5) whose right-hand sides are defined in a sufficiently broad neighborhood of the point $(0, 0, 0)$.

19.2. **The main lemma.** In neighborhoods of the points $\xi = -q$ and $\xi = q$, $q > 0$, system (19.5) has [17] invariant integral manifolds $\eta_1 = H_1(\xi, \eta_2, \varepsilon, \mu)$ and $\eta_1 = H_2(\xi, \eta_2, \varepsilon, \mu)$, respectively, where $H_1|_{\varepsilon=0} \equiv H_2|_{\varepsilon=0} = -\xi^2$, of which the first is exponentially stable with exponent of order ε^{-1}, and the second is unstable with the same exponent. These manifolds (as was shown in §16) are closely connected with the duck trajectories of system (19.5)—the trajectories of it with zero approximation $(-\xi^2, \eta_{20}(\xi))$—because a solution of the boundary value problem

$$\frac{d\eta_1}{d\xi} = \frac{\varepsilon g_1(\xi, \eta_1, \eta_2, \varepsilon, \mu)}{\xi^2 + \eta_1}, \qquad \frac{d\eta_2}{d\xi} = \frac{\varepsilon g_2(\xi, \eta_1, \eta_2, \varepsilon, \mu)}{\xi^2 + \eta_1}, \tag{19.8}$$

$$\eta_1|_{\xi=-q} = H_1|_{\xi=-q}, \qquad \eta_1|_{\xi=q} = H_2|_{\xi=q}, \tag{19.9}$$

is obviously a duck trajectory.

Lemma 19.1. *There is a smooth function* $\mu_{\text{cr}}(\alpha, \varepsilon)$ *such that for* $\mu = \mu_{\text{cr}}(\alpha, \varepsilon)$ *the solution of system* (19.8) *with initial conditions*

$$\eta_1|_{\xi=-q} = H_1|_{\xi=-q}, \qquad \eta_2|_{\xi=-q} = \eta_{20}(-q) + \varepsilon\alpha \tag{19.10}$$

satisfies the boundary conditions (19.9). *Moreover, the following asymptotic expansions are valid*:

$$\mu_{\text{cr}}(\alpha, \varepsilon) = \sum_{k=1}^{\infty} \mu_k(\alpha)\varepsilon^k, \tag{19.11}$$

$$\eta_1 = -\xi^2 + \sum_{k=1}^{\infty} a_k(\xi, \alpha)\varepsilon^k, \qquad \eta_2 = \eta_{20}(\xi) + \sum_{k=1}^{\infty} b_k(\xi, \alpha)\varepsilon^k. \tag{19.12}$$

We dwell first on a formalism for determining the coefficients of the series (19.11) and (19.12). Substituting them in (19.8) and equating the expressions for the same powers of ε, we get the recursive sequence of equations

$$\begin{aligned}\frac{g'_{1\mu}(\xi, -\xi^2, \eta_{20}, 0, 0)\mu_k}{a_1} &- \frac{g_1(\xi, -\xi^2, \eta_{20}, 0, 0)}{a_1^2}a_{k+1} \\ &+ \frac{g'_1\eta_2(\xi, -\xi^2, \eta_{20}, 0, 0)}{a_1}b_k = \varphi_k(\xi, \alpha),\end{aligned} \tag{19.13}$$

$$\begin{aligned}\frac{db_k}{d\xi} = &\frac{g'_{2\mu}(\xi, -\xi^2, \eta_{20}, 0, 0)\mu_k}{a_1} + \frac{g'_{2\eta_2}(\xi, -\xi^2, \eta_{20}, 0, 0)}{a_1}b_k \\ &- \frac{g_2(\xi, -\xi^2, \eta_{20}, 0, 0)}{a_1^2}a_{k+1} + \psi_k(\xi, \alpha),\end{aligned} \tag{19.14}$$

where $k = 1, 2, \dots$, $a_1 = -g_1(\xi, -\xi^2, \eta_{20}, 0, 0)/(2\xi) = 1/2 + \cdots$, and the nonhomogeneities φ_k and ψ_k depend only on the a_r, b_s, and μ_j with indices $r \le k$ and $s, j \le k-1$. Expressing a_{k+1} from (19.13) and substituting in (19.14), we get for determining the functions b_k differential equations of the form

$$\frac{db_k}{d\xi} = \kappa(\xi)b_k + F_k(\xi), \qquad b_k(-q) = \begin{cases} \alpha & \text{for } k = 1, \\ 0 & \text{for } k > 1, \end{cases}$$

where the matrix $\kappa(\xi)$ and the vector-valued function F_k are smooth in a neighborhood of $\xi = 0$. Returning to (19.13), we conclude that to find from it a smooth function $a_{k+1}(\xi, \alpha)$ it is necessary that

$$2g'_{1\mu}(0, 0, 0, 0)\mu_k = \varphi_k(0, \alpha), \qquad b_k|_{\xi=0} = 0, \tag{19.15}$$

from which we determine the unknown vector μ_k (the system of linear equations (19.15) is uniquely solvable in view of (19.3)).

Accordingly, we have described a formalism for finding the series (19.11) and (19.12). After this the proof of the lemma is completed by the standard scheme: let $\mu = \sum_{k=1}^{k_0} \mu_k(\alpha)\varepsilon^k + \delta_{k_0}\varepsilon^{k_0+1}$ in (19.8) and (19.9); show that the solutions of (19.8) with the initial condition (19.10) and the initial condition

$$\eta_1|_{\xi=q} = H_2|_{\xi=q}, \qquad \eta_2|_{\xi=q} = \eta_{20}(q) + \sum_{k=1}^{k_0} \varepsilon^k b_k(q, \alpha),$$

respectively, can be extended to the point $\xi = 0$; and find the unknown function $\delta_{k_0}(\alpha, \varepsilon)$ from the condition that these solutions coincide at $\xi = 0$. The only difficulty arising in the implementation of this scheme is the analysis of the linear system

$$\varepsilon\, dh_1/d\xi = a(\xi, \varepsilon)h_1 + b(\xi, \varepsilon)h_2, \qquad dh_2/d\xi = c(\xi, \varepsilon)h_1 + d(\xi, \varepsilon)h_2, \tag{19.16}$$

which is the linearization of (19.8) on the partial sums of (19.12). In (19.16) all the coefficients depend smoothly on their arguments, and

$$\begin{aligned} a(\xi, 0) &= 4\xi + \cdots, & b(0, 0) &= 2g'_{1\eta_2}(0, 0, 0, 0, 0), \\ c(0, 0) &= 0, & d(0, 0) &= 2g'_{2\eta_2}(0, 0, 0, 0, 0). \end{aligned} \tag{19.17}$$

Therefore, the properties of system (19.16) are analogous in large respect to those of system (16.18), studied in §16. In particular, here also it is necessary to construct for $-q \le \xi \le -\varepsilon^\lambda$ and $-\varepsilon^\lambda \le \xi \le 0$, $0 < \lambda < 1/2$, the asymptotics of the integral manifold $h_1 = \omega(\xi, \varepsilon)h_2$, where the row matrix $\omega(\xi, \varepsilon)$ is found from the equation

$$\varepsilon\, d\omega/d\xi + \varepsilon\omega c(\xi, \varepsilon)\omega = a\omega + b.$$

Without dwelling on the computations, we note that the solutions of (19.16) with initial conditions independent of ε behave for $\xi = -q$ as follows: the component $h_2(\xi, \varepsilon)$ remains bounded uniformly with respect to ε on the whole interval $[-q, 0]$, and the component h_1 is bounded on any interval $[-q, -q_1]$, $q_1 > 0$, but has order $\varepsilon^{-1/2}$ at the point $\xi = 0$.

19.3. Instability of duck cycles. Assume that the degenerate system (19.7) has a duck trajectory such that $\eta_{20}(-q_0) = \eta_{20}(q_0)$, where $q_0 > q$, and

$$-\int_{-q}^{q_0} 4\xi^2 \, d\xi / g_1(\xi, -\xi^2, \eta_{20}(\xi), 0, 0) < 0. \tag{19.18}$$

Under analogous conditions a relaxation system on the plane has a stable duck cycle, while if the inequality opposite to that in (19.18) holds for $q = q_0$, then any cycle with the zero approximation under consideration is unstable.

Suppose that in (19.5) $\mu = \mu_{\mathrm{cr}}(\alpha, \varepsilon) + \Delta\mu$, $\Delta\mu = \mu_0 \exp(-\delta_0/\varepsilon)$, where

$$\begin{gathered} g'_{1\mu}(0, 0, 0, 0, 0)\mu_0 < 0, \\ \delta_0 = -\int_0^{q_0} 4\xi^2 \, d\xi / g_1(\xi, -\xi^2, \eta_{20}(\xi), 0, 0) > 0. \end{gathered} \tag{19.19}$$

It follows from Lemma 19.1 and conditions (19.18) and (19.19) that the trajectories of system (19.5) with initial conditions $\eta_1|_{\xi=-q} = H_1|_{\xi=-q} + \varepsilon\beta$ and $\eta_2|_{\xi=-q} = \eta_{20}(-q) + \varepsilon\alpha$ are at a distance of order ε from the above closed duck trajectory of the degenerate system. Therefore, the operator

$$\Pi \colon \begin{array}{l} \beta \to \Delta(\alpha, \beta, \varepsilon) \exp(-\gamma/\varepsilon), \\ \alpha \to \Phi(\alpha, \beta, \varepsilon) \end{array} \tag{19.20}$$

is defined in the plane $\xi = -q$, where $\gamma > 0$, Δ and Φ are smooth functions, and

$$\Phi(\alpha, 0, \varepsilon) = B\alpha + b + O(\varepsilon), \qquad \Phi'_\alpha(\alpha, 0, \varepsilon) = B + O(\varepsilon). \tag{19.21}$$

Using the expansions (19.11) and (19.12) in (19.8), we can get for the matrix B in (19.21) an explicit formula from which it follows that 1 is not an eigenvalue when

$$\det[b'_{1\alpha}(-q_0, \alpha) - b'_{1\alpha}(q_0, \alpha)] \neq 0. \tag{19.22}$$

Since $b_1(\xi, \alpha)$ depends linearly on α, a check of condition (19.22) reduces to a computation of the determinant of a constant matrix.

Theorem 19.1. *Assume the inequality* (19.22). *Then for* $\mu = \mu_{\mathrm{cr}}(\alpha(\varepsilon), \varepsilon) + \Delta\mu$, *where* $\alpha(\varepsilon) = (I - B)^1 b + O(\varepsilon)$ *is a component of a fixed point of the operator* (19.20), *the system* (19.5) *has a unique duck cycle with the zero approximation described above, which is unstable if* $a_{12} \cdot a_{21} \neq 0$.

The instability assertion of the theorem follows from the properties of system (19.16): its solutions with initial conditions independent of ε and given for $\xi = -q$ become of order $\exp(\delta/\varepsilon)$, $\delta > 0$, for $\xi = q$.

It can be shown that an a priori existing duck cycle of arbitrary structure is unstable in the general case. Thus, in contrast to systems with a single slow variable, the creation of an ordinary relaxation cycle in system (19.1) is not connected with duck cycles.

BIBLIOGRAPHY

1. E. F. Mishchenko and L. S. Pontryagin, *Periodic solutions of systems of differential equations close to discontinuous solutions*, Dokl. Akad. Nauk SSSR **102** (1955), 889–891. (Russian)

2. L. S. Pontryagin, *The asymptotic behavior of solutions of systems of differential equations with a small parameter multiplying the higher derivatives*, Izv. Akad. Nauk SSSR Ser. Mat. **21** (1957), 605–626; English transl. in Amer. Math. Soc. Transl. (2) **18** (1961).

3. E. F. Mishchenko and L. S. Pontryagin, *Proof of some asymptotic formulas for solutions of differential equations with a small parameter*, Dokl. Akad. Nauk SSSR **120** (1958), 967–969. (Russian)

4. ——, *Derivation of some asymptotic estimates for solutions of differential equations with a small parameter multiplying the derivatives*, Izv. Akad. Nauk SSSR Ser. Mat. **23** (1959), 643–660. (Russian)

5. E. F. Mishchenko and N. Kh. Rozov, *Differential equations with a small parameters and relaxation oscillations*, "Nauka", Moscow, 1975; English transl., Plenum Press, New York, 1980.

6. A. A. Andronov, A. A. Vitt, and S. E. Khaĭkin, *Theory of oscillations*, 2nd ed., Fizmatgiz, Moscow, 1959; English transl., Pergmamon Press, Oxford, and Addison-Wesley, Reading, Mass., 1966.

7. A. K. Zvonkin and M. A. Shubin, *Nonstandard analysis and singular perturbations of ordinary differential equations*, Uspekhi Mat. Nauk **39** (1984), no. 2 (236), 77–129; English transl. in Russian Math. Surveys **39** (1984).

8. A. B. Vasil′eva and V. F. Butuzov, *Asymptotic expansions of solutions of singularly perturbed equations*, "Nauka", Moscow, 1973. (Russian)

9. A. Yu. Kolesov et al., *Asymptotic integration of the variational system of a multidimensional relaxation cycle.* I, II, Differentsial′nye Uravneniya **23** (1987), 1881–1889, 2036–2047; English transl. in Differental Equations **23** (1987).

10. ——, *A relaxation system in a neighborhood of a breakaway point: reduction to the regular case*, Uspekhi Mat. Nauk **43** (1988), no. 2 (260), 141–142; English transl. in Russian Math. Surveys **43** (1988).

11. A. Yu. Kolesov and E. F. Mishchenko, *Asymptotics of relaxation oscillations*, Mat. Sb. **137 (179)** (1988), 3–18; English transl. in Math. USSR Sb. **65** (1990).

12. A. Yu. Kolesov et al., *On chaos phenomena in three-dimensional relaxation systems*, Mat. Zametki **46** (1989), no. 2, 153–155. (Russian) *

13. A. Yu. Kolesov and E. F. Mishchenko, *On a problem of Pontryagin*, Differentsial′nye Uravneniya **25** (1989), 1888–1891; English transl. in Differential Equations **25** (1989).

14. ——, *Existence and stability of a relaxation torus*, Uspekhi Mat. Nauk **44** (1989), no. 3 (267), 161–162; English transl. in Russian Math. Surveys **44** (1989).

15. Yu. S. Kolesov, *Periodic solutions of relay systems with distributed parameters*, Mat. Sb. **83 (125)** (1970), 349–371; English transl. in Math. USSR Sb. **12** (1970).

16. Carles Bonet, *Singular perturbation of relaxed periodic orbits*, J. Differential Equations **66** (1987), 301–339.

17. Yu. A. Mitropol′skiĭ and O. B. Lykova, *Integral manifolds in nonlinear mechanics*, "Nauka", Moscow, 1973. (Russian)

18. L. P. Shil′nikov, *On a problem of Poincaré and Birkhoff*, Mat. Sb. **74 (116)** (1967), 378–397; English transl. in Math. USSR Sb. **3** (1967).

19. A. N. Sharkovskiĭ, Yu. L. Maĭsterenko, and E. Yu. Romanenko, *Difference equations and their applications*, "Naukova Dumka", Kiev, 1986. (Russian)

20. L. S. Pontryagin and L. V. Rodygin, *Approximation solution of a system of ordinary differential equations with a small parameter multiplying the derivatives*, Dokl. Akad. Nauk SSSR **131** (1960), 255–258; English transl. in Dokl. Akad. Nauk SSSR **1** (1960).

21. ——, *A periodic solution of a system of ordinary differential equations with a small parameter multiplying the derivatives*, Dokl. Akad. Nauk SSSR **132** (1960), 537–540; English transl. in Soviet Math. Dokl. **1** (1960).

22. L. S. Pontryagin and E. F. Mishchenko, *Some questions in the theory of differential equations with a small parameter*, Trudy Mat. Inst. Steklov. **169** (1985), 99–118; English transl. in Proc. Steklov Inst. Math. **1986**, no. 4 (169).

23. N. N. Bogolyubov and Yu. A. Mitropol′skiĭ, *Asymptotic methods in the theory of nonlinear oscillations*, 4th ed., "Nauka", Moscow, 1974; English transl. of 2nd ed., Gordon and Breach, New York, 1962.

24. D. V. Anosov, *On limit cycles of systems of differential equations with a small parameter multiplying the highest derivatives*, Mat. Sb. **50 (92)** (1960), 299–334; English transl. in Amer. Math. Soc. Transl. (2) **33** (1963).

---

* *Editor's note.* According to MR **91a**:58162, the content of this note is the same as their paper, *A mixing attractor in relaxation systems*, Dokl. Akad. Nauk SSSR **306** (1989), 38–40; English transl. in Soviet Math. Dokl. **39** (1989).

25. A. Yu. Kolesov and Yu. S. Kolesov, *Construction of the normal form in a neighborhood of a cycle with the help of the Krylov-Bogolyubov-Mitropol'skiĭ asymptotic method*, Differentsial'nye Uravneniya **24** (1988), 891–894. (Russian)
26. M. T. Grekhova, editor, *Auto-wave processes in systems with diffusion*, Gor'kov. Gos. Univ., Gorky, 1981. (Russian)
27. G. R. Ivanitskiĭ, V. I. Krinskiĭ, and E. E. Sel'kov, *Mathematical biophysics of a cell*, "Nauka", Moscow, 1978. (Russian)
28. A. Yu. Kolesov, *Parabolic relaxation systems*, Dokl. Akad. Nauk SSSR **306** (1989), 1297–1300; English transl. in Soviet Math. Phys. Dokl. **39** (1989).
29. A. M. Turing, *The chemical basis of morphogenesis*, Philos. Trans. Roy. Soc. London Ser. B. **237** (1952/54), 37–72.
30. A. Yu. Kolesov, *Stability of relaxation auto-oscillations in systems with diffusion*, Dokl. Akad. Nauk SSSR **294** (1987), 575–578; English transl. in Soviet Phys. Dokl. **32** (1987).
31. ——, *Properties of a certain linear system connected with the stability of a relaxation cycle in media with diffusion*, Uspekhi Mat. Nauk **43** (1988), no. 3 (261), 185–186; English transl. in Russian Math. Surveys **43** (1988).
32. A. Yu. Kolesov, et al., *Multidimensional relaxation oscillations in media with diffusion*, Dokl. Akad. Nauk SSSR **302** (1988), 1312–1315; English transl. in Soviet Math. Dokl. **38** (1989).
33. Yu. S. Kolesov, *The Schauder principle and stability of periodic solutions*, Dokl. Akad. Nauk SSSR **188** (1969), 1234–1236; English transl. in Soviet Math. Dokl. **10** (1969).
34. ——, *An investigation of stability of solutions of parabolic second-order equations in the critical case*, Izv. Akad. Nauk SSSR Ser. Mat. **33** (1969), 1356–1372; English transl. in Math. USSR Izv. **3** (1969).
35. A. M. Il'in, A. S. Kalashnikov, and O. A. Oleĭnik, *Second-order linear equations of parabolic type*, Uspekhi Mat. Nauk **17** (1962), no. 3 (105), 3–146; English transl. in Russian Math. Surveys **17** (1962).
36. O. A. Ladyzhenskaya, V. A. Solonnikov, and N. N. Ural'tseva, *Linear and quasilinear equations of parabolic type*, "Nauka", Moscow, 1967; English transl., Amer. Math. Soc., Providence, RI, 1968.
37. V. V. Strygin and V. A. Sobolev, *Separation of motions by the method of integral manifolds*, "Nauka", Moscow, 1988. (Russian)
38. A. Yu. Kolesov, *Migration effects in a single-species biocenosis*, Nonlinear Oscillations and Ecology (Yu. S. Kolesov, editor), Yaroslav. Gos. Univ., Yaroslavl, 1984, pp. 34–61. (Russian)
39. Yu. S. Kolesov, *The host-parasite problem*, Dynamics of Biological Populations, Gor'kov. Gos. Univ., Gorky, 1984, pp. 16–29. (Russian)
40. Philip Hartman, *Ordinary differential equations*, Wiley, New York, 1964.
41. A. N. Kulikov, *On smooth invariant manifolds of a semigroup of nonlinear operators acting in a Banach space*, Studies in Stability and the Theory of Oscillations (Yu. S. Kolesov and P. P. Zabreiko, editors), Yaroslav. Gos. Univ., Yaroslavl, 1976, pp. 67–85. (Russian)
42. Yu. S. Kolesov, *The problem of adequacy of ecological equations*, Preprint, No. 1901–85, Yaroslav. Gos. Univ., Yaroslavl, 1985 = Manuscript deposited at VINITI, 1985. (Russian)
43. Yu. S. Kolesov and A. Yu. Kolesov, *Biological and physical parabolic systems*, Mathematical Models in Biology and Medicine, vyp. 2 (D. I. Shvitra, editor), Inst. Mat. i Kibernet. Akad. Nauk Litovsk. SSR, Vilnius, 1987, pp. 5–23. (Russian)
44. ——, *Properties of biological and physical parabolic systems*, Mathematical Models in Problems in the Rational Use of Nature, Rostov-on-Don, 1986, pp. 106–107. (Russian)
45. Yu. S. Kolesov and V. V. Maĭorov, *Spatial and temporal self-organization in single-species biocenosis*, Dynamics of Biological Populations, Gor'kov. Univ., Gorky, 1986, pp. 3–18. (Russian)
46. A. Yu. Kolesov, *The structure of a neighborhood of a homogeneous cycle in a medium with diffusion*, Izv. Akad. Nauk SSSR Ser. Mat. **53** (1989), 345–362; English transl. in Math. USSR Izv. **34** (1990).
47. Yu. S. Kolesov, *Mathematical models of ecology*, Studies in Stability and the Theory of Oscillations (Yu. S. Kolesov, editor), Yaroslav. Gos. Univ., Yaroslavl, 1979, pp. 3–40. (Russian)
48. A. N. Kulikov, *On bifurcations of the creation of invariant tori*, Studies in Stability and the Theory of Oscillations (Yu. S. Kolesov, editor), Yaroslav. Gos. Univ., Yaroslavl, 1983, pp. 112–117. (Russian)
49. V. I. Arnol'd, *Supplementary chapters in the theory of ordinary differential equations*, "Nauka", Moscow, 1978; English transl., *Geometric methods in the theory of ordinary differential equations*, Springer-Verlag, Berlin, and New York, 1982.

50. N. K. Gavrikov, *On bifurcations of periodic motions close to an internal resonance* 1:3, Studies in Stability and the Theory of Oscillations (Yu. S. Kolesov and P. P. Zabreĭko, editors), Yaroslav. Gos. Univ., Yaroslavl, 1977, pp. 192–199. (Russian)

51. Johan Grasman, *Asymptotic methods for relaxation oscillations and applications*, Springer-Verlag, Berlin, 1987.

52. N. N. Bautin and E. A. Leontovich [-Andronova], *Methods and rules for the qualitative study of dynamical systems on the plane*, "Nauka", Moscow, 1976. (Russian)

Translated by H. H. McFADEN

Труды Матем. Инст. Стеклова
197 (1991)

Proceedings of the Steklov
Institute of Mathematics
1993, Issue 1

REGIMES WITH INCREASINGLY MORE FREQUENT SWITCHINGS IN OPTIMAL CONTROL PROBLEMS

UDC 517.977

M. I. ZELIKIN AND V. F. BORISOV

INTRODUCTION

In 1960 at the First International Congress of IFAC in Moscow A. T. Fuller [1] gave an example of a two-dimensional linear control system with quadratic performance criterion in which an optimal control has infinitely many switching points on a finite time interval. This behavior later came to be called the "Fuller phenomenon", and the corresponding control regime became known in the English-language literature as a "chattering" regime. As the Russian equivalent of this term we propose using the term "regime with increasingly more frequent switchings".

An explicit solution of the Fuller problem was obtained only due to the fact that it has a one-parameter group of symmetries. The work of Fuller met with great interest, but was perceived rather as curious pathology. It was forgotten soon after publication. Interest in the Fuller phenomenon was reawakened thirty years later when other examples with optimal chattering regimes began to appear in publications (see [2]–[14]). These were mainly two-dimensional problems having a symmetry group like thc Fullcr problcm.

Interest in the Fuller phenomenon has recently become even more heightened due, in particular, to the intensive efforts to realize the true nature of such a phenomenon as regular synthesis. The concept of regular synthesis—a system of axioms that must be satisfied by "typical" synthesis of optimal trajectories—was introduced by Boltyanskiĭ [15] for proving sufficient conditions for optimality. However, typical examples of problems were later published (in [16]–[18], and elsewhere) in which different ones of the Boltyanskiĭ axioms failed. One of the essential requirements of the Boltyanskiĭ axiomatics is that on any finite segment of a synthesis trajectory there are finitely many control switching points. This axiom leads immediately to the question of how typical the Fuller phenomenon is. This question was answered in the now well-known papers [19] and [20] of Kupka, where it is proved that for the Hamiltonian systems of the Pontryagin maximum principle with a one-dimensional control and with phase space of dimension ≥ 8 there exists at least one trajectory with infinitely many switchings on a finite time interval in the general position case. Borisov [21] later showed that the trajectorywise equivalence of synthesis of optimal trajectories to the synthesis in the Fuller problem is a situation of finite codimension in the set of two-dimensional control systems with an integral performance criterion.

There is an important connection between the existence of a chattering regime for a control system and the existence of singular regimes for it. A singular regime is understood to be a trajectory at whose points the Pontryagin maximum condition does

1991 *Mathematics Subject Classification*. Primary 49K15, 93B50, 93C15; Secondary 58F05, 93C85.

not uniquely determine a value of the controlling variable; in other words, the maximum of the Pontryagin function is attained at more than one point. In the present article we consider problems affinely generated by a scalar control $u \in [-1, 1]$. Therefore, the Pontryagin function depends linearly on u, and the coefficient of u, which is a function H_1 of the phase and the conjugate variables, is identically zero along a singular trajectory. The control on such a trajectory is not determined from the condition that the Pontryagin function have a maximum, but it can be obtained by successive differentiation of the identity $H_1 = 0$. The order of the derivative for which the coefficient of the control first turns out to be nonzero is necessarily an even number $2q$. The number q is called the order of the singular trajectory. According to the Kelley-Kopp-Moyer theorem ([22]; see also [23]) the joining of a singular trajectory of second (or any even) order with a nonsingular piecewise smooth arc is not optimal. Therefore, for optimal trajectories a singular part is joined not with a piecewise smooth arc, but with a chattering trajectory. It is for this reason that we investigate here optimal trajectories joined with singular trajectories of second order.

This article consists of four chapters. In Chapter 1 we present a complete formal solution of the classical Fuller problem. Besides the properties of this solution known from the literature we investigate the asymptotic behavior as $t \to +\infty$ and as $t \to -\infty$ of all the trajectories of a discontinuous Hamiltonian system of the Pontryagin maximum principle for Fuller problems (Lemmas 5 and 6).

Chapter 2 is fundamental. In it we show how to reduce a system of equations of the Pontryagin maximum principle to a form convenient for study when there are singular regimes of second order. Then we investigate this system. It is proved that a neighborhood of the manifold S of singular regimes of second order can be represented as a direct sum of two bundles over the base S, $p^+ : \mathfrak{N}^+ \to S$ and $p^- : \mathfrak{N}^- \to S$, with two-dimensional piecewise smooth fibers $\mathfrak{N}^+_\sigma$ and $\mathfrak{N}^-_\sigma$ ($\operatorname{codim} S = 4$ in the general position situation). Trajectories beginning in $\mathfrak{N}^+_\sigma$ tend to the point $\sigma \in S$ in a finite time with infinitely many switchings that become increasingly more frequent as σ is approached. Trajectories beginning in $\mathfrak{N}^-_\sigma$ tend to the point σ in reverse time also in a finite time and with infinitely many switchings that become increasingly more frequent as σ is approached (Theorem 1). A description is then given of Lagrange manifolds made up of chattering trajectories. Theorem 2 says that the complete inverse image of a Lagrangian manifold $\mathfrak{M}$ belonging to the singular manifold S under the mapping p^+ is also a Lagrangian manifold. Knowledge of the Lagrangian manifolds allows us to construct an optimal synthesis containing chattering trajectories. Sufficient conditions for the optimality of such a synthesis are given by Theorem 3 and the assertions following it.

Chapter 3 is devoted to certain generalizations of the results in Chapter 2 and to conjectures relating to the case of singular trajectories of order higher than 2. Here, in particular, a solution is given for a three-dimensional analogue of the Fuller problem, and self-similar solutions of a four-dimensional analogue of it are found.

Chapter 4 consists entirely of concrete examples that can be solved by the technique developed in this paper. In addition to model examples, we investigate here nontrivial nonlinear problems of applied content. Among them are the problem of optimal allocation of resources in a model of controlled technological progress, the problem of controlling a two-component robot, and Lawden's problem of the motion of a rocket in the Newtonian gravity field of a single attracting center. For all these problems we construct an optimal synthesis containing singular trajectories of second order and chattering trajectories.

It should be mentioned that the problem of optimal allocation of resources is considered for the first time in the formulation presented here. The problem of optimal

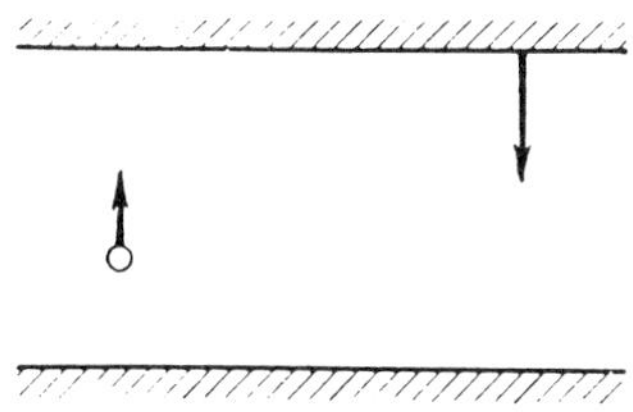

FIGURE 1

control of a robot has been treated in [24]–[26] and elsewhere. However, a chattering solution of it was first constructed by Zelikin and Borisov in [27]. Most famous is Lawden's problem. Not only papers but even whole monographs have been devoted to its investigation (see, for example, [28]–[32]). For a long time the main attention was directed to the question of optimal trajectories of intermediate thrust, i.e., singular trajectories, in Lawden's problems. It would not be an exaggeration to state that it was this question that gave rise in its time to the theory of necessary conditions for optimality of higher order for singular trajectories. Breakwell and Dixon's article [33] was the first step in the investigation of optimal synthesis for Lawden's problem. There it was observed that quadratic approximation of the problem along a concrete singular optimal trajectory leads to the nonsymmetric Fuller problem investigated by Marchal [3]. Considerations on the form of an optimal synthesis for quadratic approximation of Lawden's problem are presented in [33] (though without rigorous proof). In this paper we describe a class of optimal syntheses that contain chattering trajectories for Lawden's problem itself.

At first glance it may seem that the optimal regimes found below are practically inapplicable to the original applied problems because they cannot be physically realized. To clarify our point of view we give the following simple example. Imagine two walls coming closer together with constant velocity, and a perfectly elastic ball which bounces off one of these walls and then off the other (Figure 1).

It will be assumed that at the moment the ball bounces its velocity changes sign instantaneously while staying the same in magnitude. Then as the walls approach each other, the number of bounces increases without limit, tending to infinity at the moment when the ball finds itself squeezed between the walls. Whoever has tried to press a bouncing ping-pong ball against the table with a paddle has encountered a similar situation. Of course, a real ball bounces only a finite, though very large, number of times. But this happens only because the impact of a real ball is not ideal—it requires nonzero time. The less this time, the greater will be the number of bounces. In the ideal case of instantaneous impact there will be infinitely many bounces in a finite interval of time.

Therefore, we regard the examples of chattering regimes constructed here as describing completely natural situations, to within the idealization used in the problem, of course. The practical value of knowing these optimal regimes consists in the fact that they can help one find ways of improving the value of the functional, and they can enable one to estimate the loss in the functional for the suboptimal regimes actually used.

CHAPTER 1. THE FULLER PROBLEM

We consider the following problem.

Problem 1. *Minimize the integral*

(1.1a)
$$\int_0^\infty |x|^q \, dt \to \inf, \qquad q > 1,$$

on the trajectories of the control system

(1.1b) $$\dot{x} = y, \qquad \dot{y} = u$$

with the initial conditions $x(0) = x_0$ *and* $y(0) = y_0$.

Locally measurable functions $u(t) \in [-1, 1]$ bounded in modulus are taken as admissible controls; the admissible trajectories $x(t)$, $y(t)$ are absolutely continuous.

Problem 1 has been investigated by several authors (see [1], [3], and [12]). However, since we found no exposition convenient for our purposes in the literature, we first give a complete solution of Problem 1.

It follows from Lemmas 12 and 13 in Chapter 3 that for arbitrary initial conditions there exists a unique optimal trajectory in Problem 1. The solution due to Fuller is based on the existence of a one-parameter group of symmetries in Problem 1. Consider the one-parameter group $\mathfrak{G} = \{g_\lambda\}$, $\lambda > 0$, of diffeomorphisms of $\mathbf{R}^2$ given by

$$g_\lambda(x, y) = (\lambda^2 x, \lambda y).$$

Denote by $v(x, y, u) = (y, u)$ the right-hand side of (1.1b). Then for any $\lambda > 0$

$$\mathscr{D} g_\lambda|_{(x,y)} v(x, y, u) = \lambda v(g_\lambda(x, y), u)$$

for the differential $\mathscr{D} g_\lambda$, i.e., the system (1.1b) is homogeneous with respect to the action of the group $\mathfrak{G}$. Consequently, if $x(t)$, $y(t)$ is an admissible trajectory of the system (1.1b) with the control $u(t)$, then

(1.2) $$(x_\lambda(t), y_\lambda(t)) = g_\lambda(x(t/\lambda), y(t/\lambda))$$

is an admissible trajectory of (1.1b) with the control $u_\lambda(t) = u(t/\lambda)$.

Lemma 1. *Suppose that* $x(t)$, $y(t)$ *is the optimal trajectory of Problem* 1 *with the control* $u(t)$ *and initial conditions* x_0 *and* y_0. *Then* $x_\lambda(t)$, $y_\lambda(t)$ *is an optimal trajectory in Problem* 1 *with the control* $u_\lambda(t)$ *and initial conditions* $g_\lambda(x_0, y_0)$.

Proof. The optimality of the trajectory $x_\lambda(t)$, $y_\lambda(t)$ follows from the invertibility of the mapping (1.2):

$$(x(t), y(t)) = g_{1/\lambda}(x_\lambda(\lambda t), y_\lambda(\lambda t))$$

and the relation

$$\int_0^\infty |x_\lambda|^q \, dt = \lambda^{2q+1} \int_0^\infty |x|^q \, dt.$$

The lemma is proved.

The orbits of the action of $\mathfrak{G}$ are the branches of the parabola

(1.3) $$x = Cy^2,$$

where either $y > 0$ or $y < 0$.

It follows from Lemma 1 that in Problem 1 an optimal control, as a function of the initial state, can be chosen to be invariant under the action of $\mathfrak{G}$. In particular, the set of control switching points consists of the branches of the parabolas of the form (1.3).

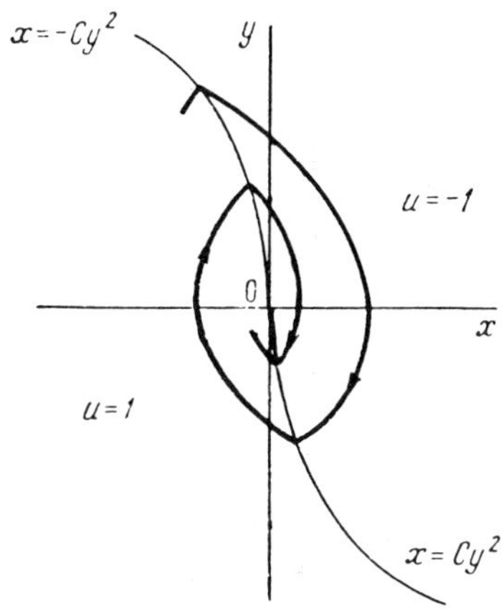

FIGURE 2

Lemma 2. *For an arbitrary* $q > 1$ *the switching curve in Problem* 1 *has the form*

$$x = -Cy^2 \operatorname{sgn} y, \tag{1.4}$$

where C *is a constant depending on* q, *and* $0 < C < 1/2$. *The optimal trajectories of Problem* 1 *have the form of spirals twisting into the origin of coordinates and intersecting the curve* (1.4) *infinitely many times* (*Figure* 2). *The control is equal to* -1 *above the switching curve and* $+1$ *below it.*

Proof. Let $x(t)$, $y(t)$ be an optimal trajectory in Problem 1 with control $u(t)$, and suppose that it is in the half-plane $y > 0$ for $t \in (t_0, t_1)$. Then on (t_0, t_1) the control u either is constant and equal to $+1$ or -1, or has a single switching from $+1$ to -1. Indeed, choosing $x(t)$ as a parameter on the curve $x(t)$, $y(t)$, we get

$$\int_{t_0}^{t_1} |x|^q \, dt = \int_{x(t_0)}^{x(t_1)} \frac{|x|^q \, dx}{y(x)}. \tag{1.5}$$

The curvilinear integral on the right-hand side of (1.5) is the smaller, the larger $y(x)$ is.

The trajectories of (1.1b) with constant control $u \neq 0$ have the form

$$x(t) = x_0 + y_0 t + \tfrac{1}{2} u t^2, \qquad y = y_0 + ut. \tag{1.6}$$

Eliminating the parameter t, we get that

$$x = \frac{1}{2u} y^2 + x_0 - \frac{1}{2u} y_0^2. \tag{1.7}$$

If the points $A(x(t_0), y(t_0))$ and $B(x(t_1), y(t_1))$ lie on the curve (1.7) with $u = 1$ or $u = -1$, then there are no other trajectories of (1.1b) with control $|u| \leq 1$ that join the points A and B and belong to the half-plane $y > 0$. If A and B do not lie on the curve (1.7) with $u = 1$ or $u = -1$, then among all the admissible trajectories of (1.1b) going through A and B and belonging to the half-plane $y > 0$, the "highest" trajectory and hence the one minimizing the functional (1.5) is the trajectory with a single switching of the control from $u = 1$ to $u = -1$ (Figure 3; see next page).

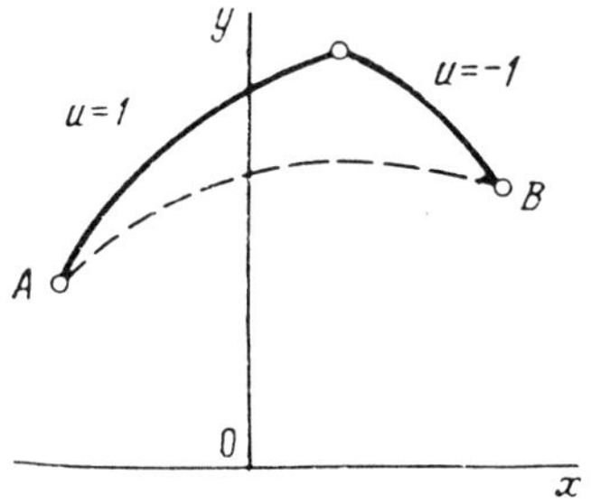

FIGURE 3

It follows from symmetry considerations that the part of the optimal trajectory lying entirely in the half-plane $y < 0$ either does not have control switchings or has a single switching from $u = -1$ to $u = 1$.

Consequently, the switching curve in Problem 1 consists of two branches of parabolas: $x = C_1 y^2$ for $y \geq 0$ and $x = C_2 y^2$ for $y \leq 0$. The mapping $(x, y) \mapsto (-x, -y)$ clearly carries optimal trajectories of Problem 1 into optimal trajectories. Therefore, the switching curve is symmetric with respect to the origin: $C_1 = -C_2$. For the integral (1.1a) to converge on trajectories of the system (1.1b) with the switching curve (1.4) it is necessary that each trajectory tend to the origin. This gives the estimate $0 < C \leq 1/2$. We show that $C \neq 1/2$. For $C = 1/2$ the motion along the switching curve (1.4) itself must be optimal in Problem 1. However, as we now show, the curve (1.4) does not satisfy the conditions of the Pontryagin maximum principle for Problem 1.

Let $x(t), y(t)$ be an optimal trajectory in Problem 1, and let $u(t)$ be a control on it. Then by the Pontryagin maximum principle [34], there exist an absolutely continuous vector-valued function $\psi_1(t), \psi_2(t)$ and a number $\lambda_0 \geq 0$, not all zero, such that

$$\dot{\psi}_1 = \lambda_0 q|x|^{q-1} \operatorname{sgn} x, \quad \dot{\psi}_2 = -\psi_1, \quad \dot{x} = y, \quad \dot{y} = u(t); \tag{1.8a}$$

$$\max_{|u|\leq 1} u\psi_2(t) = u(t)\psi_2(t) \text{ (a.e.)}, \tag{1.8b}$$

where a.e. means almost everywhere.

Assume that $\lambda_0 = 0$. Then it follows from (1.8a) that $\psi_2(t) = -\psi_{10}t + \psi_{20}$ is a linear function not identically equal to zero, and hence the control on the optimal trajectory has at most one switching. In this case the integral (1.1a) diverges. Therefore, $\lambda_0 \neq 0$. We set $\lambda_0 = 1/q$ everywhere below.

Assume that the optimal trajectory $x(t), y(t)$ satisfies (1.4) with $C = 1/2$. For definiteness let $y(0) = y_0 > 0$. Then it follows from (1.6) and (1.7) that

$$(x(t), y(t)) = \begin{cases} (-\frac{1}{2}y_0^2 + y_0 t - \frac{1}{2}t^2, y_0 - t), & 0 \leq t < y_0, \\ (0, 0), & t \geq y_0. \end{cases}$$

Let $\psi_2(t), \psi_1(t)$ be the corresponding solution of (1.8). For $t \geq y_0$ we have $u(t) = 0$, which implies that $\psi_2(t) = \psi_1(t) = 0$. For $0 \leq t < y_0$ we have $x(t) < 0$. Then it follows from (1.8a) that $\psi_1(t) < 0$ and $\psi_2(t) > 0$. But $u(t) = \operatorname{sgn} \psi_2(t) = -1$ must hold for $0 \leq t < y_0$ in view of (1.8b). Contradiction. The lemma is proved.

We show that the moments of time between successive control switchings on optimal trajectories form a decreasing geometric progression. Let A_0, A_1, and A_2 be three successive switching points on some optimal trajectory. For definiteness suppose that A_0 lies above the x-axis. It follows from (1.6) that the coordinates of the points $A_0(-Cy_0^2, y_0)$, $y_0 > 0$, $A_1(Cy_1^2, y_1)$, $y_1 < 0$, and $A_2(-Cy_2^2, y_2)$, $y_2 > 0$ are connected by the relations

$$(1.9)\qquad \begin{aligned} y_1 &= y_0 - t_0, & Cy_1^2 &= -Cy_0^2 + y_0t_0 - \tfrac{1}{2}t_0^2, \\ y_2 &= y_1 + t_1, & -Cy_2^2 &= Cy_1^2 + y_1t_1 + \tfrac{1}{2}t_1^2, \end{aligned}$$

where t_0 and t_1 are the times in moving along the optimal trajectory from A_0 to A_1 and from A_1 to A_2, respectively. Eliminating t_0 from the first system and t_1 from the second, we get

$$(C + 1/2)y_1^2 = (-C + 1/2)y_0^2, \qquad -(C + 1/2)y_2^2 = (C - 1/2)y_1^2,$$

and so

$$(1.10a)\qquad y_1/y_0 = y_2/y_1 = -[(1 - 2C)/(1 + 2C)]^{1/2}.$$

By (1.9),

$$(1.10b)\qquad t_1/t_0 = (y_2 - y_1)/(y_0 - y_1) = [(1 - 2C)/(1 + 2C)]^{1/2} < 1.$$

It is thereby proved that the time for optimal trajectories to hit the origin is finite. Therefore, the optimal trajectories in Problem 1 are chattering trajectories.

We clear up the dependence of the constant C in the equation (1.4) of the switching curve on q. The trajectories of the system (1.8) that determine optimal trajectories of Problem 1 arrive at the origin in a finite time, and hence belong to the zero level surface of the Hamiltonian:

$$(1.11)\qquad \psi_1 y + |\psi_2| - q^{-1}|x|^q = 0.$$

At the points of the switching curve (1.4) we have

$$(1.12)\qquad \psi_2 = 0, \qquad x = -Cy^2 \operatorname{sgn} y, \qquad \psi_1 = \frac{1}{q}C^q y^{2q-1}.$$

Let us integrate (1.8) with the control $u = -1$ and the initial conditions ψ_{20}, ψ_{10}, x_0, and y_0 ($y_0 > 0$) satisfying (1.12). The first control switching on this trajectory must take place at some time $\tau > 0$ at the point $(x(\tau), y(\tau))$, where $y(\tau) < 0$ and $x(\tau) = Cy^2(\tau)$. By (1.10),

$$(1.13)\qquad \tau = y_0(1 + ((1 - 2C)/(1 + 2C))^{1/2}).$$

The equation $\psi_2(\tau) = 0$ implies that

$$\frac{1}{q}C^q y_0^{2q-1}\tau + \int_0^\tau (\tau - s)| - Cy_0^2 + y_0 s - \tfrac{1}{2}s^2|^{q-1} \operatorname{sgn}(-Cy_0^2 + y_0 s - \tfrac{1}{2}s^2)\, ds = 0.$$

The trajectory $x(s) = -Cy_0^2 + y_0 s - \frac{1}{2}s^2$, $y(s) = y_0 - s$ intersects the y-axis at the time $s_0 = y_0(1 - (1 - 2C)^{1/2})$. Therefore,

$$\frac{1}{q}C^q y_0^{2q-1}\tau - \int_0^{s_0} (\tau - s)(Cy_0^2 - y_0 s + \tfrac{1}{2}s^2)^{q-1}\, ds + \int_{s_0}^{\tau} (\tau - s)(-Cy_0^2 + y_0 s - \tfrac{1}{2}s^2)^{q-1}\, ds = 0.$$

Substituting (1.13) here, we get an implicit equation on $C = C(q)$:

$$(1.14)\qquad \begin{aligned} &-\frac{1}{q}C^q \rho + \int_0^{1-(1-2C)^{1/2}} (\rho - s)(C - s + \tfrac{1}{2}s^2)^{q-1}\, ds \\ &\quad - \int_{1-(1-2C)^{1/2}}^{\rho} (\rho - s)(-C + s - \tfrac{1}{2}s^2)^{q-1}\, ds = 0, \end{aligned}$$

where $\rho = 1 + ((1 - 2C)/(1 + 2C))^{1/2}$.

Let us consider the case most important for what follows: $q = 2$. Then for $\mu = \rho - 1$ $(C = \frac{1}{2}(1-\mu^2)/(1+\mu^2))$ we get

$$\mu^4 - 3\mu^3 - 4\mu^2 - 3\mu + 1 = 0. \tag{1.15}$$

Since (1.15) is a reciprocal equation, its order can be reduced by passing to the variable $\sigma = \mu + 1/\mu$. We get that $\sigma^2 - 3\sigma - 6 = 0$, or $\sigma_1 = \frac{1}{2}(3+\sqrt{33})$ and $\sigma_2 = \frac{1}{2}(3-\sqrt{33})$. The root $\sigma_2 < 0$ does not yield positive solutions of (1.15). Corresponding to the value σ_1 are two solutions of (1.15):

$$\mu_1 = \frac{1}{4}\left(3+\sqrt{33}-\sqrt{26+6\sqrt{33}}\right)$$

and

$$\mu_2 = \frac{1}{4}\left(3+\sqrt{33}+\sqrt{26+6\sqrt{33}}\right); \qquad \mu_2 = 1/\mu_1.$$

For $\mu = \mu_1$ we get $C_1 = \frac{1}{2}(1-\mu_1^2)/(1+\mu_1^2)$; for $\mu = \mu_2$ we get $C_2 = \frac{1}{2}(1-\mu_2^2)/(1+\mu_2^2) = -C_1$. For $q = 2$ the value of the constant C in the equation (1.4) of the switching curve is equal to $C_1 = 0.444623\ldots$.

Let us now consider equation (1.14) for an arbitrary $q > 1$. We show that if C_1 is a solution of (1.14), then $C_2 = -C_1$ is also a solution of it (we have that $\mu_2 = 1/\mu_1$ for the corresponding values of $\mu = ((1-2C)/(1+2C))^{1/2}$).

Let $x(t)$, $y(t)$, $\psi_1(t)$, $\psi_2(t)$ be a solution of (1.8). Then, substituting the functions $x(-t)$, $-y(-t)$, $-\psi_1(-t)$, $\psi_2(-t)$ in (1.8), we see that they also are a solution of this system. Thus, each family of solutions of (1.8) with switching curve (1.12) determines a second family of solutions with switching curve

$$\psi_2 = 0, \quad \psi_1 = q^{-1}C^q y^{2q-1}, \quad x = Cy^2 \operatorname{sgn} y \tag{1.16}$$

with the same constant C. For $0 < C < 1/2$ the trajectories of the first family arrive at the origin in a finite time, with infinitely many control switchings. The trajectories of the second family are obtained from those of the first by reflection with respect to the plane $y = 0$, $\psi_1 = 0$ and time reversal of the motion on them. They have the form of spirals untwisting from the origin and carrying out infinitely many switchings in a finite time upon going backwards to the origin (Figure 4).

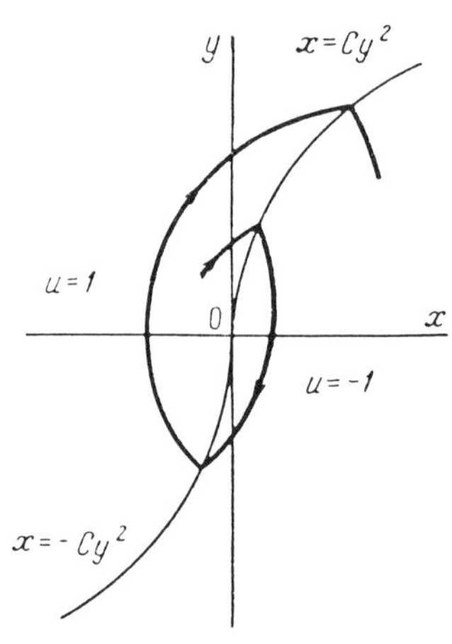

FIGURE 4

Denote by $\mathfrak{N}^+$ the family of trajectories of (1.8) corresponding to optimal synthesis in Problem 1 with the switching curve (1.12), and denote by $\mathfrak{N}^-$ the corresponding family of trajectories of (1.8) with switching curve (1.16). This family is directly related to the following problem.

Problem 2. *Minimize the integral*

$$\int_0^{T_0} |x|^q \, dt \to \inf \tag{1.17a}$$

under the constraints

$$\dot{x} = y, \qquad \dot{y} = u \in [-1, 1], \tag{1.17b}$$

$$x(0) = x_0, \quad y(0) = y_0, \quad x(T_0) = x_1, \quad y(T_0) = y_1. \tag{1.17c}$$

The time T_0 is assumed to be fixed.

Lemma 3. *For arbitrary boundary conditions* (1.17c) *there is a* $T_0 > 0$ *such that the corresponding optimal trajectory of Problem* 2 *consists of three parts abutting on each other: the trajectory of the family* $\mathfrak{N}^+$ *passing through the point* (x_0, y_0), *the trajectory coinciding with the origin of coordinates, and the trajectory of the family* $\mathfrak{N}^-$ *passing through the point* (x_1, y_1).

Proof. In view of the sufficiency theorem ([35], §6.1, Theorem 5) any solution of (1.8) satisfying the boundary conditions (1.17c) determines an optimal solution of Problem 2. The solution of Problem 2 is unique in view of the strict convexity of the functional (1.17a) on the trajectories of the linear system (1.17b).

It was shown that the trajectories of the family $\mathfrak{N}^+$ with initial conditions (x_0, y_0) hit the origin in finite time. Denote this time by $T(x_0, y_0)$. Then the time for a trajectory of the family $\mathfrak{N}^-$ to go from the origin to the point (x_1, y_1) is equal to $T(x_1, -y_1)$. It remains to take $T_0 > T(x_0, y_0) + T(x_1, -y_1)$. The lemma is proved.

We clear up the behavior of the trajectories of (1.8) that belong to the surface (1.11) and do not coincide with trajectories of the families $\mathfrak{N}^+$ and $\mathfrak{N}^-$.

The symmetry group $\mathfrak{G}$ of Problem 1 induces the symmetry group $\widetilde{\mathfrak{G}}$ of the system (1.8):

$$\widetilde{\mathfrak{G}} = \{\tilde{g}_\lambda\}, \qquad \lambda > 0,$$
$$\tilde{g}_\lambda(\psi_2, \psi_1, x, y) = (\lambda^{2q}\psi_2, \lambda^{2q-1}\psi_1, \lambda^2 x, \lambda y).$$

For brevity let $\xi = (\psi_2, \psi_1, x, y)$, and let $\tilde{v}(\xi) = (-\psi_1, |x|^{q-1}\operatorname{sgn} x, y, u)$ be the right-hand side of (1.8). By the maximum principle,

$$u(\xi) = \begin{cases} \operatorname{sgn}\psi_2 & \text{for } \psi_2 \neq 0; \\ 1 & \text{for } \psi_2 = 0, \ \psi_1 < 0; \\ & \quad \psi_2 = \psi_1 = 0, \ x < 0; \\ & \quad \psi_2 = \psi_1 = x = 0, \ y < 0; \\ 0 & \text{for } \psi_2 = \psi_1 = x = y = 0; \\ -1 & \text{otherwise.} \end{cases} \tag{1.18}$$

In this notation we have

$$\mathscr{D}\tilde{g}_\lambda|_\xi \tilde{v}(\xi) = \lambda \tilde{v}(\tilde{g}_\lambda(\xi)),$$

where $\mathscr{D}\tilde{g}_\lambda$ is the differential of the mapping $\tilde{g}_\lambda$. Consequently, if $\psi_2(t)$, $\psi_1(t)$, $x(t)$, $y(t)$ is a trajectory of (1.8), then so is $\lambda^{2q}\psi_2(t/\lambda)$, $\lambda^{2q-1}\psi_1(t/\lambda)$, $\lambda^2 x(t/\lambda)$,

$\lambda y(t/\lambda)$ with arbitrary $\lambda > 0$. We factor the system (1.8) with respect to the action of the group $\widetilde{\mathfrak{G}}$. The corresponding system with discontinuous right-hand side on the quotient space $\mathbf{R}^4\setminus\{0\}/\widetilde{\mathfrak{G}}$ is denoted by $\mathfrak{A}$. The system $\mathfrak{A}$ is uniquely determined up to trajectorywise equivalence. As the set of representatives of the cosets of $\mathbf{R}^4\setminus\{0\}/\widetilde{\mathfrak{G}}$ we take the unit sphere S given by the equation $\psi_2^2 + \psi_1^2 + x^2 + y^2 = 1$. The point where an orbit of the action of $\widetilde{\mathfrak{G}}$ intersects S is determined by the equation

$$\lambda^{4q}\psi_2^2 + \lambda^{4q-2}\psi_1^2 + \lambda^4 x^2 + \lambda^2 y^2 = 1. \tag{1.19}$$

Since for $\lambda > 0$ the left-hand side of (1.19) is a monotone function of λ, for arbitrary $\xi = (\psi_2, \psi_1, x, y)$ (1.19) has a unique solution. Denote this solution by $\Lambda(\xi)$. Consider the mapping $g\colon \mathbf{R}^4\setminus\{0\} \to S$:

$$g(\xi) = \tilde{g}_{\Lambda(\xi)}(\xi) = (\Lambda^{2q}\psi_2, \Lambda^{2q-1}\psi_1, \Lambda^2 x, \Lambda y).$$

As the vector field of the system $\mathfrak{A}$ on S we can take

$$v_S g(\xi) = \frac{1}{\Lambda(\xi)} V(\xi), \qquad V(\xi) = \mathscr{D} g|_\xi \tilde{v}(\xi).$$

The differential $\mathscr{D} g$ is given by the sum of the two matrices $M_1 + M_2$, where

$$M_1 = \begin{pmatrix} \Lambda^{2q} & 0 & 0 & 0 \\ 0 & \Lambda^{2q-1} & 0 & 0 \\ 0 & 0 & \Lambda^2 & 0 \\ 0 & 0 & 0 & \Lambda \end{pmatrix}$$

is a diagonal matrix, and

$$M_2 = \begin{pmatrix} 2q\psi_2\Lambda^{2q-1}\operatorname{grad}\Lambda \\ (2q-1)\psi_1\Lambda^{2q-2}\operatorname{grad}\Lambda \\ 2x\Lambda\operatorname{grad}\Lambda \\ y\operatorname{grad}\Lambda \end{pmatrix}$$

is a matrix of rank 1 whose rows are proportional to the vector $\operatorname{grad}\Lambda$:

$$\operatorname{grad}\Lambda = -\gamma^{-1}(\Lambda^{4q}\psi_2, \Lambda^{4q-2}\psi_1, \Lambda^4 x, \Lambda^2 y),$$
$$\gamma = 2q\Lambda^{4q-1}\psi_2^2 + (2q-1)\Lambda^{4q-3}\psi_1^2 + 2\Lambda^3 x^2 + \Lambda y^2.$$

We get

$$\begin{aligned} V(\xi) = (&-\psi_1\Lambda^{2q} + 2q\psi_2\Lambda^{2q-1}\langle\operatorname{grad}\Lambda, \tilde{v}\rangle; \\ &|x|^{q-1}\operatorname{sgn} x\Lambda^{2q-1} + (2q-1)\psi_1\Lambda^{2q-2}\langle\operatorname{grad}\Lambda, \tilde{v}\rangle; \\ &y\Lambda^2 + 2x\langle\operatorname{grad}\Lambda, \tilde{v}\rangle; u\Lambda + y\langle\operatorname{grad}\Lambda, \tilde{v}\rangle), \end{aligned}$$

where

$$\langle\operatorname{grad}\Lambda, \tilde{v}\rangle = -\gamma^{-1}(-\Lambda^{4q}\psi_2\psi_1 + \Lambda^{4q-2}\psi_1|x|^{q-1}\operatorname{sgn} x + \Lambda^4 xy + \Lambda^2 yu).$$

By (1.19), $\Lambda(\tilde{g}_\lambda(\xi)) = \lambda^{-1}\Lambda(\xi)$. This implies that

$$\langle\operatorname{grad}\Lambda(\tilde{g}_\lambda(\xi)), \tilde{v}(\tilde{g}_\lambda(\xi))\rangle = \lambda^{-2}\langle\operatorname{grad}\Lambda(\xi), \tilde{v}(\xi)\rangle; \qquad V(\tilde{g}_\lambda(\xi)) = \lambda^{-1}V(\xi).$$

Denote by $\operatorname{Orb}\xi_0 = \bigcup_{\lambda>0}\tilde{g}_\lambda(\xi_0)$ the orbit of the point ξ_0 under the action of $\widetilde{\mathfrak{G}}$. Thus, the vectors $V(\xi)/\Lambda(\xi)$ coincide for the points $\xi \in \operatorname{Orb}\xi_0$ of a single orbit.

We show that $V(\xi) \neq 0$ at all points different from the origin. Assume that $V(\xi) = 0$ at some point $\xi \neq 0$. If $y = 0$, then the last coordinate of $V(\xi)$, which is equal to $u\Lambda + y\langle \operatorname{grad}\Lambda, \tilde{v}\rangle$, is nonzero in view of (1.18). Therefore, $y \neq 0$ and $u\Lambda + y\langle \operatorname{grad}\Lambda, \tilde{v}\rangle = 0$. Then

$$V(\xi) = (\Lambda^{2q}(-\psi_1 - 2q\psi_2 u/y);\ \Lambda^{2q-1}(|x|^{q-1}\operatorname{sgn} x - (2q-1)\psi_1 u/y);\\ \Lambda^2(y - 2xu/y);\ 0) = 0.$$

From this we get successively

$$x = \frac{1}{2u}y^2; \qquad \psi_1 = \frac{y}{(2q-1)u}|x|^{q-1}\operatorname{sgn} x = \frac{y^{2q-1}}{2^{q-1}(2q-1)};$$
$$\psi_2 = -\frac{\psi_1 y}{2qu} = -\frac{y^{2q}}{2^q q(2q-1)u}.$$

Since $\psi_2 \neq 0$, it follows from the last equation that $\operatorname{sgn} u = -\operatorname{sgn}\psi_2$, while $u = \operatorname{sgn}\psi_2$ by assumption. Contradiction. This proves that the vector field v_S does not have fixed points on the surface S.

Denote by Tor the intersection of S with the zero level surface of the Hamiltonian (1.11). This manifold is given by the equations

$$\begin{aligned} &\psi_2^2 + \psi_1^2 + x^2 + y^2 = 1, \\ &\psi_1 y + |\psi_2| - q^{-1}|x|^q = 0. \end{aligned} \tag{1.20}$$

Lemma 4. *The manifold* Tor *determined by the system of equations* (1.20) *is homeomorphic to the two-dimensional torus.*

Proof. We construct a homotopy of Tor by considering the family of manifolds $T^{\alpha,\beta}$ $(\alpha, \beta > 1)$ determined by the equations

$$\psi_2^2 + \psi_1^2 + x^2 + y^2 = 1, \tag{1.21a}$$
$$\alpha^{-1}|\psi_2|^\alpha + \psi_1 y - \beta^{-1}|x|^\beta = 0. \tag{1.21b}$$

Let us show that the gradients of the left-hand sides of (1.21),

$$\tfrac{1}{2}\nabla_1 = (\psi_2, \psi_1, x, y),$$
$$\nabla_2 = (|\psi_2|^{\alpha-1}\operatorname{sgn}\psi_2, y, -|x|^{\beta-1}\operatorname{sgn} x, \psi_1),$$

are linearly independent at the points of the manifold $T^{\alpha,\beta}$. Assume not: $\frac{1}{2}\nabla_1 = \lambda\nabla_2$. We have

$$\begin{aligned} &\psi_2 = \lambda|\psi_2|^{\alpha-1}\operatorname{sgn}\psi_2, \qquad \psi_1 = \lambda y, \\ &x = -\lambda|x|^{\beta-1}\operatorname{sgn} x, \qquad y = \lambda\psi_1. \end{aligned} \tag{1.22}$$

Since the vectors $\frac{1}{2}\nabla_1$ and ∇_2 are nonzero at the points of $T^{\alpha,\beta}$, it follows that $\lambda \neq 0$. Then it follows from the first equation in (1.22) that $\lambda > 0$ for $\psi_2 \neq 0$, and from the third that $\lambda < 0$ for $x \neq 0$. If $\psi_2 = 0$ and $\lambda < 0$, then $\psi_1 y \leq 0$ by the second and third equations in (1.22). Then $\psi_1 y - \beta^{-1}|x|^\beta = 0$ by (1.21b), i.e., $x = 0$ and $\psi_1 y = 0$, and hence $x = y = \psi_1 = \psi_2 = 0$, which contradicts (1.21a). But if $x = 0$ and $\lambda > 0$, then $\psi_1 y \geq 0$ by the second and third equations in (1.22). Then $\alpha^{-1}|\psi_2|^\alpha + \psi_1 y = 0$, by (1.21b), i.e., $\psi_2 = 0$ and $\psi_1 y = 0$, and hence $x = y = \psi_1 = \psi_2 = 0$, which contradicts (1.21a). Thus, the systems (1.21) and (1.22) are incompatible. Consequently, the manifolds determined by (1.21a) and (1.21b) intersect transversally for all $\alpha > 1$ and $\beta > 1$. From this it follows that we have a homotopy of the manifold $T^{\alpha,\beta}$ as α and β vary in the domain $\alpha > 1$, $\beta > 1$.

For $\alpha = 1$ and $\beta > 1$ the above argument is applicable to each of the two smooth parts:

$$Q_+ = \{\psi_1 y + |\psi_2| - \beta^{-1}|x|^\beta = 0 | \psi_2 \geq 0\},$$
$$Q_- = \{\psi_1 y + |\psi_2| - \beta^{-1}|x|^\beta = 0 | \psi_2 \leq 0\},$$

which make up the surface $\psi_1 y + |\psi_2| - \beta^{-1}|x|^\beta = 0$. The manifolds Q_+ and Q_- intersect the sphere transversally in smooth two-dimensional manifolds without singularities and with a common smooth boundary corresponding to the section $\psi_2 = 0$. Consequently, the homotopy of $T^{\alpha,\beta}$ extends (with loss of smoothness) to the value $\alpha = 1$.

For $\alpha = 1$ and $\beta = q$ the manifold $T^{1,q}$ coincides with Tor. For $\alpha = \beta = 2$, (1.21b) takes the form $\psi_1 y + \frac{1}{2}\psi_2^2 - \frac{1}{2}x^2 = 0$, where the left-hand side is a quadratic form of signature zero. Such a quadratic form describes a toric cone intersecting the unit sphere in a two-dimensional torus. The lemma is proved.

As already shown, the dynamic system on Tor does not have singular points. The trajectories of the families $\mathfrak{N}^+$ and $\mathfrak{N}^-$ determine two piecewise smooth closed cycles of the system $\mathfrak{A}$. Denote them by Z^+ and Z^-, respectively.

Lemma 5. *The system* $\mathfrak{A}$ *on* Tor *does not have closed cycles other than* Z^+ *and* Z^-.

Proof. Assume the opposite: $\mathfrak{A}$ has a third cycle Z different from Z^+ and Z^-. Consider the trajectory $W(t)$ of (1.8) with initial conditions $W(0) \in \operatorname{Orb}\xi_0$, where $\xi_0 \in Z$ is a switching point on Z. Since Z is a cycle, we again have $W(t_0) = \tilde{g}_{\lambda_0}(W(0)) \in \operatorname{Orb}\xi_0$ at some time $t_0 > 0$. The system (1.8) is homogeneous with respect to the action of the group $\tilde{\mathfrak{G}}$. Consequently, the trajectory $\tilde{g}_{\lambda_0} W(t/\lambda_0)$ is the solution of (1.8) with initial conditions $W(t_0)$. It is easy to see that at points different from the origin the system (1.8) has the property of right- and left-sided uniqueness. Therefore,

$$\tilde{g}_{\lambda_0} W(t/\lambda_0) = W(t + t_0). \tag{1.23}$$

Trajectories satisfying a relation of the type (1.23) are usually called self-similar trajectories. The relation (1.23) means that the trajectory of (1.8) with initial conditions $W(t_0)$ is obtained from the original trajectory $W(t)$ by contraction by a factor of λ_0 along the orbits of the action of $\tilde{\mathfrak{G}}$. In particular, the next return of the trajectory $W(t + t_0)$ to the orbit $\operatorname{Orb}\xi_0$ takes place at the time $\lambda_0 t_0$, and $W(\lambda_0 t_0 + t_0) = \tilde{g}_{\lambda_0} W(t_0) = \tilde{g}_{\lambda_0^2} W(0)$. From this it follows that there is a sequence of points of intersection of the trajectory $W(t)$ with $\operatorname{Orb} W(0)$:

$$W(0),\ W(t_0) = \tilde{g}_{\lambda_0} W(0),\ W(t_0 + \lambda_0 t_0) = \tilde{g}_{\lambda_0^2} W(0), \dots,$$
$$W(t_0 + \lambda_0 t_0 + \dots + \lambda_0^{n-1} t_0) = \tilde{g}_{\lambda_0^n} W(0), \dots.$$

There are three possibilities: $0 < \lambda_0 < 1$, $\lambda_0 = 1$, and $\lambda_0 > 1$. In the case $0 < \lambda_0 < 1$ we have

$$\lim_{n\to\infty} \tilde{g}_{\lambda_0^n} W(0) = 0.$$

The time T at which the trajectory $W(t)$ hits the origin can be computed as follows:

$$T = t_0 + \lambda_0 t_0 + \dots + \lambda_0^n t_0 + \dots = t_0/(1 - \lambda_0).$$

By the sufficiency theorem ([35], §6.1, Theorem 5), $W(t)$ is optimal in Problem 2 with $T_0 = T$ if we set $x_1 = y_1 = 0$ in (1.17c) and take the point (x_0, y_0) to be the projection of the point $W(0)$ on the space of x, y. This contradicts Lemma 3.

For $\lambda_0 > 1$ we have

$$\lim_{n\to-\infty} \tilde{g}_{\lambda_0^n} W(0) = 0.$$

In this case the trajectory $W(t)$ arrives at the origin at the time $-T<0$, where

$$T=t_0/\lambda_0+t_0/\lambda_0^2+\cdots+t_0/\lambda_0^n+\cdots=\lambda_0 t_0/(\lambda_0-1).$$

The trajectory $W(t-T)$ is optimal in Problem 2 with $T_0=T$ and with the boundary conditions $x_0=y_0=0$ if (x_1, y_1) is taken to be the projection of the point $W(0)$. This also contradicts Lemma 3.

If $\lambda_0=1$, then the trajectory $W(t)$ is optimal in Problem 2, with the boundary conditions $(x_0, y_0)=(x_1, y_1)$ chosen to be the projection of $W(0)$ and with $T_0=nt_0$, where n is any natural number. For sufficiently large n this also contradicts Lemma 3. The lemma is proved.

Thus, the system $\mathfrak{A}$ has on the surface Tor only two cycles: Z^+ and Z^-. The manifold Tor is smooth away from the points of the switching curve $l=\mathrm{Tor}\cap\{\psi_2=0\}$ (l is homeomorphic to S^1); the vector field v_S is also smooth at the points of Tor l. For the system (1.8) the manifold of singular trajectories (that lie entirely on the switching surface $\psi_2=0$) coincides with the origin of coordinates; therefore, the intersection of an arbitrary trajectory of the system $\mathfrak{A}$ with l is discrete. By the Poincaré-Bendixson theorem ([36], Chapter 16, Theorem 2.1), the α- and ω-limit sets of the trajectories of $\mathfrak{A}$ consist of the curves Z^+ and Z^-.

Lemma 6. *The cycle Z^+ is unstable, while Z^- is stable.*

Proof. The idea of the proof is as follows. By Lemma 3, precisely those trajectories lying on the surface Orb Z^+ arrive in a finite time at the origin of coordinates. We show here that if the cycle Z^+ is the ω-limit set of some trajectory Z of the system $\mathfrak{A}$ on the manifold Tor, then the trajectories of (1.8) lying in Orb Z also arrive at the origin in a finite time, contrary to Lemma 3.

Let M_1 and M_2 be the points where Z^+ intersects the switching curve l. The trajectory Z^+ consists of two smooth arcs, of which one corresponds to the control $u=1$ and the other to the control $u=-1$. Since $\psi_1|_{M_i}\neq 0$, $i=1, 2$, the trajectory Z^+ arrives at l and leaves it at the points M_i at a nonzero angle and with nonzero phase velocity (the surface $\Sigma=\{\psi_2=0\}$ is invariant under the action of $\widetilde{\mathfrak{G}}$, and hence $(\mathrm{Ker}\,\mathscr{D}g)|_\Sigma\subset T\Sigma$, while $d\psi_2/dt|_{(1.8)}=-\psi_1$). By the theorem on smooth dependence of the solutions of differential equations on the initial data, there is a small neighborhood U of M_1 on l such that the trajectories of $\mathfrak{A}$ with initial conditions in U also intersect l transversally at the initial time and at the times of the first and second control switchings.

Let $W(t)$ be a trajectory of (1.8) with initial conditions $W(0)=W_0\in\mathrm{Orb}\,U$. Denote by $t(W_0)$ the second switching time on $W(t)$. The function $t(\cdot)$ is continuous on Orb U, and hence is bounded on any bounded subset of Orb U. In particular,

$$\sup_{W_0\in U} t(W_0)\le T<\infty.$$

From the homogeneity of the system (1.8) with respect to the action of $\widetilde{\mathfrak{G}}$ it follows that $t(\tilde g_\lambda(W_0))=\lambda t(W_0)$. Setting $\lambda=\Lambda(W_0)$, we get $t(W_0)=t(g(W_0))/\Lambda(W_0)$. From this we get the estimate

$$t(W_0)\le T/\Lambda(W_0) \tag{1.24}$$

for an arbitrary $W_0\in\mathrm{Orb}\,U$.

Let $\rho(W_0)=\Lambda(W_0)/\Lambda(W(t(W_0)))$. It is clear that the function $\rho(\cdot)$ is continuous on Orb U and constant on the orbits of $\widetilde{\mathfrak{G}}$. It follows from (1.10) that $\rho(M_1)=(1-2C)/(1+2C)\in(0, 1)$. We shrink the neighborhood U if necessary so that $\rho(W_0)\le\rho_0<1$ for any $W_0\in U$.

Assume that there is a trajectory $\widetilde{Z}$ of the system $\mathfrak{A}$ ($\widetilde{Z}(0) \in U$) whose ω-limit set coincides with Z^+. Let $\tau_1, \dots, \tau_n, \dots$ be the successive switching times on $\widetilde{Z}$. Choosing the neighborhood U sufficiently small, in view of Lemma 5 we have

$$\lim_{k\to\infty} \widetilde{Z}(\tau_{2k}) = M_1, \qquad \lim_{k\to\infty} \widetilde{Z}(\tau_{2k+1}) = M_2.$$

Denote by $\widetilde{W}(t)$ the trajectory of (1.8) with initial conditions $\widetilde{W}(0) = \widetilde{Z}(0)$. Let $t_2, t_4, \dots$ be the successive switching times with even indices on the trajectory $\widetilde{W}(t)$. We get

$$\rho(\widetilde{W}(t_2)) \cdots \rho(\widetilde{W}(t_{2n})) = \frac{\Lambda(\widetilde{W}(t_2))}{\Lambda(\widetilde{W}((t_4))} \cdots \frac{\Lambda(\widetilde{W}(t_{2n}))}{\Lambda(\widetilde{W}(t_{2n+2}))},$$

and hence

$$\Lambda(\widetilde{W}(t_{2n+2})) = \frac{\Lambda(\widetilde{W}(t_2))}{\rho(\widetilde{W}(t_2)) \cdot \dots \cdot \rho(\widetilde{W}(t_{2n}))} \geq \frac{\Lambda(\widetilde{W}(t_2))}{\rho_0^n}.$$

Consequently, $\lim_{n\to\infty} \Lambda(\widetilde{W}(t_{2n})) = \infty$ and $\lim_{n\to\infty} \widetilde{W}(t_{2n}) = 0$.

We estimate the time at which $\widetilde{W}(t)$ hits the origin. It follows from (1.24) that

$$t_{2n+2} - t_{2n} \leq \frac{T}{\Lambda(\widetilde{W}(t_{2n}))} \leq \frac{T\rho_0^{n-1}}{\Lambda(\widetilde{W}(t_2))}.$$

Therefore,

$$\tau = \sum_{n=1}^{\infty} (t_{2n+2} - t_{2n}) \leq \frac{T}{\Lambda(\widetilde{W}(t_2))} \sum_{n=1}^{\infty} \rho_0^{n-1} < \infty.$$

As we have already seen, this means that $\widetilde{W}(t)$ is optimal in Problem 2 with $T_0 = \tau$ if (x_0, y_0) is taken to be the projection of the point $\widetilde{W}(t_2)$ and if $x_1 = y_1 = 0$. This contradicts Lemma 3. Thus, the cycle Z^+ is the α-limit set of the trajectories of the system $\mathfrak{A}$ on Tor and is therefore unstable. Correspondingly, Z^- is the unique stable cycle of $\mathfrak{A}$. The lemma is proved.

Consequently, (1.8) has two two-dimensional manifolds $\mathfrak{N}^+$ and $\mathfrak{N}^-$ made up of trajectories hitting the origin as the time increases and decreases, respectively, in finite time and with infinitely many control switchings. Any trajectory of the system (1.8) belonging to the zero level surface of the Hamiltonian tends as $t \to +\infty$ (in the "angle" metric, in which the orbits of $\widetilde{\mathfrak{G}}$ play the role of rays) to the two-dimensional manifold formed by the trajectories in $\mathfrak{N}^-$, and to the manifold $\mathfrak{N}^+$ as $t \to -\infty$. The assertion just proved gives a complete description of all the solutions of (1.8).

Chapter 2. Chattering trajectories in a neighborhood of the manifold of singular trajectories of second order

The main content of this chapter was published in [37]–[39].

1

We consider the following problem

Problem 3. *Minimize the integral*

$$J(x(\cdot)) = \int_0^T (f_0(x) + uf_1(x))\,dt \to \inf$$

on the trajectories of a control system

$$\dot{x} = \varphi_0(x) + u\varphi_1(x)$$

that is affinely generated by a scalar control $u \in [-1, 1]$.

Here $x \in \mathbf{R}^n$, $f_i \colon \mathbf{R}^n \to \mathbf{R}$, and $\varphi_i \colon \mathbf{R}^n \to \mathbf{R}^n$, $i = 0, 1$. The functions f_i and φ_i are assumed to be sufficiently smooth. Problem 3 is determined by the choice of an initial state x_0 and a smooth target manifold $\mathcal{M} \subset \mathbf{R}^n$. The admissible controls are measurable functions, and the admissible trajectories are absolutely continuous.

If the target manifold is fixed, and the initial conditions are chosen from some open neighborhood U of $\mathcal{M}$, then we get the problem of constructing an optimal synthesis in U when an optimal control is regarded as a function of the phase state.

The system of equations of the Pontryagin maximum principle for Problem 3 can be written in the form

$$\dot{y} = I \operatorname{grad}(H_0(y) + uH_1(y)), \qquad u = \operatorname{sgn} H_1(y). \tag{2.1}$$

Here $y = (\psi, x) \in \mathbf{R}^{2n}$, $I = \left(\begin{smallmatrix} 0 & -E \\ E & 0 \end{smallmatrix}\right)$, E being the $n \times n$ identity matrix, and $H_i(y) = \psi\varphi_i(x) - \lambda_0 f_i(x)$, $i = 0, 1$. A solution of (2.1) is understood in the sense of Filippov ([40], Chapter 2, §4.2a).

Since the time T in Problem 3 is not fixed, the optimal trajectories belong to the zero level surface of the Hamiltonian: $H_0(y(t)) + uH_1(y(t)) = 0$.

A trajectory $y(t)$ of (2.1) is said to be singular on the interval (t_0, t_1) if $H_1(y(t)) = 0$ for $t \in (t_0, t_1)$. The number q is called the *order* (*intrinsic order*) of the singular trajectory [41] if

$$\begin{gathered} \left.\frac{\partial}{\partial u}\frac{d^k}{dt^k}\right|_{(2.1)} H_1(y) = 0, \qquad k = 0, 1, \dots, 2q-1, \\ \left.\frac{\partial}{\partial u}\frac{d^{2q}}{dt^{2q}}\right|_{(2.1)} H_1(y) \neq 0 \end{gathered} \tag{2.2}$$

for all y belonging to some open neighborhood of the trajectory $y(t)$. If the relations (2.2) hold only at the points of $y(t)$ itself, then q is called the *local order* of the singular trajectory [41]. It can be shown [22] that q is an integer. The following necessary condition for optimality of singular trajectories is proved in [22] (the Kelley condition):

$$K(y(t)) = (-1)^q \left.\frac{\partial}{\partial u}\frac{d^{2q}}{dt^{2q}}\right|_{(2.1)} H_1(y(t)) \leq 0.$$

Theorem A (Kelley, Kopp, and Moyer [22]). *Suppose that the singular trajectory* $y(t)$ *of the system* (2.1) *has second* (*or any even*) *order* (*intrinsic order*) *for* $t \in (t_0, t_1)$ *and that the Kelley condition in strict form holds*: $K(y(t)) < 0$. *If the control* $u(t)$ *on the trajectory* $y(t)$ *is a* C^∞*-smooth function, then* $y(t)$ *cannot be joined with a nonsingular piecewise smooth trajectory of* (2.1) *if the control is discontinuous at the joining point of the singular and nonsingular parts.*

Consequently, if such a joining takes place, then the nonsingular part must contain infinitely many control switching points $u = \operatorname{sgn} H_1$.

Denote by

$$\operatorname{ad}_{H_0} H_1 = \sum_{i=1}^{n} \left(\frac{\partial H_0}{\partial \psi_i}\frac{\partial H_1}{\partial x_i} - \frac{\partial H_0}{\partial x_i}\frac{\partial H_1}{\partial \psi_i} \right)$$

the Poisson bracket of the functions H_0 and H_1. Assume that the system (2.1) has a singular trajectory $y(t)$ of second order in a neighborhood of which the functions $z_i = \mathrm{ad}_{H_0}^{i-1} H_1$, $i = 1, 2, 3, 4$, are independent. We supplement $z = (z_1, z_2, z_3, z_4)$ by functions $w = (w_1, \dots, w_{2n-4})$ in such a way that $\det(D(z, w)/D(y))|_{y(t)} \neq 0$. As a result of the substitution $y \mapsto (z, w)$, (2.1) has the form

$$\begin{gathered} \dot{z}_1 = z_2, \quad \dot{z}_2 = z_3, \quad \dot{z}_3 = z_4, \quad \dot{z}_4 = \alpha(z, w) + u\beta(z, w), \\ \dot{w} = \gamma(z, w, u), \quad u = \operatorname{sgn} z_1. \end{gathered} \tag{2.3}$$

It follows from the definition of the order of a singular trajectory that $\beta(y(t)) \neq 0$. We assume that the Kelley condition holds in strict form in a neighborhood of $y(t)$: $\beta < 0$. In order that the motion along a singular trajectory satisfy the constraint $|u| \leq 1$ we require that $|\alpha| < -\beta$.

It will be shown below that when the Kelley condition holds and $|\alpha| < -\beta$, then the one-parameter family $\mathfrak{N}_\sigma^+$ of nonsingular chattering trajectories of the system (2.3) hits each point $z = 0$, $w = \sigma$, and the one-parameter family N_σ^- of reverse-time chattering trajectories of (2.3) leaves each point $z = 0$, $w = \sigma$.

2. Chattering trajectories

There are many interesting problems in which a chattering regime is used in going to the manifold of singular trajectories of local order 2.

We remark that the necessity of using a chattering regime in this case does not follow from Theorem A. In order to include in our consideration also this broader class of problems we consider a certain generalization of (2.3), the system of equations with discontinuous right-hand side

$$\begin{gathered} \dot{z}_1 = z_2 + f_1(z, w, u), \quad \dot{z}_2 = z_3 + f_2(z, w, u), \quad \dot{z}_3 = z_4 + f_3(z, w, u), \\ \dot{z}_4 = \alpha(w) + u\beta(w) + f_4(z, w, u), \quad \dot{w} = F(z, w, u), \quad u = \operatorname{sgn} z_1. \end{gathered} \tag{2.4}$$

Here $z = (z_1, z_2, z_3, z_4) \in \mathbf{R}^4$ and $w = (w_1, \dots, w_m) \in \mathbf{R}^m$. It will be assumed that $|z| < \delta$ and $|w - w^0| < \delta$, where δ is sufficiently small. The functions f_i $(i = 1, \dots, 4)$ are of class C^{k_0+5-i}, and the functions α, β, and F of class C^{k_0} jointly in their variables, $1 \leq k_0 \leq \infty$, $\beta < 0$, $|\alpha| < -\beta$. The dependence of α and β on z is included in (2.4) in the term $f_4(z, w, u)$.

Let $g_\kappa(z) = (\kappa^4 z_1, \kappa^3 z_2, \kappa^2 z_3, \kappa z_4)$, $\kappa > 0$. Everywhere below it is assumed that

$$\varlimsup_{\kappa \to +0} \kappa^{-(5-i)} |f_i(g_\kappa(z), w, u)| < C, \qquad i = 1, \dots, 4, \tag{2.5}$$

where C is a constant. The system of equations (2.4) satisfying the condition (2.5) can be regarded as a small perturbation of the system (2.3) with respect to the action of the group of g_κ.

Remark 1. To clear up the possibility of reducing the system (2.1) to the form (2.4), (2.5) we should proceed as in the determination of the system (2.3). Namely, the function $H_1(\psi, x)$ must be taken as z_1. By virtue of (2.1) its total derivative does not contain the control and can be taken as z_2. However, it is sometimes convenient to denote by z_2 a function differing from dH_1/dt by a quantity of greater order of smallness than z_1, and so on. A control system in which the control appears only in the expression for the second derivative (and most problems in mechanics reduce to such systems) often has a singular regime of second order and can be investigated by the method proposed below (see Chapter 4, Problems 16 and 17).

It follows from (2.5) that $f_i(0, w, u) = 0$, $i = 1, \dots, 4$. Consequently, there is a family of singular trajectories of (2.4) that fills the surface $S_0 = \{z, w | z = 0\}$. By the definition of a solution in the Filippov sense for systems with discontinuous right-hand side, the control on singular trajectories is $u = -\alpha(w)/\beta(w)$, and

$$\dot{w} = \frac{\alpha+\beta}{2\beta} F(0, w, -1) + \frac{\beta-\alpha}{2\beta} F(0, w, 1).$$

The standard method of solving the Cauchy problem for an equation with discontinuous right-hand side is to integrate the system on a region of continuity of the control. This method cannot be used to find chattering trajectories, because for such trajectories there is no region of continuity of the control that abuts on the point of joining with the singular part. To find a chattering trajectory of the system (2.4) we proceed as follows. Let $S = \{z, w | z_1 = 0\}$. Consider the Poincaré mapping $\Phi: S \to S$ assigning to a point $(z^0, w^0) \in S$ the first intersection point $(z^1, w^1) \in S$ with S of the trajectory of (2.4) that leaves (z^0, w^0). We find a family of chattering trajectories by starting from the fact that the switching surface of these trajectories is invariant under Φ. The singularity of Φ is resolved by passing in $S \backslash \{z_4 = 0\}$ to the new coordinates

(2.6) $$z_4 = \kappa, \quad z_3 = \lambda\kappa^2, \quad z_2 = \mu\kappa^3, \quad w = \sigma + \kappa\nu.$$

On the surface S we introduce the coordinates σ: $z = 0$, $w = \sigma$. Under the substitution (2.6) the $(m+2)$-dimensional surface $\kappa = 0$ is "glued in" in place of the point $z = 0$, $w = \sigma$. We show that the mapping Φ^2 has exactly two fixed points on the surface $\kappa = 0$ for each fixed σ. The theorem on an invariant manifold of a diffeomorphism can be used to prove that in a neighborhood of each of these fixed points there is a Φ^2-invariant manifold that is the switching surface of some family of chattering trajectories.

Remark 2. We consider the system of equations of the Pontryagin maximum principle in Problem 1 with $q = 2$:

(2.7) $$\dot{\psi}_2 = -\psi_1, \quad \dot{\psi}_1 = x, \quad \dot{x} = y, \quad \dot{y} = \operatorname{sgn} \psi_2.$$

In Chapter 1 it was shown that equation (2.7) is homogeneous with respect to the action of the group $\{\tilde{g}_\kappa\}$, where $\tilde{g}_\kappa(\psi_2, \psi_1, x, y) = (\kappa^4\psi_2, \kappa^3\psi_1, \kappa^2 x, \kappa y)$. The first return mapping $\Phi: S \to S$ of the switching surface carries the orbit

$$\operatorname{Orb}(\lambda_0, \mu_0) = \bigcup_{\kappa>0} (0, \mu_0\kappa^3, \lambda_0\kappa^2, \kappa)$$

into the different orbit $\operatorname{Orb}(\lambda_1, \mu_1)$. Here the quotient mapping $(\lambda_0, \mu_0) \mapsto (\lambda_1, \mu_1)$ arises. The substitution $(\psi_1, x, y) \to (\lambda, \mu, \kappa)$ assigns the disjoint lines $\lambda = \lambda_i$, $\mu = \mu_i$ $(i = 0, 1)$ to the points of the mutually tangent curves $\operatorname{Orb}(\lambda_0, \mu_0)$ and $\operatorname{Orb}(\lambda_1, \mu_1)$. Up to a change of notation the system (2.4) is a small perturbation of (2.7) under the action of the group $\{\tilde{g}_\kappa\}$. Under an arbitrarily small perturbation of (2.7) the homogeneity of the system under the action of $\tilde{g}_\kappa$ is not preserved, but the first return mapping for the perturbed system coincides with the first return mapping for the unperturbed system up to terms with higher order of smallness.

Lemma 7. *Suppose that* $z(t), w(t)$, $t \in [0, \kappa T]$, *is a solution of the system* (2.4) *with initial conditions* $z(0) = g_\kappa z^0$. *Then for any* $T > 0$ *there exist* κ_0 *and* $C_1 > 0$ *such that* $\|g_{1/\kappa} z(t)\| < C_1$ *for* $0 < \kappa < \kappa_0$.

Proof. Denote by $\Lambda(\tau)$ the function $g_{1/\kappa} z(\kappa\tau)$. We differentiate Λ with respect to τ by virtue of (2.4):

$$\frac{d}{d\tau}\Lambda_i = \kappa^{-(5-i)}\frac{dz_i}{d\tau}(\kappa\tau) = \kappa^{-(4-i)}(z_{i+1}(\kappa\tau) + f_i(z(\kappa\tau), w(\kappa\tau), u(\kappa\tau))),$$

for $i = 1, 2, 3$;

$$\frac{d}{d\tau}\Lambda_4 = \kappa^{-1}\frac{dz_4}{d\tau}(\kappa\tau) = \alpha(w(\kappa\tau)) + u(\kappa\tau)\beta(w(\kappa\tau)) + f_4(z(\kappa\tau), w(\kappa\tau), u(\kappa\tau)).$$

Since $z(\kappa\tau) = g_\kappa\Lambda(\tau)$, we find by (2.5) that

$$\frac{d\Lambda_i}{d\tau} = \Lambda_{i+1} + \kappa\tilde{f}_i(\Lambda(\tau), w(\kappa\tau), u(\kappa\tau)), \qquad i = 1, 2, 3,$$

$$\frac{d\Lambda_4}{d\tau} = \alpha(w(\kappa\tau)) + u(\kappa\tau)\beta(w(\kappa\tau)) + \kappa\tilde{f}_4(\Lambda(\tau), w(\kappa\tau), u(\kappa\tau)),$$

where the functions $\tilde{f}_i(\Lambda, w, u) = \kappa^{-(4-i)}f_i(g_\kappa\Lambda, w, u)$ are continuous and uniformly bounded with respect to κ. Let

$$\Lambda(\tau) = e^{A\tau}\left(\xi(\tau) + \int_0^\tau e^{-As}v(s)\,ds + z^0\right),$$

$$A = \begin{pmatrix} 0 & 1 & 0 & 0 \\ 0 & 0 & 1 & 0 \\ 0 & 0 & 0 & 1 \\ 0 & 0 & 0 & 0 \end{pmatrix}, \qquad v = \begin{pmatrix} 0 \\ 0 \\ 0 \\ \alpha(w(\kappa s)) + u(\kappa s)\beta(w(\kappa s)) \end{pmatrix}.$$

Then we have

$$\dot{\xi} = \kappa e^{-A\tau}\tilde{f}\left(e^{A\tau}\left(\xi(\tau) + \int_0^\tau e^{-As}v(s)\,ds + z^0\right), w(\kappa\tau), u(\kappa\tau)\right), \tag{2.8}$$
$$\tilde{f} = (\tilde{f}_1, \dots, \tilde{f}_4), \qquad \xi(0) = 0.$$

Denote by C_2 the maximum of the absolute values of the functions $(e^{-A\tau}\tilde{f})_i$, $i = 1, \dots, 4$, for $\tau \in [0, T]$, $|\xi_i| < \xi_0$, $i = 1, \dots, 4$. Then the solution $\xi(\tau)$ of (2.8) with the initial conditions $\xi(0) = 0$ is determined and does not leave the domain $|\xi_i(\tau)| < \xi_0$, at least not in time $\tau > \xi_0/C_2\kappa_0$. It remains to choose $\kappa_0 < \xi_0/C_2T$. The lemma is proved.

Suppose that $(z^0, w^0) \in S$, $(z^1, w^1) = \Phi(z^0, w^0)$, and $T(z^0, w^0) > 0$ is the time of first intersection with S of the trajectory of (2.4) leaving the point (z^0, w^0). We write the mapping Φ in the coordinates $(\mu, \lambda, \kappa, \nu)$. Let $\Phi(\mu_0, \lambda_0, \kappa_0, \nu_0) = (\mu_1, \lambda_1, \kappa_1, \nu_1)$ and $T(z_0, w_0) = \kappa_0\tau_0$. Let us rewrite (2.4) as a system of integral equations. We have

$$\kappa_1 = \kappa_0 + \int_0^{\kappa_0\tau_0}\left(\alpha\left(\sigma + \kappa_0\nu_0 + \int_0^\tau F(z(s), w(s), u)\,ds\right)\right.$$
$$\left. + u\beta\left(\sigma + \kappa_0\nu_0 + \int_0^\tau F(z(s), w(s), u)\,ds\right) + uf_4(z(\tau), w(\tau), u)\right)d\tau$$
$$= \kappa_0(1 + (\alpha(\sigma) + u\beta(\sigma))\tau_0) + \kappa_0 Q_1,$$

where Q_1 denotes the function

$$Q_1 = \int_0^{\tau_0}\left(\left(\alpha\left(\sigma + \kappa_0\nu_0 + \int_0^{\tau\kappa_0} F(s)\,ds\right) - \alpha(\sigma)\right)\right.$$
$$\left. + u\left(\beta\left(\sigma + \kappa_0\nu_0 + \int_0^{\tau\kappa_0} F(s)\,ds\right) - \beta(\sigma)\right) + uf_4(\kappa_0\tau)\right)d\tau,$$
$$F(s) = F(z(s), w(s), u), \qquad f_4(\tau) = f_4(z(\tau), w(\tau), u).$$

By the theorem on the smooth dependence of the solutions of differential equations on the initial data, the functions $z(s)$ and $w(s)$ are k_0 times continuously differentiable with respect to μ_0, λ_0, κ_0, ν_0, τ_0, and u, where the number k_0 is determined by the smoothness of the right-hand sides of (2.4). Therefore, the function Q_1 is of class C^{k_0} jointly in its variables. We show that

$$\overline{\lim_{\kappa\to+0}}\, \kappa^{-1}|Q_1| < C_3,$$

where the constant C_3 does not depend on the remaining arguments of Q_1. It follows from Lemma 7 that for any T_0

$$\overline{\lim_{\kappa\to+0}}\, \kappa^{-1}|f_4(z(\kappa\tau), w(\kappa\tau), u)| < C_4$$

for $0 \le \tau \le T_0$, where C_4 can be chosen to depend only on T_0. By the mean-value theorem,

$$\left|\alpha\left(\sigma + \kappa_0\nu_0 + \int_0^{\tau\kappa_0} F(s)\,ds\right) - \alpha(\sigma)\right| \le \|\alpha'\|\kappa_0 \left\|\nu_0 + \int_0^{\tau} F(s\kappa_0)\,ds\right\|,$$
$$\left|\beta\left(\sigma + \kappa_0\nu_0 + \int_0^{\tau\kappa_0} F(s)\,ds\right) - \beta(\sigma)\right| \le \|\beta'\|\kappa_0 \left\|\nu_0 + \int_0^{\tau} F(s\kappa_0)\,ds\right\|.$$

From now on, $\Theta(\kappa^N)$ denotes a function of class C^{k_0+N-1} in μ, λ, κ, τ, ν, and u such that

$$\overline{\lim_{\kappa\to+0}}\, \kappa^{-N}|\Theta(\kappa^N)| < C,$$

where C is some constant. In this notation, for example, $\Theta(1)$ is an arbitrary function of class C^{k_0-1}. Thus, it has been proved that $Q_1 = \Theta(\kappa)$.

Integrating the equation $\dot z_3 = z_4 + f_3(z, w, u)$, we get

$$\lambda_1\kappa_1^2 = (\lambda_0 + \tau_0 + \tfrac{1}{2}(\alpha(\sigma) + u\beta(\sigma))\tau_0^2)\kappa_0^2 + \kappa_0^2 Q_2 + \kappa_0 Q_3,$$

where

$$Q_2 = \int_0^{\tau_0}\int_0^{\tau}\left(\left(\alpha\left(\sigma + \nu_0\kappa_0 + \int_0^{\tau'\kappa_0} F(s)\,ds\right) - \alpha(\sigma)\right)\right.$$
$$\left. + u\left(\beta\left(\sigma + \nu_0\kappa_0 + \int_0^{\tau'\kappa_0} F(s)\,ds\right) - \beta(\sigma)\right) + f_4(\kappa_0\tau')\right) d\tau'\,d\tau,$$
$$Q_3 = \int_0^{\tau_0} f_3(z(\kappa_0\tau), w(\kappa_0\tau), u)\,d\tau.$$

It is not hard to see that $Q_2 = \Theta(\kappa_0)$ and $Q_3 = \Theta(\kappa_0^2)$. It can be proved similarly that

$$\mu_1\kappa_1^3 = (\mu_0 + \lambda_0\tau_0 + \tfrac{1}{2}\tau_0^2 + \tfrac{1}{6}(\alpha(\sigma) + u\beta(\sigma))\tau_0^3)\kappa_0^3 + \kappa_0^3 Q_4 + \kappa_0^2 Q_5 + \kappa_0 Q_6,$$
$$\nu_1\kappa_1 = (\nu_0 + \tau_0 F(0, \sigma, u))\kappa_0 + \kappa_0 Q_7,$$
$$\kappa_0^4\tau(\mu_0 + \tfrac{1}{2}\lambda_0\tau_0 + \tfrac{1}{6}\tau_0^2 + \tfrac{1}{24}(\alpha(\sigma) + u\beta(\sigma))\tau_0^3) + \kappa_0^4 Q_8 + \kappa_0^3 Q_9 + \kappa_0^2 Q_{10} + \kappa_0 Q_{11} = 0,$$

where the functions Q_1, Q_2, Q_4, Q_7, and Q_8 are of class $\Theta(\kappa_0)$, the functions Q_3, Q_5, and Q_9 are of class $\Theta(\kappa_0^2)$, Q_6 and Q_{10} are of class $\Theta(\kappa_0^3)$, and Q_{11} is of class $\Theta(\kappa_0^4)$.

Lemma 8. *Let* $G(\kappa, \Sigma) = \sum_{i=1}^{m} \kappa^{m-i+1}\varphi_i(\kappa, \Sigma)$, $\varphi_i = \Theta(\kappa^i)$; *here* Σ *stands for the collection of arguments* $\mu, \lambda, \nu, \tau, u$. *Then* $\kappa^{-m}G(\kappa, \Sigma) = \Theta(\kappa)$.

Proof. It is clear that

$$\overline{\lim_{\kappa\to 0}}\, \kappa^{-(m+1)}|G| < \infty.$$

We show that if $\varphi = \Theta(\kappa^{l+1})$, then $\kappa^{-l}\varphi \in C^{k_0}$. Since $\varphi \in C^{k_0+l}$, it follows from Taylor's formula that

$$\varphi(\kappa, \Sigma) = \sum_{i=0}^{l-1} \frac{\partial^i}{\partial \kappa^i}\varphi(0, \Sigma)\frac{\kappa^i}{i!} + \int_0^\kappa \frac{(\kappa-\Theta)^{l-1}}{(l-1)!}\frac{\partial^l}{\partial\kappa^l}\varphi(\theta, \Sigma)\,d\theta$$
$$= \kappa^l \int_0^1 \frac{(1-\theta)^{l-1}}{(l-1)!}\frac{\partial^l}{\partial\kappa^l}\varphi(\theta\kappa, \Sigma)\,d\theta.$$

By assumption, the first $l-1$ partial derivatives of φ with respect to κ vanish for $\kappa = 0$. The function $\partial^l\varphi/\partial\kappa^l$ is of class C^{k_0}. Thus, $\varphi(\kappa, \Sigma) = \kappa^l\varphi_1(\kappa, \Sigma)$, where φ_1 is a function of class C^{k_0}. The lemma is proved.

By Lemma 8, Φ can be written in the form

$$\kappa_1 = \kappa_0(1 + (\alpha(\sigma) + u\beta(\sigma))\tau_0) + \kappa_0\Theta(\kappa_0),$$
$$\lambda_1 = \frac{\lambda_0 + \tau_0 + \frac{1}{2}(\alpha(\sigma) + u\beta(\sigma))\tau_0^2}{(1 + (\alpha(\sigma) + u\beta(\sigma))\tau_0)^2} + \Theta(\kappa_0),$$
$$\mu_1 = \frac{\mu_0 + \lambda_0\tau_0 + \frac{1}{2}\tau_0^2 + \frac{1}{6}(\alpha(\sigma) + u\beta(\sigma))\tau_0^3}{(1 + (\alpha(\sigma) + u\beta(\sigma))\tau_0)^3} + \Theta(\kappa_0),$$
$$\nu_1 = \frac{\nu_0 + \tau_0 F(0, \sigma, u)}{1 + (\alpha(\sigma) + u\beta(\sigma))\tau_0} + \Theta(\kappa_0),$$
$$\mu_0 + \tfrac{1}{2}\lambda_0\tau_0 + \tfrac{1}{6}\tau_0^2 + \tfrac{1}{24}(\alpha(\sigma) + u\beta(\sigma))\tau_0^3 + \Theta(\kappa_0) = 0.$$

Let $\Phi(\mu_1, \lambda_1, \kappa_1, \nu_1) = (\mu_2, \lambda_2, \kappa_2, \nu_2)$ and $T(z_1, w_1) = \kappa_1\tau_1$. We consider initial conditions μ_0, λ_0, κ_0, and ν_0 such that $u = -1$ for $t \in (0, \kappa_0\tau_0)$ and $u = 1$ for $t \in (\kappa_0\tau_0, \kappa_0\tau_0 + \kappa_1\tau_1)$. For such a sequence of switchings it suffices to require that, first, $T(z_0, w_0) > 0$ and $T(z_1, w_1) > 0$, and, second, $\dot z_1(+0) < 0$ and $\dot z_1(\tau_0\kappa_0 + 0) > 0$. From this we get

$$\tau_0\kappa_0 > 0, \quad \tau_1\kappa_1 > 0, \quad \mu_0\tau_0 < 0, \quad \mu_1\tau_1 > 0. \tag{2.9}$$

The mapping Φ^2 is now determined by the system of equations

$$\begin{aligned} \kappa_{i+1} &= \kappa_i(1 + A_i\tau_i) + \kappa_i\Theta(\kappa_i), \\ \lambda_{i+1} &= (\lambda_i + \tau_i + \tfrac{1}{2}A_i\tau_i^2)(1 + A_i\tau_i)^{-2} + \Theta(\kappa_i), \\ \mu_{i+1} &= (\mu_i + \lambda_i\tau_i + \tfrac{1}{2}\tau_i^2 + \tfrac{1}{6}A_i\tau_i^3)(1 + A_i\tau_i)^{-3} + \Theta(\kappa_i), \\ \nu_{i+1} &= (\nu_i + \tau_i F(0, \sigma, u_i))(1 + A_i\tau_i)^{-1} + \Theta(\kappa_i), \end{aligned} \tag{2.10a}$$

where the functions $\tau_i = \tau_i(\mu_i, \lambda_i, \kappa_i, \nu_i)$ are implicitly given by the equations

$$\mu_i + \tfrac{1}{2}\lambda_i\tau_i + \tfrac{1}{6}\tau_i^2 + \tfrac{1}{24}A_i\tau_i^3 + \Theta(\kappa_i) = 0. \tag{2.10b}$$

Here $i = 0, 1$, $u_0 = -1$, $u_1 = 1$, $A_0 = \alpha(\sigma) - \beta(\sigma)$, $A_1 = \alpha(\sigma) + \beta(\sigma)$, $A_0 > 0$, and $A_1 < 0$. In (2.10) we can set $A_0 = 1$ and $A_1 = b < 0$ without loss of generality, because the substitution

$$\mu' = A_0^2\mu, \quad \lambda' = A_0\lambda, \quad \nu' = A_0\nu, \quad \tau' = A_0\tau, \quad b = A_1A_0^{-1} \tag{2.11}$$

leads to this.

The mapping Φ^2 is piecewise smooth. In its place we introduce the mapping $\widetilde{\Phi}^2$ determined by (2.10). The mapping $\widetilde{\Phi}^2$ coincides with Φ^2 for all points at which (2.9) holds under the following additional conditions: each of equations (2.10b) does not have a root of odd multiplicity lying respectively between 0 and τ_0 and between 0 and τ_1. Let us find the coordinates of the fixed points of $\widetilde{\Phi}^2$ on the surface $\kappa = 0$. We get the system of equations

$$\lambda_1 = \frac{\lambda_0 + \tau_0 + \frac{1}{2}\tau_0^2}{(1+\tau_0)^2}, \qquad \lambda_0 = \frac{\lambda_1 + \tau_1 + \frac{1}{2}b\tau_1^2}{(1+b\tau_1)^2},$$

$$\text{(2.12a)} \qquad \mu_1 = \frac{\mu_0 + \lambda_0\tau_0 + \frac{1}{2}\tau_0^2 + \frac{1}{6}\tau_0^3}{(1+\tau_0)^3}, \qquad \mu_0 = \frac{\mu_1 + \lambda_1\tau_1 + \frac{1}{2}\tau_1^2 + \frac{1}{6}b\tau_1^3}{(1+b\tau_1)^3},$$

$$\nu = \frac{\nu_0 + \tau_0 F(0, \sigma, -1)}{1+\tau_0}, \qquad \nu_0 = \frac{\nu_1 + \tau_1 F(0, \sigma, 1)}{1 + b\tau_1}.$$

The functions $\tau_0 = \tau_0(\lambda, \mu)$ and $\tau_1 = \tau_1(\lambda, \mu)$ are solutions of the system

$$\text{(2.12b)} \qquad \begin{aligned} \mu_0 + \tfrac{1}{2}\lambda_0\tau_0 + \tfrac{1}{6}\tau_0^2 + \tfrac{1}{24}\tau_0^3 &= 0, \\ \mu_1 + \tfrac{1}{2}\lambda_1\tau_1 + \tfrac{1}{6}\tau_1^2 + \tfrac{1}{24}b\tau_1^3 &= 0. \end{aligned}$$

Since the variable ν does not appear in (2.12b), the last two equations in (2.12a) for finding the coordinates λ and μ of a fixed point are not needed and can be excluded from consideration for the time being.

Remark 3. Solution of (2.12) involves finding the switching curve in the following problem of Fuller and Marchal.

Problem 4.

$$\int_0^\infty x^2\,dt \to \inf,$$

$$\dot{x} = y, \quad \dot{y} = u, \quad x(0) = x_0, \quad y(0) = y_0,$$
$$u \in [b, 1], \qquad b < 0.$$

The optimal trajectories in Problem 4 were found in [3]. It turns out that Problem 4 is a model problem for the system (2.4) in the sense that its solutions determine the asymptotic behavior of the chattering trajectories of (2.4) upon approaching the singular manifold S_0 of this system. Most important for us will be to clear up the dependence of the solutions of Problem 4 on the parameter b, and to compute the cyclic points of order two of the first-return mapping for the system of equations of the Pontryagin maximum principle after resolution of the singularity of this mapping.

The system of equations for the Pontryagin maximum principle for Problem 4 can be written in the form

$$\text{(2.13)} \qquad \dot{\psi}_1 = x, \quad \dot{\psi}_2 = -\psi_1, \quad \dot{x} = y, \quad \dot{y} = u,$$

where $u = 1$ for $\psi_2 > 0$ and $u = b$ for $\psi_2 < 0$.

We let $z_1 = \psi_2$, $z_2 = -\psi_1$, $z_0 = -x$, and $z_4 = -y$, and then make the substitution (2.6). Just as in Lemma 2 it can be shown that the switching curve of the optimal trajectories of Problem 4 has the form

$$\psi_2 = 0, \qquad \psi_1 = \begin{cases} \mu_0 y^3, & y < 0, \\ \mu_1 y^3, & y > 0, \end{cases} \qquad x = \begin{cases} \lambda_0 y^2, & y < 0, \\ \lambda_1 y^2, & y > 0. \end{cases}$$

To find (λ_0, μ_0) and (λ_1, μ_1) we get precisely the system (2.12), with the equations containing the coordinate ν omitted.

Denote by φ the mapping $(\lambda_0, \mu_0) \mapsto (\lambda_2, \mu_2)$ given by (2.10) with $\kappa = 0$, and then discard the equations containing ν. The mapping φ is not uniquely determined. Denote by φ_+ the mapping corresponding to the first positive root of (2.12b), and by φ_- the mapping corresponding to its first negative root. The following assertion is valid for the system (2.13). If $\psi_2(t), \psi_1(t), x(t), y(t)$ is a solution of (2.13), then $\psi_2(-t), -\psi_1(-t), x(-t), -y(-t)$ is clearly also a solution of (2.13) $((d^4/d\tau^4)\psi_2(-t) = (d^4/dt^4)\psi_2(\tau)|_{\tau=-t})$. Therefore, if $\varphi_+(\lambda_0, \mu_0) = (\lambda_2, \mu_2)$, then $\varphi_-(\lambda_2, \mu_2) = (\lambda_0, \mu_0)$, i.e.,

$$\varphi_-(\lambda, \mu) = (\varphi_+)^{-1}(\lambda, \mu). \tag{2.14}$$

Lemma 9. *The system* (2.12) *has two collections of solutions,* $(\lambda_i^+, \mu_i^+, \nu_i^+)$ *and* $(\lambda_i^-, \mu_i^-, \nu_i^-)$, $i = 0, 1$, *with* $\lambda_0^+ \in (0, \frac{1}{2})$, $\lambda_1^+ \in (1/(2b), 0)$, $\lambda_0^- \in (1/(2b), 0)$, $\lambda_1^- \in (0, \frac{1}{2})$, $\mu_i^+ = \frac{1}{2}(\lambda_i^+)^2$, *and* $\mu_i^- = \frac{1}{2}(\lambda_i^-)^2$. *The functions* $\tau_i^\varepsilon = \tau_i(\lambda, \mu)|_{\lambda=\lambda_i^\varepsilon, \mu=\mu_i^\varepsilon}$, $i = 0, 1$, $\varepsilon =$ " + ", " − ", *are* C^{k_0}*-continuously differentiable at the point* $(\lambda_i^\varepsilon, \mu_i^\varepsilon)$ *and satisfy the following estimate*:

$$\begin{aligned} &\tau_0^+ \in (-2, -1), \qquad \tau_1^+ \in (-1/b, -2/b), \\ &\tau_0^- \in (-\infty, -2), \qquad \tau_1^- \in (-2/b, \infty). \end{aligned}$$

Proof. We find the solutions of (2.11) corresponding to the zero level surface of the Hamiltonian of (2.13):

$$\psi_1 y + u\psi_2 - \tfrac{1}{2}x^2 = 0.$$

In this case we can set $\mu_0 = \frac{1}{2}\lambda_0^2$ and $\mu_1 = \frac{1}{2}\lambda_1^2$ in (2.12). Then the coordinates of a fixed point of the mapping $\widetilde{\Phi}^2$ are determined by the system of equations

$$\begin{aligned} &\lambda_1 = (\lambda_0 + \tau_0 + \tfrac{1}{2}\tau_0^2)(1 + \tau_0)^{-2}, \\ &12\lambda_0^2 + 12\lambda_0\tau_0 + 4\tau_0^2 + \tau_0^3 = 0; \end{aligned} \tag{2.15a}$$

$$\begin{aligned} &\lambda_0 = (\lambda_1 + \tau_1 + \tfrac{1}{2}b\tau_1^2)(1 + b\tau_1)^{-2}, \\ &12\lambda_1^2 + 12\lambda_1\tau_1 + 4\tau_1^2 + b\tau_1^3 = 0. \end{aligned} \tag{2.15b}$$

Let

$$\sigma = 1 + b\tau_1, \qquad 2b\lambda_0 - 1 = \alpha, \qquad 2b\lambda_1 - 1 = \beta. \tag{2.16}$$

It follows from (2.15b) that

$$\begin{aligned} &\sigma^2 = \beta/\alpha, \\ &\sigma^3 + \sigma^2 + (1 + 6\beta)\sigma + 3\beta^2 = 0. \end{aligned}$$

Eliminating the parameter σ, we get

$$\alpha\beta(1 + 3\alpha\beta)^2 = (6\alpha\beta + \alpha + \beta)^2. \tag{2.17}$$

Letting $b = 1$, $\tau_1 = \tau_0'$, $\lambda_1 = \lambda_0'$, and $\lambda_0 = \lambda_1'$ in (2.16), we write the system (2.15a) after eliminating the variable σ in the form

$$\alpha'\beta'(1 + 3\alpha'\beta')^2 = (6\alpha'\beta' + \alpha' + \beta')^2,$$

where

$$\begin{aligned} &\alpha' = 2\lambda_1 - 1 = b^{-1}(\beta + 1 - b), \\ &\beta' = 2\lambda_0 - 1 = b^{-1}(\alpha + 1 - b). \end{aligned} \tag{2.18}$$

Let

$$X = \alpha\beta, \qquad Y = \alpha + \beta. \tag{2.19}$$

The curve (2.17) is given on the (X, Y)-plane by the equation

$$X(1+3X)^2 = (6X+Y)^2. \tag{2.20}$$

The system (2.19) is solvable with respect to α and β if

$$4X \leq Y^2. \tag{2.21}$$

The transformation (2.18) can be written in the coordinates X, Y in the form

$$X' = \frac{1}{b^2}X + \frac{1-b}{b^2}Y + \frac{(1-b)^2}{b^2}, \qquad Y' = \frac{1}{b}Y + \frac{2(1-b)}{b}. \tag{2.22}$$

Thus, the problem has been reduced to the determination of the points of the curve (2.20) belonging to the domain (2.21) that are carried by the mapping (2.22) into points of the same curve. The mapping (2.22) is affine. Making the change of variables

$$\xi = X + Y + 1, \qquad \eta = Y + 2, \tag{2.23}$$

we write the mapping (2.22) in the form

$$\xi' = \frac{1}{b^2}\xi, \qquad \eta' = \frac{1}{b}\eta. \tag{2.24}$$

In the coordinates ξ, η the curve (2.20) has the form

$$(\xi - \eta + 1)(3(\xi - \eta + 1) + 1)^2 = (6(\xi - \eta + 1) + \eta - 2)^2.$$

Setting $\xi - \eta + 1 = t^2$, we get the following parametric representation of it:

$$\xi = 3t^3 - 5t^2 + t + 1, \qquad \eta = 3t^3 - 6t^2 + t + 2. \tag{2.25}$$

The transformation (2.24) carrics points of the parabola $\xi = C\eta^2$ into points of the same parabola. Consequently, the solution of (2.15) is determined by a pair of points of the curve (2.25) that lie on the intersection of this curve with some parabola $\xi = C\eta^2$ on different sides of the x-axis. Here the parameter C of the parabola and the number b in (2.24) become functions of t. We remark that the inverse images of these points under the substitution (2.23) must satisfy (2.21), which is equivalent to the inequality

$$4\xi \leq \eta^2.$$

We investigate the curve (2.25). The graphs of $\xi(t)$ and $\eta(t)$ are pictured in Figures 5 and 6. Note that

$$\frac{d\eta}{d\xi} = \frac{9t^2 - 12t + 1}{(9t-1)(t-1)}, \qquad \frac{d^2\eta}{d\xi^2} = \frac{2(9t^2-1)}{(9t-1)^3(t-1)^3}.$$

The graph of $\eta(\xi)$ is pictured in Figure 7 (see next page).

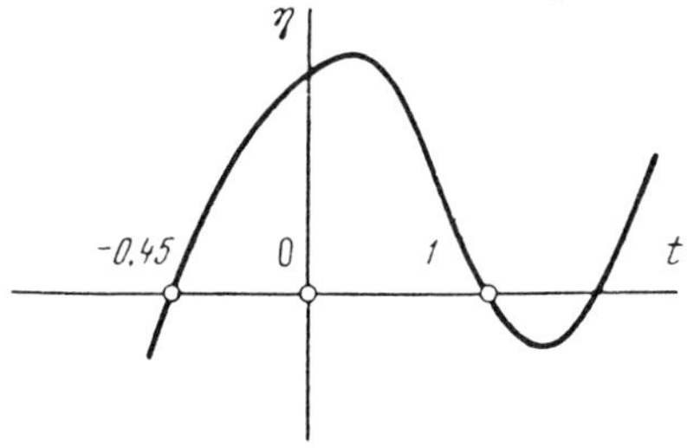

FIGURE 5

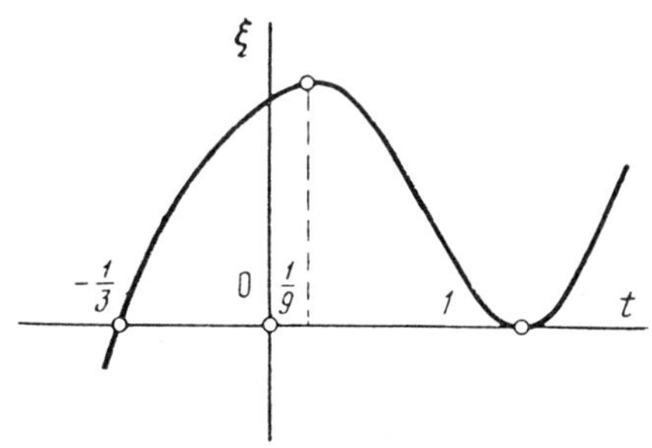

FIGURE 6

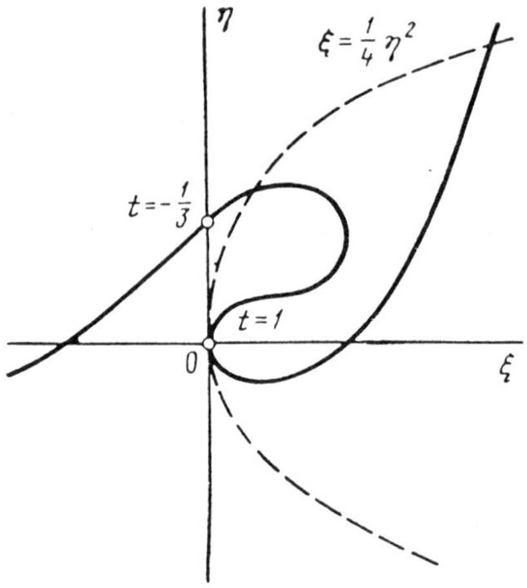

FIGURE 7

The values of the parameter t such that the curve (2.25) intersects the parabola $\xi = \frac{1}{4}\eta^2$ are determined from the equation $3t^2(t-1)^2(3t^2-6t-1)=0$. Its roots are 0, 1, $-0.1547\ldots$, and $2.1541\ldots$. Consequently, the parabolas $\xi = C\eta^2$ with $C < 1/4$, $C \neq 0$, intersect the curve (2.25) at exactly two points, not counting the origin. For $0 < C < 1/4$ these points are on the same side of the x-axis, and for $C < 0$ they are on different sides. Thus, the points of interest to us belong to the half-plane $\xi < 0$. We show that for $C < 0$ the intersection of the curve (2.25) with the parabolas $\xi = C\eta^2$ is transversal (except for the point at the origin). To the parabolas $\xi = C\eta^2$ intersecting the curve (2.25) tangentially there correspond the multiple roots of the equation

$$3t^3 - 5t^2 + t + 1 = C(3t^3 - 6t^2 + t + 2)^2,$$

i.e., the roots for which, simultaneously,

$$9t^2 - 10t + 1 = 2C(3t^3 - 6t^2 + t + 2)(9t^2 - 12t + 1).$$

Eliminating the parameter C from our system of equations, we get $t(t-1)^3(9t+1)=0$. Consequently, for $t < -1/3$ (i.e., in the half-plane $\xi < 0$) there are no multiple points of intersection. By this and the implicit function theorem, the functions $C(t)$, $b(t)$, and $\tau(t)$ uniquely determined for $t < -1/3$ by the relations

$$\eta(\tau) = \eta(t)b^{-1}(t), \qquad \xi(\tau) = \xi(t)b^{-2}(t), \qquad C(t) = \xi(t)\eta^{-2}(t) \tag{2.26}$$

are continuously differentiable. Here $(\xi(t), \eta(t))$ and $(\xi(\tau), \eta(\tau))$ satisfy (2.25).

We show that the mapping $t \to b(t)$ is one-to-one for $t < -1/3$. Note that the function $\eta(\xi)$ determined by (2.25) is monotonically increasing for $\xi < 0$.

Assume that $b(t_1) = b(t_2)$ for some values of t_1 and t_2 ($t_1 < t_2 < -1/3$). It suffices to consider the case $\eta(t_1) > 0$. Then, by the definition of the mapping (2.24),

$$\xi(t_1) < \xi(t_2) < 0, \qquad 0 < \eta(t_1) < \eta(t_2),$$
$$\xi(\tau(t_1)) = b^{-2}\xi(t_1) < b^{-2}\xi(t_2) = \xi(\tau(t_2)),$$
$$\eta(\tau(t_1)) = b^{-1}\eta(t_1) > b^{-1}\eta(t_2) = \eta(\tau(t_2)),$$

where the function $\tau(t)$ is determined by (2.26). The last two inequalities contradict the monotone increase of $\eta(\xi)$ in the half-plane $\xi < 0$. Thus, it is proved that for

$t < -1/3$ the function $b(t)$ is smooth (even analytic) and monotonically increasing from $-\infty$ to 0. To the value of the parameter t for which $\eta(t) = 0$ there corresponds the value $b = -1$ $(t = -0.4574\ldots)$.

We show that the solutions of (2.15) satisfy one of the constraints

$$0 < \lambda_0 < 1/2, \qquad 1/(2b) < \lambda_1 < 0 \tag{2.27a}$$

or

$$1/(2b) < \lambda_0 < 0, \qquad 0 < \lambda_1 < 1/2. \tag{2.27b}$$

The parameters α and β appear in (2.17) symmetrically. Therefore, if the pair (λ_0, λ_1) is a solution of (2.15), so is the pair (λ_1, λ_0) for the same value of b.

The system (2.27) is equivalent to

$$b - 1 < \alpha < -1, \qquad -1 < \beta < 0; \tag{2.28a}$$

$$-1 < \alpha < 0, \qquad b - 1 < \beta < -1. \tag{2.28b}$$

For $b = -1$ the relations (2.28) are satisfied. As shown in Chapter 1, $\lambda_0 = -\lambda_1 = \pm 0.4446\ldots$ in this case. Since $b(t)$ is a continuous function, it remains to show that the system (2.19) cannot have any of the following three values of α as a solution: $\alpha = 0$, $\alpha = -1$, or $\alpha = b - 1$. For $\alpha = 0$ we get $X = 0$. But by (2.23) we have $X = \xi - \eta + 1 = t^2 > 0$. For $\alpha = -1$ it follows from (2.19) that $\beta = -X$ and $X + Y + 1 = 0$. But in view of (2.23) for $t < -1/3$ we have $X + Y + 1 = \xi < 0$. It remains to consider the case $\alpha = b - 1$. By (2.23), substitution of $\alpha = b - 1$ in (2.19) gives us

$$(b - 1)^2 + (2 - \eta)(b - 1) + \xi - \eta + 1 = 0.$$

Since $b < 0$ and $\xi < 0$, we get

$$b = \eta/2 - (\eta^2/4 - \xi)^{1/2}. \tag{2.29}$$

Assume that there exist two values $t_2 < t_1 < -\frac{1}{3}$ of the parameter t such that $t_2 = \tau(t_1)$ and such that (2.29) holds with $\xi = \xi(t_1)$ and $\eta = \eta(t_1)$ for $b = b(t_1)$ (the functions $\tau(t)$ and $b(t)$ are determined by (2.26)). By the definition of the mapping (2.26), $\eta(t_2) = \eta(t_1)/b(t_1)$. By (2.29),

$$b(t_1) = \tfrac{1}{2}\eta(t_1)(1 - (1 - 4C(t_1))^{1/2}),$$

and so

$$\eta(t_2) = \eta(t_1)/b(t_1) = 2/(1 - (1 - 4C(t_1))^{1/2}).$$

Let $\rho = \eta(t_2)$. Then $C(t_1) = (\rho - 1)\rho^{-2}$. Since $\xi(t_2)\eta^{-2}(t_2) = \xi(t_1)\eta^{-2}(t_1) = C(t_1)$, we get from (2.25) that

$$3t_2^3 - 5t_2^2 + t_2 + 1 = \rho - 1, \qquad 3t_2^3 - 6t_2^2 + t_2 + 2 = \rho.$$

Eliminating ρ, we get that $t_2^2 = 0$. Thus, we have shown that the function $b(t)$ cannot satisfy (2.29), whatever the value of the parameter $t < -1/3$. The estimate (2.28) is thereby proved.

Let $\lambda_0, \lambda_1, \tau_0, \tau_1$ be the solutions of (2.15). We show that then either

$$\begin{array}{ll} 0 < \lambda_0 < 1/2, & 1/(2b) < \lambda_1 < 0, \\ -2 < \tau_0 < -1, & -1/b < \tau_1 < -2/b, \end{array} \tag{2.30a}$$

or

$$\begin{array}{ll} 1/(2b) < \lambda_0 < 0, & 0 < \lambda_1 < 1/2, \\ \tau_0 < -2, & \tau_1 > -2/b. \end{array} \tag{2.30b}$$

Suppose first that $0 < \lambda_0 < 1/2$ and $1/(2b) < \lambda_1 < 0$. It follows from (2.15a) that $\lambda_0+\tau_0+\frac{1}{2}\tau_0^2 < 0$, from which $\tau_0+\frac{1}{2}\tau_0^2 < 0$, or (since $\tau_0 \neq -1$) $\tau_0 \in (-2, -1) \cup (-1, 0)$. If $\tau_0 > -1$, we have from (2.15a) that

$$0 = 12\lambda_0^2 + 12\lambda_0\tau_0 + 4\tau_0^2 + \tau_0^3 > 12\lambda_0^2 + 12\lambda_0\tau_0 + 3\tau_0^2 = 3(2\lambda_0 + \tau_0)^2 \geq 0.$$

Contradiction. Thus, $-2 < \tau_0 < -1$.

We show that $-1/b < \tau_1 < -2/b$. It follows from (2.15b) that $\lambda_1+\tau_1+\frac{1}{2}b\tau_1^2 > 0$, and so $\tau_1 + \frac{1}{2}b\tau_1^2 > 0$. Since $\tau_1 \neq -1/b$, we have $\tau_1 \in (0, -1/b) \cup (-1/b, -2/b)$. If $\tau_1 < -1/b$, then from (2.15b) we have

$$0 = 12\lambda_1^2 + 12\lambda_1\tau_1 + 4\tau_1^2 + b\tau_1^3 > 12\lambda_1^2 + 12\lambda_1\tau_1 + 3\tau_1^2 = 3(2\lambda_1 + \tau_1)^2 \geq 0.$$

Contradiction. Therefore, $-1/b < \tau_1 < -2/b$.

We prove (2.30b). In this case it follows from (2.15a) that $\lambda_0 + \tau_0 + \frac{1}{2}\tau_0^2 > 0$, and so $\tau_0 + \frac{1}{2}\tau_0^2 > 0$, or $\tau_0 \in (-\infty, -2) \cup (0, \infty)$. If $\tau_0 > 0$, then

$$0 = 12\lambda_0^2 + 12\lambda_0\tau_0 + 4\tau_0^2 + \tau_0^3 = 3(2\lambda_0 + \tau_0)^2 + \tau_0^2 + \tau_0^3 > 0.$$

Contradiction. Thus, $\tau_0 < -2$. Further, from (2.15b) we get $\lambda_1 + \tau_1 + \frac{1}{2}b\tau_1^2 < 0$, and so $\tau_1 + \frac{1}{2}b\tau_1^2 < 0$, or $\tau_1 \in (-\infty, 0) \cup (-2/b, \infty)$. For $\tau_1 < 0$ we have

$$0 = 12\lambda_1^2 + 12\lambda_1\tau_1 + 4\tau_1^2 + b\tau_1^3 = 3(2\lambda_1 + \tau_1)^2 + \tau_1^2 + b\tau_1^3 > 0.$$

Contradiction. Therefore, $\tau_1 > -2/b$. The lemma is proved.

It is now not hard to write the component ν of the solution of (2.12):

$$\nu_0^\varepsilon = \frac{\tau_0^\varepsilon F(0, \sigma, -1) + \tau_1^\varepsilon(1 + \tau_0^\varepsilon)F(0, \sigma, 1)}{(1 + \tau_0^\varepsilon)(1 + b\tau_1^\varepsilon) - 1},$$
$$\nu_1^\varepsilon = \frac{\tau_1^\varepsilon F(0, \sigma, 1) + \tau_0^\varepsilon(1 + b\tau_1^\varepsilon)F(0, \sigma, -1)}{(1 + \tau_0^\varepsilon)(1 + b\tau_1^\varepsilon) - 1}. \tag{2.31}$$

In view of Lemma 9 the mapping $\widetilde{\Phi}^2$ has two pairs of fixed points; we write $M^+(0, \mu_0^+, \lambda_0^+, \nu_0^+)$ and $M^-(0, \mu_0^-, \lambda_0^-, \nu_0^-)$.

Lemma 10. *The mapping $\widetilde{\Phi}^2$ is hyperbolic at the points M^+ and M^-. The differential $\mathscr{D}\widetilde{\Phi}^2(M^+)$ has one eigenvalue less than 1 in modulus, and $m+2$ eigenvalues greater than 1 in modulus. The differential $\mathscr{D}\widetilde{\Phi}^2(M^-)$ has one eigenvalue greater than 1 in modulus, and $m+2$ eigenvalues less than 1 in modulus.*

Proof. We first consider the point M^+. Let us compute the differential $\mathscr{D}^+ = \mathscr{D}\widetilde{\Phi}^2(M^+)$ of the mapping $\widetilde{\Phi}^2$. Differentiating (2.10) and using the normalization (2.11), we get that

$$\left.\frac{\partial\kappa_2}{\partial\kappa_0}\right|_{M^+} = \left.\frac{\partial\kappa_2}{\partial\kappa_1}\right|_{\Phi(M^+)} \left.\frac{\partial\kappa_1}{\partial\kappa_0}\right|_{M^+} = (1 + b\tau_1^+)(1 + \tau_0^+).$$

Let $\rho_0^+ = (1 + b\tau_1^+)(1 + \tau_0^+)$. It follows from (2.30a) that $\tau_0^+ \in (-2, -1)$ and $\tau_1^+ \in (-1/b, -2/b)$, which implies that $\rho_0^+ \in (0, 1)$. Moreover,

$$\frac{\partial\kappa_2^+}{\partial\mu_0} = \frac{\partial\kappa_2^+}{\partial\lambda_0} = \frac{\partial\kappa_2^+}{\partial\nu_0} = 0$$

at the point M^+.

Everywhere below, the superscript "+" will mean that the corresponding quantity is computed for the point M^+, and "$-$" for the point M^-.

Let

$$Q^+ = \begin{pmatrix} \left(\dfrac{\partial \lambda_2}{\partial \lambda_0}\right)^+ & \left(\dfrac{\partial \lambda_2}{\partial \mu_0}\right)^+ \\ \left(\dfrac{\partial \mu_2}{\partial \lambda_0}\right)^+ & \left(\dfrac{\partial \mu_2}{\partial \mu_0}\right)^+ \end{pmatrix}.$$

Then $Q^+ = I_2^+ I_1^+$, where

$$I_2^+ = \begin{pmatrix} \left(\dfrac{\partial \lambda_2}{\partial \lambda_1}\right)^+ & \left(\dfrac{\partial \lambda_2}{\partial \mu_1}\right)^+ \\ \left(\dfrac{\partial \mu_2}{\partial \lambda_1}\right)^+ & \left(\dfrac{\partial \mu_2}{\partial \mu_1}\right)^+ \end{pmatrix}, \qquad I_1^+ = \begin{pmatrix} \left(\dfrac{\partial \lambda_1}{\partial \lambda_0}\right)^+ & \left(\dfrac{\partial \lambda_1}{\partial \mu_0}\right)^+ \\ \left(\dfrac{\partial \mu_1}{\partial \lambda_0}\right)^+ & \left(\dfrac{\partial \mu_1}{\partial \mu_0}\right)^+ \end{pmatrix}.$$

From (2.10) it follows that $(\partial \lambda_2/\partial \nu_0)^+ = (\partial \mu_2/\partial \nu_0)^+ = 0$. Finally, $(\partial \nu_2/\partial \nu_0)^+ = (\rho_0^+)^{-1} E_m$, where E_m is the $m \times m$ identity matrix. We can now write

$$\mathscr{D}^+ = \begin{pmatrix} \rho_0^+ & 0 & 0 & 0 \\ * & \left(\dfrac{\partial \lambda_2}{\partial \lambda_0}\right)^+ & \left(\dfrac{\partial \lambda_2}{\partial \mu_0}\right)^+ & 0 \\ * & \left(\dfrac{\partial \mu_2}{\partial \lambda_0}\right)^+ & \left(\dfrac{\partial \mu_2}{\partial \mu_0}\right)^+ & 0 \\ * & * & * & (\rho_0^+)^{-1} E_m \end{pmatrix},$$

where the asterisks denote quantities whose values are not essential.

We show that the eigenvalues of the matrix $Q^+ = I_2^+ I_1^+$ are greater than 1 in modulus. By the implicit function theorem, we have from (2.10) that

$$\left(\frac{\partial \lambda_1}{\partial \lambda_0}\right)^+ = \frac{1}{(1+\tau_0^+)^2} + \frac{1-2\lambda_1^+}{1+\tau_0^+}\left(\frac{\partial \tau_0}{\partial \lambda_0}\right)^+; \qquad \left(\frac{\partial \lambda_1}{\partial \mu_0}\right)^+ = \frac{1-2\lambda_1^+}{1+\tau_0^+}\left(\frac{\partial \tau_0}{\partial \mu_0}\right)^+;$$

$$\left(\frac{\partial \mu_1}{\partial \mu_0}\right)^+ = \frac{1}{(1+\tau_0^+)^3} + \frac{\lambda_1^+ - 3\mu_1^+}{1+\tau_0^+}\left(\frac{\partial \tau_0}{\partial \mu_0}\right)^+;$$

$$\left(\frac{\partial \mu_1}{\partial \lambda_0}\right)^+ = \frac{1}{(1+\tau_0^+)^3} + \frac{\lambda_1^+ - 3\mu_1^+}{1+\tau_0^+}\left(\frac{\partial \tau_0}{\partial \mu_0}\right)^+; \qquad \left(\frac{\partial \tau_0}{\partial \lambda_0}\right)^+ = \frac{\tau_0^+}{2}\left(\frac{\partial \tau_0}{\partial \mu_0}\right)^+;$$

$$\left(\frac{\partial \tau_0}{\partial \mu_0}\right)^+ = -\left(\frac{1}{2}\lambda_0^+ + \frac{1}{3}\tau_0^+ + \frac{1}{8}(\tau_0^+)^2\right)^{-1};$$

$$\left(\frac{\partial \lambda_2}{\partial \lambda_1}\right)^+ = \frac{1}{(1+b\tau_1^+)^2} + \frac{1-2b\lambda_0^+}{1+b\tau_1^+}\left(\frac{\partial \tau_1}{\partial \lambda_1}\right)^+; \qquad \left(\frac{\partial \lambda_2}{\partial \mu_1}\right)^+ = \frac{1-2b\lambda_0^+}{1+b\tau_1^+}\left(\frac{\partial \tau_1}{\partial \mu_1}\right)^+;$$

$$\left(\frac{\partial \mu_2}{\partial \lambda_1}\right)^+ = \frac{\tau_1^+}{(1+b\tau_1^+)^3} + \frac{\lambda_0^+ - 3b\mu_0^+}{1+b\tau_1^+}\left(\frac{\partial \tau_1}{\partial \lambda_1}\right)^+;$$

$$\left(\frac{\partial \mu_2}{\partial \mu_1}\right)^+ = \frac{1}{(1+b\tau_1^+)^3} + \frac{\lambda_0^+ - 3b\mu_0^+}{1+b\tau_1^+}\left(\frac{\partial \tau_1}{\partial \mu_1}\right)^+; \qquad \left(\frac{\partial \tau_1}{\partial \lambda_1}\right)^+ = \frac{\tau_1^+}{2}\left(\frac{\partial \tau_1}{\partial \mu_1}\right)^+;$$

$$\left(\frac{\partial \tau_1}{\partial \mu_1}\right)^+ = -\left(\frac{1}{2}\lambda_1^+ + \frac{1}{3}\tau_1^+ + \frac{1}{8}b(\tau_1^+)^2\right)^{-1}.$$

We show that $\det I_1^+ < 0$ and $\det I_2^+ < 0$, and hence the matrices I_1^+ and I_2^+ have real simple eigenvalues. A direct computation gives us that

$$\begin{aligned}\det I_1^+ &= \frac{1}{(1+\tau_0^+)^5} + \frac{\lambda_1^+ - 3\mu_1^+}{(1+\tau_0^+)^3}\left(\frac{\partial\tau_0}{\partial\mu_0}\right)^+ \\ &+ \frac{1-2\lambda_1^+}{(1+\tau_0^+)^4}\left(\frac{\partial\tau_0}{\partial\lambda_0}\right)^+ - \frac{1-2\lambda_1^+}{(1+\tau_0^+)^4}\tau_0^+\left(\frac{\partial\tau_0}{\partial\mu_0}\right)^+ \\ &= \frac{1}{(1+\tau_0^+)^5} + \left(\frac{\lambda_1^+ - 3\mu_1^+}{(1+\tau_0^+)^3} - \frac{1-2\lambda_1^+}{(1+\tau_0^+)^4}\frac{\tau_0^+}{2}\right)\left(\frac{\partial\tau_0}{\partial\mu_0}\right)^+; \\ \det I_2^+ &= \frac{1}{(1+b\tau_1^+)^5} + \frac{\lambda_0^+ - 3\beta\mu_0^+}{(1+b\tau_1^+)^3}\left(\frac{\partial\tau_1}{\partial\mu_1}\right)^+ \\ &+ \frac{1-2b\lambda_0^+}{(1+b\tau_1^+)^4}\left(\frac{\partial\tau_1}{\partial\lambda_1}\right)^+ - \frac{1-2b\lambda_0^+}{(1+b\tau_1^+)^4}\tau_1^+\left(\frac{\partial\tau_1}{\partial\mu_1}\right)^+ \\ &= \frac{1}{(1+b\tau_1^+)^5} + \left(\frac{\lambda_0^+ - 3b\mu_0^+}{(1+b\tau_1^+)^3} - \frac{1-2b\lambda_0^+}{(1+b\tau_1^+)^4}\frac{\tau_1^+}{2}\right)\left(\frac{\partial\tau_1}{\partial\mu_1}\right)^+.\end{aligned} \tag{2.32}$$

From (2.30a) and the relation $\mu_i^+ = \frac{1}{2}(\lambda_i^+)^2 > 0$ it follows that, first, $1+\tau_0^+ < 0$, and, second, each term in the parentheses multiplying $(\partial\tau_0/\partial\mu_0)^+$ in the expression (2.32) for $\det I_1^+$ is positive:

$$\frac{\lambda_1^+ - 3\mu_1^+}{(1+\tau_0^+)^3} \geq 0, \qquad -\frac{1-2\lambda_1^+}{(1+\tau_0^+)^4}\frac{\tau_0^+}{2} \geq 0.$$

It remains to show that

$$(\partial\tau_0/\partial\mu_0)^+ = -\left(\tfrac{1}{2}\lambda_0^+ + \tfrac{1}{3}\tau_0^+ + \tfrac{1}{8}(\tau_0^+)^2\right)^{-1} < 0$$

for all values $b < 0$.

It was shown earlier that the expression $\frac{1}{2}\lambda_0^+ + \frac{1}{3}\tau_0^+ + \frac{1}{8}(\tau_0^+)^2$ depends continuously on b. A direct computation gives us that the expression $\frac{1}{2}\lambda_0^+ + \frac{1}{3}\tau_0^+ + \frac{1}{8}(\tau_0^+)^2 = 0.00011\ldots$ is not equal to zero for $b = -1$. We show that $\frac{1}{2}\lambda_0^+ + \frac{1}{3}\tau_0^+ + \frac{1}{8}(\tau_0^+)^2$ is nonvanishing for all $b < 0$. Assume the opposite. Then for some $b < 0$ the following two expressions are simultaneously equal to zero:

$$\tfrac{1}{2}(\lambda_0^+)^2 + \tfrac{1}{2}\lambda_0^+\tau_0^+ + \tfrac{1}{6}(\tau_0^+)^2 + \tfrac{1}{24}(\tau_0^+)^3 = 0;$$
$$\tfrac{1}{2}\lambda_0^+ + \tfrac{1}{3}\tau_0^+ + \tfrac{1}{8}(\tau_0^+)^2 = 0.$$

Multiplying the second equation by $-\frac{1}{3}\tau_0^+$ and adding it to the first, we get the system

$$\tfrac{1}{2}(\lambda_0^+)^2 + \tfrac{1}{3}\lambda_0^+\tau_0^+ + \tfrac{1}{18}(\tau_0^+)^2 = 0, \qquad \tfrac{1}{2}\lambda_0^+ + \tfrac{1}{3}\tau_0^+ + \tfrac{1}{8}(\tau_0^+)^2 = 0.$$

Multiplying the second equation by $-\lambda_0^+$ and adding it to the first, we get $(\frac{1}{18} - \frac{1}{8}\lambda_0^+)(\tau_0^+)^2 = 0$. Hence $\lambda_0^+ = 4/9$. The value of τ_0^+ can be found from the quadric equation $\frac{2}{9} + \frac{1}{3}\tau_0^+ + \frac{1}{8}(\tau_0^+)^2 = 0$. We get that $\tau_0^+ = -4/3$. Then $\lambda_1^+ = (\lambda_0^+ + \tau_0^+ + \frac{1}{8}(\tau_0^+)^2)(1+\tau_0^+)^{-2} = 0$, which does not satisfy (2.30a). Thus, it has been proved that $\det I_1^+ < 0$. We prove that $\det I_2^+ < 0$.

It follows from (2.30a) and the relation $\mu_i^+ = \frac{1}{2}(\lambda_i^+)^2 > 0$ that $1 + b\tau_1^+ < 0$, and each term in the parentheses multiplying $(\partial\tau_1/\partial\mu_1)^+$ in the expression (2.32) for $\det I_2$ is negative:

$$\frac{\lambda_0^+ - 3b\mu_0^+}{(1+b\tau_1^+)^3} \leq 0, \qquad -\frac{1-2b\lambda_0^+}{(1+b\tau_1^+)^4}\frac{\tau_1^+}{2} \leq 0.$$

We show that

$$\left(\frac{\partial \tau_1}{\partial \mu_1}\right)^+ = -(\tfrac{1}{2}\lambda_1^+ + \tfrac{1}{3}\tau_1^+ + \tfrac{1}{8}b(\tau_1^+)^2)^{-1} > 0.$$

For $b = -1$ we have that $\frac{1}{2}\lambda_1^+ + \frac{1}{3}\tau_1^+ + \frac{1}{8}b(\tau_1^+)^2 = -1.802\ldots$. Suppose that the following hold simultaneously for some $b < 0$:

$$\tfrac{1}{2}(\lambda_1^+)^2 + \tfrac{1}{2}\lambda_1^+\tau_1^+ + \tfrac{1}{6}(\tau_1^+)^2 + \tfrac{1}{24}b(\tau_1^+)^3 = 0,$$
$$\tfrac{1}{2}\lambda_1^+ + \tfrac{1}{2}\tau_1^+ + \tfrac{1}{8}b(\tau_1^+)^2 = 0.$$

Multiplying the first equation by $-\frac{1}{3}\tau_1^+$ and adding it to the second we get

$$\tfrac{1}{2}(\lambda_1^+)^2 + \tfrac{1}{3}\lambda_1^+\tau_1^+ + \tfrac{1}{18}(\tau_1^+)^2 = 0, \qquad \tfrac{1}{2}\lambda_1^+ + \tfrac{1}{3}\tau_1^+ + \tfrac{1}{8}b(\tau_1^+)^2 = 0.$$

From this it follows that

$$\lambda_1^+ = 4/(9b), \qquad \tau_1^+ = -4/(3b),$$
$$\lambda_0^+ = (\lambda_1^+ + \tau_1^+ + \tfrac{1}{2}b(\tau_1^+)^2)(1 + b\tau_1^+)^{-2} = 0,$$

which does not satisfy (2.30a). The inequality $\det I_2^+ < 0$ is thereby proved.

We now show that the matrices I_1^+ and I_2^+ do not have eigenvalues ≤ 1 in modulus. For this we verify that $\xi = \pm 1$ does not satisfy the characteristic equations

$$\xi^2 - \operatorname{tr} I_j^+ \xi + \det I_j^+ = 0, \qquad j = 1, 2.$$

Let us first consider the case $j = 1$:

$$\operatorname{tr} I_1^+ = \frac{1}{(1+\tau_0^+)^2} + \frac{1}{(1+\tau_0^+)^3} + \frac{1-2\lambda_1^+}{1+\tau_0^+}\left(\frac{\partial \tau_0}{\partial \lambda_0}\right)^+ + \frac{\lambda_1^+ - 3\mu_1^+}{1+\tau_0^+}\left(\frac{\partial \tau_0}{\partial \mu_0}\right)^+$$
$$(2.33) \qquad = \frac{1}{(1+\tau_0^+)^2} + \frac{1}{(1+\tau_0^+)^3} + \left(\frac{\lambda_1^+ - 3\mu_1^+}{1+\tau_0^+} + \frac{1-2\lambda_1^+}{1+\tau_0^+}\frac{\tau_0^+}{2}\right)\left(\frac{\partial \tau_0}{\partial \mu_0}\right)^+.$$

We show that $|\det I_1| > 1 + |\operatorname{tr} I_1|$. Since $|1 + \tau_0^+| < 1$, in (2.33) each term in the parentheses multiplying $(\partial\tau_0/\partial\mu_0)^+$ in the expression for $\operatorname{tr} I_1^+$ is majorized in modulus by the corresponding term in the parentheses multiplying $(\partial\tau_0/\partial\mu_0)^+$ in the expression (2.32) for $\det I_1^+$:

$$\left|\frac{\lambda_1^+ - 3\mu_1^+}{1+\tau_0^+}\right| \leq \frac{\lambda_1^+ - 3\mu_1^+}{(1+\tau_0^+)^3}, \qquad \left|\frac{1-2\lambda_1^+}{1+\tau_0^+}\frac{\tau_0^+}{2}\right| \leq -\frac{1-2\lambda_1^+}{(1+\tau_0^+)^4}\frac{\tau_0^+}{2}.$$

It remains to show that

$$-\frac{1}{(1+\tau_0^+)^5} > 1 + \left|\frac{1}{(1+\tau_0^+)^2} + \frac{1}{(1+\tau_0^+)^3}\right|.$$

Let $p = (1 + \tau_0^+)^{-1}$. Since $p < -1$, we get that $-p^5 > 1 - p^2 - p^3$, or the equivalent inequality $(p^3 - 1)(p^2 - 1) < 0$. It is thereby proved that the matrix I_1^+ cannot have eigenvalues equal to 1 in modulus. Since I_1^+ depends continuously on $b < 0$, this implies that the arrangement of eigenvalues of I_1^+ with respect to the unit circle is the same for all $b < 0$. For $b = -1$ a direct computation gives us that

$$I_1^+ = \begin{pmatrix} 268.57\ldots & -404.97\ldots \\ 16638.2\ldots & -26719.5\ldots \end{pmatrix},$$

which implies that $\xi^2 - 26450.9\ldots\xi - 438067.7\ldots = 0$, $\xi_1 = 26467.5\ldots$, $\xi_2 = -16.55\ldots$. It is thereby proved that the eigenvalues of I_1^+ are strictly greater than 1 in modulus. The arguments are analogous for $j = 2$. We have

$$\operatorname{tr} I_2^+ = \frac{1}{(1+b\tau_1^+)^2} + \frac{1}{(1+b\tau_1^+)^3} + \left(\frac{\lambda_0^+ - 3b\mu_0^+}{1+b\tau_1^+} + \frac{1-2b\lambda_0^+}{1+b\tau_1^+}\frac{\tau_1^+}{2}\right)\left(\frac{\partial \tau_1}{\partial \mu_1}\right)^+.$$

We need to show that $|\det I_2^+| > 1 + |\operatorname{tr} I_2^+|$. Since $|1 + b\tau_1^+| < 1$, each term in the parentheses multiplying $(\partial\tau_1/\partial\mu_1)^+$ in the expression for $\operatorname{tr} I_2^+$ can be estimated from above in modulus by the corresponding terms in the parentheses multiplying $(\partial\tau_1/\partial\mu_1)^-$ in the expression (2.32) for $\det I_2^+$ (taken with reverse sign, since both these terms are negative):

$$\left|\frac{\lambda_0^+ - 3b\mu_0^+}{1 + b\tau_1^+}\right| \leq -\frac{\lambda_0^+ - 3b\mu_0^+}{(1 + b\tau_1^+)^3},$$

$$\left|\frac{1 - 2b\lambda_0^+}{1 + b\tau_1^+}\frac{\tau_1^+}{2}\right| \leq \frac{1 - 2b\lambda_0^+}{(1 + b\tau_1^+)^4}\frac{\tau_1^+}{2}.$$

Moreover,

$$-(1 + b\tau_1^+)^{-5} > 1 + |(1 + b\tau_1^+)^{-2} + (1 + b\tau_1^+)^{-3}|,$$

since $(1 + b\tau_1^+)^{-1} < -1$.

For $b = -1$ it is not hard to compute that the matrix I_2^+ coincides with the matrix

$$\begin{pmatrix} \left(\dfrac{\partial\lambda_1}{\partial\lambda_0}\right)^+ & -\left(\dfrac{\partial\lambda_1}{\partial\mu_0}\right)^+ \\ -\left(\dfrac{\partial\mu_1}{\partial\lambda_0}\right)^+ & \left(\dfrac{\partial\mu_1}{\partial\mu_0}\right)^+ \end{pmatrix},$$

computed at the point (λ_0^+, μ_0^+), and thus has the same eigenvalues as I_1^+ for $b = -1$.

We have shown that the eigenvalues of the matrices I_1^+ and I_2^+ are real, distinct (of different signs), and strictly greater than 1 in modulus. Consider their product $I_2^+ I_1^+$. Note first that if over the field $\mathbf{R}$ the norm of some real matrix A satisfies the estimate $\|A\| > C$, then the same estimate is true also over the field $\mathbf{C}$. This implies that the spectrum of A lies outside the disk of radius C. Since $\|I_2^+ I_1^+ r\| > \|I_1^+ r\| > \|r\|$, the spectrum of the matrix $Q^+ = I_2^+ I_1^+$ lies outside the unit disk.

We now return to the original notation, getting rid of the normalization condition (2.11). For this the notation for the solutions λ_i^ε, μ_i^ε, ν_i^ε, τ_i^ε, $i = 0, 1$, $\varepsilon =$ "+", "−", of (2.12) is changed to $\lambda_i^{\varepsilon'}$, $\mu_i^{\varepsilon'}$, $\nu_i^{\varepsilon'}$, $\tau_i^{\varepsilon'}$. The functions computed earlier for the mapping (2.10) after the substitution (2.11) will also be indicated by a prime. In this notation the coordinates of the fixed point of the mapping (2.10) have the form

$$M^\varepsilon(0,\ \lambda_0^\varepsilon = A_0^{-1}\lambda_0^{\varepsilon'},\ \mu_0^\varepsilon = A_0^{-2}\mu_0^{\varepsilon'},\ \nu_0^\varepsilon = A_0^{-1}\nu_0^{\varepsilon'}).$$

Further, $\tau_i^\varepsilon = A_0^{-1}\tau_i^{\varepsilon'}$, $A_0 = \alpha(\sigma) - \beta(\sigma)$, and $A_1 = \alpha(\sigma) + \beta(\sigma)$. In (2.30) the quantity b should be set equal to $A_0^{-1}A_1$ in estimating $\lambda_i^{\varepsilon'}$, $\mu_i^{\varepsilon'}$, and $\tau_i^{\varepsilon'}$. We get

$$0 < \lambda_0^+ < 1/(2A_0), \qquad 1/(2A_1) < \lambda_1^+ < 0, \qquad -2/A_0 < \tau_0^+ < -1/A_0;$$
$$-1/A_1 < \tau_1^+ < -2/A_1; \qquad 1/(2A_1) < \lambda_0^- < 0, \qquad 0 < \lambda_1^- < 1/(2A_0),$$
$$\tau_0^- < -2/A_0, \qquad \tau_1^- > -2/A_1.$$

For the coordinate ν of the points M_i^ε ($M_0^\varepsilon = M^\varepsilon$, $M_1^\varepsilon = \Phi M_0^\varepsilon$) in (2.31) we get

$$\nu_i^\varepsilon = \frac{\tau_i^\varepsilon F(0, \sigma, u_i) + \tau_{1-i}^\varepsilon(1 + A_i\tau_i^\varepsilon)F(0, \sigma, u_{1-i})}{(1 + A_i\tau_i^\varepsilon)(1 + A_{1-i}\tau_{1-i}^\varepsilon) - 1}. \tag{2.34}$$

Let us show that the matrix $\mathscr{D}^+$ does not change under the substitution inverse to (2.11). We have that

$$\frac{\partial \kappa_2}{\partial \kappa_0} = (1 + A_1\tau_1^+)(1 + A_0\tau_0^+) = \left(1 + \frac{A_1}{A_0}\tau_1^{+'}\right)(1 + \tau_0^{+'}) = \frac{\partial \kappa_2'}{\partial \kappa_0},$$

$$\frac{\partial \lambda_1}{\partial \lambda_0} = \frac{1}{(1 + A_0\tau_0^+)^2} + \frac{1 - 2\lambda_1^+ A_0}{1 + A_0\tau_0^+}\frac{\partial \tau_0}{\partial \lambda_0} = \frac{1}{(1 + \tau_0^{+'})^2} + \frac{1 - 2\lambda_1^{+'}}{1 + \tau_0^{+'}}\frac{\partial \tau_0}{\partial \lambda_0},$$

$$\frac{\partial \tau_0}{\partial \lambda_0} = \frac{-\frac{1}{2}\tau_0^+}{\frac{1}{2}\lambda_0^+ + \frac{1}{3}\tau_0^+ + \frac{1}{8}A_0(\tau_0^+)^2} = \frac{-\frac{1}{2}\tau_1^{+'}}{\frac{1}{2}\lambda_0^{+'} + \frac{1}{3}\tau_0^{+'} - \frac{1}{8}(\tau_0^{+'})^2} = \frac{\partial \tau_0'}{\partial \lambda_0},$$

from which it follows that $\partial\lambda_1/\partial\lambda_0 = \partial\lambda_1'/\partial\lambda_0$. The coincidence of the remaining elements of the matrices $\mathscr{D}^+$ and $\mathscr{D}^{+'}$ is proved in exactly the same way. Since $\mathscr{D}^+$ has block-triangular form, its spectrum does not intersect the unit circle.

We now consider the point M^-. The matrix $\mathscr{D}^- = D\widetilde{\Phi}^2(M^-)$ has the form

$$\mathscr{D}^- = \begin{pmatrix} \rho_0^- & 0 & 0 & 0 \\ * & \left(\frac{\partial\lambda_2}{\partial\lambda_0}\right)^- & \left(\frac{\partial\lambda_2}{\partial\mu_0}\right)^- & 0 \\ * & \left(\frac{\partial\mu_2}{\partial\lambda_0}\right)^- & \left(\frac{\partial\mu_2}{\partial\mu_0}\right)^- & 0 \\ * & * & * & (\rho_0^-)^{-1}E_m \end{pmatrix},$$

where the matrix

$$Q^- = \begin{pmatrix} \left(\frac{\partial\lambda_2}{\partial\lambda_0}\right)^- & \left(\frac{\partial\lambda_2}{\partial\mu_0}\right)^- \\ \left(\frac{\partial\mu_2}{\partial\lambda_0}\right)^- & \left(\frac{\partial\mu_2}{\partial\mu_0}\right)^- \end{pmatrix}$$

is computed at the point M^-, and $\rho_0^- = (1 + \tau_0^-)(1 + b\tau_0^-)$. It follows from (2.14) that Q^- coincides with the matrix $(Q^+)^{-1}$ inverse to Q^+. Therefore, $\operatorname{Spec} Q^-$ lies interior to the unit disk. In view of (2.30), $\rho_0^- > 1$. Consequently, $\operatorname{Spec}\mathscr{D}^-$ does not intersect the unit circle. Lemma 10 is proved

Since the mapping $\widetilde{\Phi}^2$ is hyperbolic at the points M^ε, $\varepsilon =$ "+" or "−", it follows from the theorem on an invariant manifold of a diffeomorphism in a neighborhood of a fixed point ([42], §5.2, Theorem 2) that through M^ε there passes a C^{k_0}-curve which is invariant under the mapping $\widetilde{\Phi}^2$ and tangent to the eigenvector of $\mathscr{D}^\varepsilon$ with eigenvalue ρ_0^ε. At M^ε we have $\mu_0^\varepsilon > 0$ and $\tau_0^\varepsilon < 0$; therefore, the branch of the invariant curve lying in the half-plane $\kappa < 0$ corresponds to the condition (2.8). Denote it by $\gamma_0^\varepsilon(\sigma)$. Let $\gamma_1^\varepsilon(\sigma) = \Phi\gamma_0^\varepsilon(\sigma)$ and $\gamma^\varepsilon = \gamma_0^\varepsilon \cup \gamma_1^\varepsilon$. The mapping $\widetilde{\Phi}^2$ depends smoothly on σ as a parameter. Consequently, the curve γ^ε has the form

$$\lambda = \lambda_i^\varepsilon(\kappa, \sigma), \qquad \mu = \mu_i^\varepsilon(\kappa, \sigma), \qquad \nu = \nu_i^\varepsilon(\kappa, \sigma),$$

where $i = 0$ for $\kappa < 0$, and $i = 1$ for $\kappa > 0$; $\lambda_i^\varepsilon(0, \sigma) = \lambda_i^\varepsilon$, $\mu_i^\varepsilon(0, \sigma) = \mu_i^\varepsilon$, and $\nu_i^\varepsilon(0, \sigma) = \nu_i^\varepsilon$. The functions $\lambda_i^\varepsilon(\cdot)$, $\mu_i^\varepsilon(\cdot)$, and $\nu_i^\varepsilon(\cdot)$ depend C^{k_0}-smoothly on κ and σ. (The smooth dependence on σ follows from [43].)

Denote by $\Gamma_\sigma^\varepsilon$ the inverse image of the curve $\gamma^\varepsilon(\sigma)$ under the substitution (2.6), i.e., the curve

(2.35)
$$z_1 = 0, \qquad z_2 = \mu_i^\varepsilon(\kappa, \sigma)\kappa^3, \qquad z_3 = \lambda_i^\varepsilon(\kappa, \sigma)\kappa^2,$$
$$z_4 = \kappa, \qquad w = \sigma + \nu_i^\varepsilon(\kappa, \sigma)\kappa$$

($i = 0$ for $\kappa < 0$, and $i = 1$ for $\kappa > 0$), and by Γ^ε the $(m+1)$-dimensional switching surface $\Gamma^\varepsilon = \bigcup_{\sigma\in S_0}\Gamma_\sigma^\varepsilon$. Denote by $\mathfrak{N}_\sigma^\varepsilon$ the family of trajectories of (2.4) with initial conditions belonging to $\Gamma^\varepsilon_\sigma$, and by $\Sigma^\varepsilon_\sigma$ the manifold swept out by the trajectories of the family $\mathfrak{N}^\varepsilon_\sigma$. Let $\Sigma^\varepsilon = \bigcup_{\sigma\in S_0}\Sigma^\varepsilon_\sigma$. We define the mapping $p^\varepsilon\colon \Sigma^\varepsilon \to S_0$ by setting $p^\varepsilon(M) = \sigma$ for any $M \in \Sigma^\varepsilon_\sigma$.

Definition 1. Let Σ be a manifold with $\dim\Sigma = r$, let $B \subset \Sigma$ be a submanifold of it with $\dim B = l < r$, and let $p\colon \Sigma \to B$ be a continuous surjective mapping. The triple (p, Σ, B) is called the fiber bundle Σ with base B and C^{k_0}-piecewise smooth fibers $\mathscr{F} \cong \mathbf{R}^{r-l}$ if for any point $\xi \in \Sigma$ there exist a neighborhood $U \subset \Sigma$ and a homeomorphism $\tau\colon U \to \mathbf{R}^r$ such that U can be represented as the union $U = U_0 \cup U_+ \cup U_-$, where $U_+ = \tau^{-1}\{\xi|\xi_1 > 0\}$, $U_0 = \tau^{-1}\{\xi|\xi_1 = 0\}$, and $U_- = \tau^{-1}\{\xi|\xi_1 < 0\}$. Further:

1) $\tau(B\cap U) = \{\xi|\xi_1 = \cdots = \xi_{r-l} = 0\}$ and for any point $M \in B$

$$\tau p^{-1}(M) = \{\xi|\xi_j = (\tau(M))_j,\ j = r-l+1, \dots, r\}.$$

2) The restrictions $\tau|_{U_+}$ and $\tau|_{U_-}$ are diffeomorphisms of class C^{k_0} that can be extended as C^{k_0}-diffeomorphisms to the points of $U_0\backslash B$, the restriction $\tau|_{U_0\backslash B}$ is a diffeomorphism of class C^{k_0} that can be extended as a C^{k_0}-diffeomorphism to the points of B, and the restriction $\tau|_B$ is a C^{k_0}-diffeomorphism.

Theorem 1. *The triple $(p^\varepsilon, \Sigma^\varepsilon, S_0)$ is an $(m+2)$-dimensional fiber bundle Σ^ε with base S_0 and piecewise smooth (of class C^{k_0}) fibers $\Sigma^\varepsilon_\sigma \cong \mathbf{R}^2$.*

Proof. It follows from (2.35) that, up to terms of higher order of smallness with respect to κ,

$$\frac{\mathscr{D}(z_4, w)}{\mathscr{D}(\kappa, \sigma)} = \begin{pmatrix} 1 & 0 \\ \nu^\varepsilon_i(\sigma) & E_m \end{pmatrix}$$

for the surface Γ^ε.

By the implicit function theorem, this implies that Γ^ε is C^{k_0}-smooth everywhere away from $\kappa = 0$. As $\kappa \to \pm 0$ the tangent planes to Γ^ε have limits. These limits coincide and are determined by the basis $\partial/\partial z_4$, $\partial/\partial w$. Let $\Gamma^\varepsilon_{\sigma 0} = \Gamma^\varepsilon_\sigma \cap \{\kappa < 0\}$ and $\Gamma^\varepsilon_{\sigma 1} = \Gamma^\varepsilon_\sigma \cap \{\kappa > 0\}$. It follows from (2.35) that the surfaces $\Gamma^\varepsilon_{\sigma i}$ are located on different sides of the plane $z_4 = 0$, and they intersect $z_4 = 0$ at a nonzero angle. This implies that the surface Γ^ε is of class C^{k_0} at the points of $\Gamma^\varepsilon\backslash S_0$ and of class C^1 everywhere, including points of the singular manifold S_0.

It remains to show that the manifold Σ^ε is smooth at the points of $\Sigma^\varepsilon\backslash\Gamma^\varepsilon$ and can be smoothly extended from both sides to $\Gamma^\varepsilon\backslash S_0$. The coordinates on Σ^ε will be specified as follows. Let the system (2.4) be integrated with the initial conditions $(z^0(\kappa, \sigma), w^0(\kappa, \sigma)) \in \Gamma^\varepsilon$ on the interval $[0, \kappa\tau_i(\kappa, \sigma))$. The function $\tau_i(\kappa, \sigma) = \tau_i(z^0(\kappa, \sigma), w^0(\kappa, \sigma))$ is determined by (2.10b) ($i = 0$ for $\kappa < 0$, $i = 1$ for $\kappa > 0$). Using Lemma 7 in the same way as in the derivation of (2.10), we can show that

$$\begin{gathered} z_1 = \kappa^3\Theta(\kappa), \qquad z_2 = \kappa^2\Theta(\kappa), \\ z_3 = (\lambda^\varepsilon_i + \tau + \tfrac{1}{2}A_i\tau^2)\kappa^2 + \kappa^2\Theta(\kappa), \\ z_4 = (1 + A_i\tau_i)\kappa + \kappa\Theta(\kappa), \\ w = \sigma + (\nu^\varepsilon_i + \tau F(0, \sigma, u_i))\kappa + \kappa\Theta(\kappa). \end{gathered} \tag{2.36}$$

Here $\tau = t/\kappa$, $A_i = \alpha(\sigma) + u_i\beta(\sigma)$, $u_i = (-1)^{i+1}$, $i = 0$ for $\kappa < 0$, and $i = 1$ for $\kappa > 0$.

Up to terms of higher order of smallness we can write

$$\frac{\mathscr{D}(z_3, z_4, w)}{\mathscr{D}(t, \kappa, \sigma)} = \begin{pmatrix} (1 + A_i\tau)\kappa & 2(\lambda^\varepsilon_i + \tau + \frac{1}{2}A_i\tau^2)\kappa & \kappa\Theta(\kappa) \\ A_i & 1 + A_i\tau & \Theta(\kappa) \\ F(0, \sigma, u_i) & \nu^\varepsilon_i + \tau F(0, \sigma, u_i) & E_m \end{pmatrix}.$$

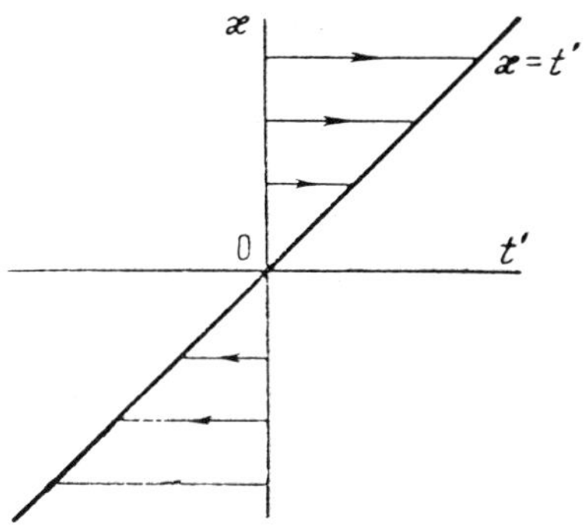

FIGURE 8

From this it follows that for sufficiently small $\kappa \neq 0$ and for $0 < t < \kappa\tau_i(\kappa, \sigma)$ the manifolds $\Sigma_\sigma^\varepsilon$ and Σ^ε are C^{k_0}-smooth manifolds of dimension 2 and $m+2$, respectively, that can be smoothly extended to the points of $\Gamma^\varepsilon \backslash S_0$.

We consider the mapping $\tilde{\tau}: \Sigma^\varepsilon \to \mathbf{R}^{m+2}$, $\tilde{\tau}(z(t), w(t)) = (\kappa, \sigma, t' = t\tau_i^{-1}(\kappa, \sigma))$.

The image $\tilde{\tau}(\Sigma_\sigma^\varepsilon)$ of the manifold $\Sigma_\sigma^\varepsilon$ on the (κ, t')-plane under the mapping $\tilde{\tau}$ is the union of two triangular sectors, $\kappa > t' > 0$ and $\kappa < t' < 0$, and the line $t' = 0$ (Figure 8). The image of the arc of a trajectory of the system (2.4) with initial conditions on $\Gamma_{\sigma i}^\varepsilon$ to the next switching point on the (κ, t')-plane is the line segment $\kappa = \kappa_0$, $0 \leq t' < \kappa_0$ for $i = 1$, and the line segment $\kappa = \kappa_0$, $-\kappa_0 < t' \leq 0$ for $i = 0$.

To construct the mapping τ determining a fiber bundle structure with piecewise smooth fibers in Definition 1 it is necessary to identify the half-line $\kappa = t'$, $\kappa < 0$ with $t' = 0$, $\kappa > 0$, and the half-line $\kappa = t'$, $\kappa > 0$ with $t' = 0$, $\kappa < 0$. For this we take the composition of $\tilde{\tau}$ and a mapping carrying each fiber of $\tilde{\tau}(\Sigma_\sigma^\varepsilon)$ onto $\mathbf{R}^2$:

$$\sigma \mapsto \sigma, \qquad (\kappa, t') \mapsto (\kappa - |\varphi|(\kappa_i(\kappa, \sigma) + \kappa))(\sin \pi(\tfrac{1}{2} - |\varphi|), \cos \pi(\tfrac{1}{2} - |\varphi|)).$$

Here $\varphi = t'/\kappa$, and the functions $\kappa_i(\kappa, \sigma)$ are determined by (2.10) after substitution of the functions in (2.35) in place of z^0 and w^0. The mapping τ has the smoothness properties required in the definition of a fiber bundle with piecewise smooth fibers. The theorem is proved.

Remark 4. Let $\mathscr{M}$ be a smooth submanifold of the surface S_0. Consider the inverse image $\Sigma_{\mathscr{M}} = (p^\varepsilon)^{-1}\mathscr{M}$ of $\mathscr{M}$ in the bundle $p^\varepsilon: \Sigma^\varepsilon \to S_0$. We show that $\Sigma_{\mathscr{M}}$ is a fiber bundle (with piecewise smooth fibers) of dimension $\dim \mathscr{M} + 2$ over $\mathscr{M}$. Since the assertion is of a local character, it can be assumed without loss of generality that $\mathscr{M}$ is given by the equations $w_{k+1} = \cdots = w_m = 0$, where $k = \dim \mathscr{M}$. The switching surface of the trajectories of the system (2.4) that fill out $\Sigma_{\mathscr{M}}$ is $\Gamma_{\mathscr{M}} = \bigcup_{\sigma \in \mathscr{M}} \Gamma_\sigma^\varepsilon$. The surface $\Gamma_{\mathscr{M}}$ is given by (2.35), with $\sigma \in \mathscr{M}$. Therefore, $\Gamma_{\mathscr{M}}$ is a C^{k_0}-smooth manifold at points not in $\mathscr{M}$. The limit of the tangent planes to $\Gamma_{\mathscr{M}}$ as $\kappa \to \pm 0$ is the plane spanned by the vector $\partial/\partial z_4 + \nu_i^\varepsilon \partial/\partial w$ and the k vectors $\partial/\partial w' = \partial/\partial w_1, \ldots, \partial/\partial w_k$. It follows from (2.35) that for sufficiently small $\kappa \neq 0$ the trajectories of (2.4) leave the surface $\Gamma_{\mathscr{M}}$ at a nonzero angle with nonzero phase velocity; therefore, $\Sigma_{\mathscr{M}}$ is homeomorphic to $\mathscr{M} \times \mathbf{R}^2$. The smoothness of $\Sigma_{\mathscr{M}}$ at the points of $\Sigma_{\mathscr{M}} \backslash \Gamma_{\mathscr{M}}$ and the smooth extendibility to the points of $\Gamma_{\mathscr{M}} \backslash \mathscr{M}$ can be proved just as this was done for the whole surface Σ^ε.

3. A LOCALLY OPTIMAL SYNTHESIS CONTAINING CHATTERING TRAJECTORIES

Assume that the system (2.1) can be reduced by a nonsingular change of variables $(\psi, x) \to (z, w)$ to the form (2.4), (2.5), and let $m = 2n-4$. We call $\mathscr{M} \subset \{H = 0\}$

a Lagrangian manifold (in the broad sense) if

$$\oint_\gamma \psi\, dx = 0$$

for any piecewise smooth closed contour $\gamma \subset \mathscr{M}$. In the standard definition of a Lagrangian manifold there is the additional requirement that $\dim \mathscr{M} = n$, which we omit.

Theorem 2. *Suppose that* $\mathscr{M} \subset S_0$ *is a Lagrangian manifold. Then* $\Sigma_{\mathscr{M}} = (p^\varepsilon)^{-1}\mathscr{M}$ *is a Lagrangian manifold.*

Proof. Consider an arbitrary piecewise smooth closed contour $\gamma \subset \Sigma_{\mathscr{M}} \cap \Gamma^\varepsilon \backslash S_0$. We introduce coordinates (κ, σ) on $\Gamma^\varepsilon \backslash S_0$ by (2.35). Let $\Phi^2(\kappa, \sigma) = (\kappa_1, \sigma_1)$. Substituting (2.35) in (2.10), we get

$$\sigma_1 = \sigma, \qquad \kappa_1 = (\rho_0^\varepsilon + R(\kappa, \sigma))\kappa, \tag{2.37}$$

where $R(\kappa, \sigma) = \Theta(\kappa)$. We first consider the case $\varepsilon =$ " $+$ ".

It follows from the theorem on the Poincaré-Cartan integral invariant ([44], §4.4.3) that

$$\oint_\gamma \psi\, dx = \int_{\gamma_1 = \Phi^2 \gamma} \psi\, dx.$$

Let $\gamma_n = \Phi^2 \gamma_{n-1}$, $n = 2, 3, \ldots$. We show that $\gamma_n \to p^+\gamma$ as $n \to \infty$ in the C^1-topology. Suppose that γ_1 is given in the form

$$\kappa = K_1(s), \quad \sigma = \Sigma(s), \quad s \in [0, 1].$$

We specify γ_n by the recursion relations

$$\kappa = K_n(s), \quad \sigma = \Sigma(s), \quad s \in [0, 1],$$
$$K_n(s) = (\rho_0^+ + R(K_{n-1}(s), \Sigma(s)))K_{n-1}(s).$$

In this case the projection $p^+\gamma_n$ on S_0 is given by the equations $\kappa = 0$ and $\sigma = \Sigma(s)$. Consequently,

$$\|\gamma_n - p^+\gamma_1\|_{C^1} = \max_{s \in [0, 1]} \{|K_n(s)|, |K_n'(s)|\}.$$

We consider the system (2.4) for $|z| < \delta$, where δ is sufficiently small. By (2.37),

$$|K_{n+1}(s)| < |\rho_0^+ + O(\delta)|\, |K_n(s)|, \tag{2.38}$$

$$\begin{aligned}(K_{n+1}'(s), \Sigma'(s)) &= \mathscr{D}\Phi^2_{(K_n(s), \Sigma(s))}(K_n'(s), \Sigma'(s)) \\ &= \begin{pmatrix} \rho_0^+ + R_1, & Q \\ 0 & E_m \end{pmatrix}\Bigg|_{(K_n(s), \Sigma(s))} (K_n'(s), \Sigma'(s)) \\ &= ((\rho_0^+ + R_1)K_n'(s) + Q\Sigma'(s), \Sigma'(s)),\end{aligned} \tag{2.39}$$

where $R_1 = \kappa\Theta(1)$ and $Q = \kappa\Theta(1)$.

By (2.38) and the fact that $\rho_0^+ < 1$ we have

$$\max_{s \in [0, 1]} |K_n(s)| \to 0 \quad \text{as } n \to \infty.$$

Consequently, for any $\delta_1 > 0$ there is an n_0 such that $|Q(K_n(s), \Sigma(s))| < \delta_1$ for $n \geq n_0$. Then it follows from (2.39) that for $n \geq n_0$

$$\max_{s \in [0, 1]} |K_{n+1}'(s)| < |\rho_0^+ + Q(\delta)| \max_{s \in [0, 1]} |K_n'(s)| + \delta_1 \max_{s \in [0, 1]} |\Sigma'(s)|.$$

From this it can be shown by induction that

$$\max_{s\in[0,1]} |K'_{n_0+m}| < q^m \max_{s\in[0,1]} |K'_{n_0}(s)| + \delta_1(1+q+\cdots+q^{m-1}) \max_{s\in[0,1]} |\Sigma'(s)|,$$

where $q = |\rho_0^+ + Q(\delta)| < 1$. Consequently,

$$\lim_{n\to\infty} |K'_n(s)| \le \frac{\delta_1}{1-q} \max_{s\in[0,1]} |\Sigma'(s)|.$$

Since δ_1 is arbitrary,

$$\lim_{n\to\infty} \|K'_n(s)\|_{C^0} = 0 \quad \text{and} \quad \lim_{n\to\infty} \|\gamma_n - p^+\gamma_1\|_{C^1} = 0.$$

Therefore, as $n \to \infty$,

$$\left|\oint_{\gamma_n} \psi\,dx - \oint_{p^+\gamma_1} \psi\,dx\right| = \left|\int_0^1 (\psi_n \dot{x}_n - \psi_0 \dot{x}_0)\,dt\right|$$

$$\le \max_{t\in[0,1]} |\psi_n(t) - \psi_0(t)| \int_0^1 |\dot{x}_0|\,dt + \max_{t\in[0,1]} |\psi_n(t)|\,\|x_n(\cdot) - x_0(\cdot)\|_{C^1} \to 0.$$

Here ψ_n, x_n and ψ_0, x_0 denote the restrictions of the corresponding functions to γ_n and $p^+\gamma_1$. Thus,

$$\oint_\gamma \psi\,dx = \oint_{\gamma_n} \psi\,dx = \oint_{p^+\gamma_1} \psi\,dx = 0.$$

Since $\Sigma_{\mathscr{M}}$ was obtained from $\Sigma_{\mathscr{M}} \cap \Gamma^\varepsilon$ by extension along the trajectories of a Hamiltonian system, this implies that $\Sigma_{\mathscr{M}}$ is a Lagrangian manifold.

To prove that $(p_\varepsilon^-)^{-1}\mathscr{M}$ is a Lagrangian manifold it suffices to replace the sequence $\gamma_n - \Phi^2\gamma_{n-1}$ in the above arguments by the sequence $\gamma_n - (\Phi^2)^{-1}\gamma_{n-1}$, because the trajectories of the family $\mathfrak{N}^-$ hit the surface S_0 in reverse time. The theorem is proved.

By using our construction we can construct a broad class of fields of extremals determining a synthesis of locally optimal trajectories for different boundary conditions in Problem 3.

A trajectory $\hat{x}(\cdot)$, $\hat{x}(\widehat{T}) \in \mathscr{M}$, is said to be *locally optimal* in Problem 3 if there is a neighborhood U of $\hat{x}(\cdot)$ such that

$$\int_0^{\widehat{T}} f(\hat{x}(t))\,dt \le \int_0^T f(x(t))\,dt$$

for any admissible trajectory $x(t)$ in Problem 3 belonging to the neighborhood U $(x(0) = \hat{x}(0),\ x(T) \in \mathscr{M})$. Here $f(x)$ denotes the function $f_0(x) + uf_1(x)$.

Let $\mathscr{N}_1 \subset S_0 \cap \{H = 0\}$ be a connected smooth manifold of dimension $n-3$, and let

1) $(0, (1/2\beta)((\beta-\alpha)F(0,\sigma,1) + (\beta+\alpha)F(0,\sigma,-1))) \in T_\sigma\mathscr{N}_1$;

2) $\psi\,dx|_{\mathscr{N}_1} = 0$.

Through each point of the manifold $\mathscr{N}_1$ we draw a singular trajectory of (2.4) on a sufficiently small time interval $(-\tau, \tau)$. The manifold thus obtained is denoted by $\mathscr{N}_2$. Denote by $\mathscr{N}_2^0$ the submanifold of $\mathscr{N}_2$ from whose points it is possible to hit $\mathscr{N}_1$ by moving along a singular trajectory of (2.4) (the manifold $\mathscr{N}_2^0$ is filled by the parts of the singular trajectories for $t \in (-\tau, 0)$). Denote by $\mathfrak{N}^*$ the subset of trajectories of the family $\mathfrak{N}^+$ that fill $\Sigma^* = (p^+)^{-1}\mathscr{N}_2$. Let π be the projection of the space (ψ, x) onto the space x, i.e., $\pi(\psi, x) = x$.

Theorem 3. *Suppose that for Problem 3 the system of equations (2.1) with $\lambda_0 = 1$ can be reduced by means of a nonsingular change of variables $(\psi, x) \to (z, w)$ to the form (2.4), (2.5). Assume that the restriction $\pi|_{\Sigma^*}$ of the projection is a regular mapping at the points of $\Sigma^*\backslash\Gamma^*$, $\Gamma^* = \Sigma^* \cap \Gamma^+$, that can be regularly extended to $\Gamma^*\backslash\mathcal{N}_2$, and the restriction $\pi|_{\Gamma^*}$ is regular. Then the projections on the space of values of x of the trajectories of the family $\mathfrak{N}^*$ with initial conditions belonging to $(p^+)^{-1}\mathcal{N}_2^0$ are locally optimal in Problem 3 with $\mathcal{M} = \pi\mathcal{N}_1$ chosen as the target manifold.*

Proof. Under the conditions of the theorem there exists a continuous function $\Psi: \mathbf{R}^n \to \mathbf{R}^n$, $(\Psi(x), x) \in \Sigma^*$. The function Ψ is of class C^{k_0} at the points of the manifold $\pi\Sigma^*\backslash\pi\Gamma^*$ and can be smoothly extended to the points of $\pi\Gamma^*\backslash\pi\mathcal{N}_2$. We show that

$$\oint_\gamma \Psi(x)\,dx = 0 \tag{2.40}$$

for an arbitrary piecewise smooth closed contour $\gamma \subset \pi\Sigma^*$.

By Theorem 2, Σ^* is a Lagrangian manifold. We consider a contour $\gamma^* \subset \Sigma^*$ with $\pi\gamma^* = \gamma$. Then

$$\int_{\gamma^*} \psi\,dx = \int_\gamma \Psi(x)\,dx.$$

The contour γ^* is not piecewise smooth in general. Since $\operatorname{codim}\pi\mathcal{N}_2 = 2$ in view of the Thom transversality theorem ([45], §7, Theorem 1), the contour γ can be approximated arbitrarily closely in the C^{k_0}-topology by a piecewise smooth closed contour γ_1 disjoint from $\pi\mathcal{N}_2$ and intersecting $\pi\Gamma^*$ at finitely many points. The lifting of γ_1 to Σ^* is now piecewise smooth. Since Σ^* is a Lagrangian manifold,

$$\int_{\gamma_1} \Psi(x)\,dx = 0.$$

This yields (2.40).

Suppose now that $x(\cdot)$ is an arbitrary admissible trajectory in Problem 3 that has initial conditions x_0 and lies entirely in the domain $\pi\Sigma^*$, and let $x^*(\cdot)$ be the projection of the trajectory of the family $\mathfrak{N}^*$ with the same initial conditions. Let T and T^* be the times at which the trajectories $x(\cdot)$ and $x^*(\cdot)$ hit the target manifold $\mathcal{M}$.

By the condition of the theorem, $\mathcal{M}$ is a connected smooth manifold, and $\int_\gamma \Psi(x)\,dx = 0$ for any (not necessarily closed) piecewise smooth contour $\gamma \subset \mathcal{M}$. Therefore, by (2.40),

$$\int_{x(\cdot)} \Psi(x)\,dx = \int_{x^*(\cdot)} \Psi(x)\,dx.$$

For the trajectory $x^*(\cdot)$, which satisfies the conditions of the Pontryagin maximum principle, we have

$$\int_{x^*(\cdot)} \Psi(x)\,dx = \int_0^{T^*} f(x^*(t))\,dt.$$

On the other hand, for any trajectory belonging to the domain $\pi\Sigma^*$ (including the trajectory $x(\cdot)$),

$$\int_{x(\cdot)} \Psi(x)\,dx \le \int_0^T f(x(t))\,dt.$$

This implies that the trajectory $x^*(\cdot)$ is optimal. The theorem is proved.

Remark 5. For Problem 3 with more general boundary conditions Theorems 1 and 2 enable us to construct an optimal synthesis if we also use trajectories of the family

$\mathfrak{N}^-$. In a typical situation for this synthesis there is a singular universal manifold of codimension 2, and in a neighborhood of it there are two open domains U_1 and U_2 of phase space with the following properties. The domain U_1 is filled by trajectories of the family $\mathfrak{N}^+$, and the domain U_2, which includes the target manifold $\mathscr{M}$, is filled by trajectories of $\mathfrak{N}^-$. Each optimal trajectory of Problem 3 with initial conditions belonging to U_1 consists of three parts. The first is the part of a trajectory in $\mathfrak{N}^+$ that goes to the universal manifold in finite time with infinitely many control switchings; the second part is a singular trajectory of Problem 3 belonging to the universal manifold. The third and last part belongs to the domain U_2 and is a trajectory in $\mathfrak{N}^-$ that leaves the universal manifold with increasingly more frequent switchings in reverse time and hits the target manifold. The simplest example of such a synthesis is provided by Problem 2 considered in Chapter 1.

The assumption of Theorem 3 on regular projectibility of the manifold Σ^* on the space of x is a direct analogue of the Jacobi condition of absence of focal points on the trajectories of a field of extremals in the calculus of variations. The assertions below have the purpose of formulating explicitly verifiable sufficient conditions for the regular projectibility of Σ^*. In Lemma 11 it is shown that the surface Γ^* is of class C^1, and a basis of tangent planes $T_{(z,w)}\Gamma^*$ at the points $(z, w) \in S_0$ is found. To explicitly find the limit position of the tangent planes $T_{(z,w)}\Sigma^*$ as $z \to 0$ it is necessary to know the values $(\partial\nu/\partial\kappa)(0, w)$ and $(\partial\nu/\partial\sigma)(0, w)$ in (2.35). Nevertheless, under a certain additional condition imposed in Theorem 4 (see below) on the substitution $(\psi, x) \mapsto (z, w)$ the limit of $\mathscr{D}\pi|_{(z,w)}T\Sigma^*$ as $z \to 0$ can be explicitly computed. This enables us to formulate sufficient conditions for the regular projectibility of Σ^* on the space of x, conditions satisfied in all the problems in Chapter 4 with increasingly more frequent switchings.

Suppose that the manifold Σ^* is the same as in Theorem 3. It can be assumed without loss of generality that in a neighborhood of a point $(0, w_0) \subset \mathscr{N}_1$ the manifold $\mathscr{N}_2$ is given by the equations $w_{n-1} = w_n = \cdots = w_{2n-4} = 0$. Let

$$\partial/\partial w' = (\partial/\partial w_1, \dots, \partial/\partial w_{n-2}), \qquad \partial/\partial w'' = (\partial/\partial w_{n-1}, \dots, \partial/\partial w_{2n-4}).$$

Lemma 11. *Suppose that the mapping $\pi|_{\mathscr{N}_2}$ is regular. Assume that the vectors $\varphi_0(x)+\varphi_1(x)$ and $\varphi_0(x) - \varphi_1(x)$ at points $x \in \pi\mathscr{N}_2$ do not belong to the tangent plane $T_x\pi\mathscr{N}_2$. Then the projection of the switching surface Γ^* on the space of x is a C^1-smooth manifold that is C^{k_0}-smooth at the points of $\pi\Gamma^*\backslash\pi\mathscr{N}_2$.*

Proof. We show that the switching surface Γ^* is C^1-smooth. As follows from (2.35), Γ^* consists of two C^{k_0}-smooth pieces: $\Gamma_i^* = \bigcup_{\sigma\in\mathscr{N}_2}\Gamma_{\sigma i}^*$, $i = 0, 1$, which are located on different sides of the plane $z_4 = 0$ and intersect it at a nonzero angle. We show that the limit of the tangent planes to Γ_1^* as $\kappa \to +0$ and the limit as $\kappa \to -0$ of the tangent planes to Γ_0^* at a point $\sigma \in \mathscr{N}_2$ coincide. It follows from (2.35) that

$$T_\sigma\Gamma_i^* = \operatorname{span}(\partial/\partial z_4 + \nu_i^+(\sigma)\partial/\partial w, \partial/\partial w').$$

From (2.34) we get

(2.41)

$$F(0, \sigma, u_i) - A_i\nu_i^\varepsilon = \frac{\tau_{1-i}^\varepsilon(1 + A_i\tau_i^\varepsilon)}{(1 + A_i\tau_i^\varepsilon)(1 + A_{1-i}\tau_{1-i}^\varepsilon) - 1} \times (A_{1-i}F(0, \sigma, u_i) - A_iF(0, \sigma, u_{1-i})), \qquad i = 0, 1.$$

Thus, the vectors $(F(0, \sigma, u_i)-A_i\nu_i^\varepsilon)\partial/\partial w$, $i = 0, 1$, are proportional to the phase velocity vector of the singular trajectories of (2.4):

$$\dot{z} = 0, \quad \dot{w} = \frac{1}{2\beta}(A_1F(0, \sigma, -1) - A_0F(0, \sigma, 1)).$$

Therefore,

$$T_\sigma\Gamma_i^* = \operatorname{span}\left(A_i\frac{\partial}{\partial z_4} + A_i\nu_i^+\frac{\partial}{\partial w}, \frac{\partial}{\partial w'}\right)$$
$$= \operatorname{span}\left(A_i\frac{\partial}{\partial z_4} + F(0, \sigma, u_i)\frac{\partial}{\partial w}, \frac{\partial}{\partial w'}\right).$$

Since the linear combination of the vectors

$$A_0\frac{\partial}{\partial z_4} + F(0, \sigma, -1)\frac{\partial}{\partial w}, \qquad A_1\frac{\partial}{\partial z_4} + F(0, \sigma, 1)\frac{\partial}{\partial w}$$

with coefficients A_1 and $-A_0$ yields a vector belonging to $T_\sigma\mathscr{N}_2$, we get that $T_\sigma\Gamma_1^* = T_\sigma\Gamma_0^*$ for $\sigma \in \mathscr{N}_2$.

We show that the restriction $\pi|_{\Gamma^*}$ is a regular mapping. Let

$$\mathscr{D}_1 = \mathscr{D}\pi|_{(z, w)\in\Gamma^*}\frac{\mathscr{D}(z, w)}{\mathscr{D}(\kappa, w')}.$$

It follows from (2.35) that, up to terms of higher order of smallness,

$$\frac{\mathscr{D}(z, w)}{\mathscr{D}(\kappa, w')} = \begin{pmatrix} 0 & 0 \\ \partial z_2/\partial\kappa & \partial z_2/\partial w' \\ \dots\dots & \dots\dots \\ \partial z_4/\partial\kappa & \partial z_4/\partial w' \\ \partial w/\partial\kappa & \partial w/\partial w' \end{pmatrix} = \begin{pmatrix} 0 & 0 \\ \kappa^2\Theta(1) & \kappa^3\Theta(1) \\ \kappa\Theta(1) & \kappa^2\Theta(1) \\ 1 & 0 \\ \nu_i^+ & \partial w/\partial w' \end{pmatrix}.$$

Hence, for $\kappa = 0$

$$\mathscr{D}_1 = \mathscr{D}\pi|_{\sigma\in N_2} = \begin{pmatrix} 0 & 0 \\ 0 & 0 \\ 0 & 0 \\ 1 & 0 \\ \nu_i^+ & \partial w/\partial w' \end{pmatrix}.$$

The projection of the vector $A_i\,\partial/\partial z_4 + F(0, \sigma, u_i)\,\partial/\partial w$ on the space of x is the vector $\varphi_0(x) + u_i\varphi_1(x)$ $(x = \pi(0, \sigma))$, and by the condition of the lemma it does not belong to $T_x(\pi\mathscr{N}_2)$. Therefore, $\operatorname{rk}\mathscr{D}_1 = n - 1$. The lemma is proved.

We need the following assertion from linear algebra. Let A and B be arbitrary $n \times k$ and $k \times n$ matrices, respectively. Then the rank of AB does not change if to an arbitrary column of B we add some linear combination of the other columns of B. Indeed, let $A = (a_1, \dots, a_n)^T$, where the $a_i \in (\mathbf{R}^k)^*$ are row vectors, and let $B = (b_1, \dots, b_n)$, where the $b_i \in \mathbf{R}^k$ are column vectors. Then

$$AB = \begin{pmatrix} a_1b_1 & \cdots & a_1b_n \\ \cdots & \cdots & \cdots \\ a_nb_1 & \cdots & a_nb_n \end{pmatrix}.$$

Replacement of an arbitrary column b_{i_0} in B by the vector $b_{i_0} + \Sigma_{i\neq i_0}c_ib_i$, $c_i \in \mathbf{R}$, leads to addition of the linear combination

$$\sum_{i\neq i_0} c_i\begin{pmatrix} a_1b_i \\ \vdots \\ a_nb_i \end{pmatrix}$$

of columns of AB to the i_0th column of AB, which, of course, does not change the rank of AB. It can be proved similarly that if to an arbitrary row of A we add some linear combination of the remaining rows of A, then the rank of AB does not change.

Theorem 4. *Assume the conditions of Lemma* 11, *and assume that the restriction* $\mathscr{D}\pi|_{\sigma\in\mathscr{N}_2}$ *to the plane formed by the vectors* $(T_\sigma\Gamma^*, \partial/\partial z_3)$ *is regular. Suppose that*

$$\frac{\partial}{\partial z_3} \in \operatorname{span}\left(\frac{\partial z}{\partial x}\frac{\partial}{\partial z} + \frac{\partial z}{\partial x}\frac{\partial}{\partial w}, \frac{\partial}{\partial z_1}, \frac{\partial}{\partial z_2}, \frac{\partial}{\partial z_4}, \frac{\partial}{\partial w'}\right)$$

at the point $(0, w_0) \in \mathscr{N}_1$. *Then the condition of Theorem* 3 *on regular projectibility of the manifold* Σ^* *on the space of* x *is satisfied in a neighborhood of the point* $(0, w_0)$.

Proof. It must be shown that the rank of the matrix

$$\mathscr{D} = \mathscr{D}\pi|_{(z, w)}\frac{\mathscr{D}(z, w)}{\mathscr{D}(t, \kappa, w')}$$

at points $(z, w) \in \Sigma^*\backslash\Gamma^*$ in a small neighborhood of $(0, w_0)$ is equal to n.

Let

$$\mathscr{D}_1 = \mathscr{D}\pi|_{(0, w_0)} = \begin{pmatrix} \dfrac{\partial x_1}{\partial z} & \dfrac{\partial x_1}{\partial w'} & \dfrac{\partial x_1}{\partial w''} \\ \cdots & \cdots & \cdots \\ \dfrac{\partial x_n}{\partial z} & \dfrac{\partial x_n}{\partial w'} & \dfrac{\partial x_n}{\partial w''} \end{pmatrix}.$$

By a condition of the theorem, there is a vector

$$a_1\frac{\partial}{\partial z_1} + a_2\frac{\partial}{\partial z_2} + \frac{\partial}{\partial z_3} + a_3\frac{\partial}{\partial z_4} + a\frac{\partial}{\partial w'}$$

belonging to $\operatorname{span}(\partial x_i/\partial(z, w), i = 1, \dots, n)$. It can be assumed without loss of generality that

$$\partial x_1/\partial(z, w) = (a_1, a_2, 1, a_3, a, 0).$$

By a condition of the theorem, the rank of the matrix

$$\mathscr{D}\left(\frac{\partial}{\partial z_3}, \frac{\partial}{\partial z_4} + \nu_i^+(0, w_0)\frac{\partial}{\partial w}, \frac{\partial}{\partial w'}\right)$$

is equal to n. Therefore, $\mathscr{D}_1$ can be reduced by a change of variables in the space of x to the form

$$\mathscr{D}_1 = \begin{pmatrix} * & * & \dfrac{\partial x_1}{\partial z_3} & * & 0 & 0 \\ * & * & 0 & \dfrac{\partial x_2}{\partial z_4} & 0 & \dfrac{\partial x_2}{\partial w''} \\ * & * & 0 & * & \dfrac{\partial x'}{\partial w'} & * \end{pmatrix}, \tag{2.42}$$

where $x' = (x_3, \dots, x_n)$, $\partial x_1/\partial z_3 \neq 0$, and

$$\frac{\partial x_2}{\partial z_4} + \frac{\partial x_2}{\partial w}\nu_i^+ \neq 0, \qquad \operatorname{rk}\frac{\partial x'}{\partial w'} = n - 2.$$

It follows from (2.36) that, up to terms of higher order of smallness,

$$\mathscr{D}_2 = \frac{\mathscr{D}(z, w)}{\mathscr{D}(t, \kappa, w')} = \begin{pmatrix} \partial z/\partial t & \partial z/\partial\kappa & \partial z/\partial w' \\ \partial w/\partial t & \partial w/\partial\kappa & \partial w/\partial w' \end{pmatrix}$$

$$= \begin{pmatrix} \kappa^4\Theta(1) & \kappa^3\Theta(1) & \kappa^4\Theta(1) \\ \kappa^3\Theta(1) & \kappa^2\Theta(1) & \kappa^3\Theta(1) \\ (1 + A_i\tau)\kappa + \kappa^2\Theta(1) & 2(\lambda_i^+ + \tau + \frac{1}{2}A_i\tau^2)\kappa + \kappa^2\Theta(1) & \kappa^2\Theta(1) \\ A_i + \kappa\Theta(1) & 1 + A_i\tau + \kappa\theta(1) & \kappa\Theta(1) \\ F(0, \sigma, u_i) + \kappa\Theta(1) & \nu_i^+ + \tau F(0, \sigma, u_i) + \kappa\Theta(1) & \partial w/\partial w' \end{pmatrix}$$

at a point $(z, w) \in \Sigma^*\backslash\Gamma^*$.

After obvious algebraic transformations on the columns of $\mathscr{D}_2$ we get that $\operatorname{rk}\mathscr{D}_1\mathscr{D}_2 = \operatorname{rk}\mathscr{D}_1\mathscr{D}_3$, where

$$\mathscr{D}_3 = \begin{pmatrix} \kappa^3\Theta(1) & \kappa^3\Theta(1) & \kappa^3\Theta(1) \\ \kappa^2\Theta(1) & \kappa^2\Theta(1) & \kappa^2\Theta(1) \\ (1-2A_i\lambda_i^+)\kappa + \kappa^2\Theta(1) & (2\lambda_i^+ + \tau)\kappa + \kappa^2\Theta(1) & \kappa^2\Theta(1) \\ \kappa\Theta(1) & 1+\kappa\Theta(1) & \kappa\Theta(1) \\ F(0, \sigma, u_i) - A_i\nu_i^+ + \kappa\Theta(1) & \nu_i^+ + \kappa\Theta(1) & \partial w/\partial w' \end{pmatrix}.$$

It follows from (2.41) that $F(0, \sigma, u_i) - A_i\nu_i^+ \in T_\sigma\mathscr{N}_2$, i.e., the vector $F(0, \sigma, u_i) - A_i\nu_i^+$ can be represented as a linear combination of the vectors forming the matrix $\partial w(0, \sigma)/\partial w'$. Therefore, $\operatorname{rk}\mathscr{D}_1\mathscr{D}_3 = \operatorname{rk}\mathscr{D}_1\mathscr{D}_4$, where

$$\mathscr{D}_4 = \begin{pmatrix} \kappa^3\Theta(1) & \kappa^2\Theta(1) & \kappa^3\Theta(1) \\ \kappa^2\Theta(1) & \kappa^2\Theta(1) & \kappa^2\Theta(1) \\ (1-2A_i\lambda_i^+)\kappa + \kappa^2\Theta(1) & \kappa^2\Theta(1) & \kappa^2\Theta(1) \\ \kappa^2\Theta(1) & 1+\kappa\Theta(1) & \kappa^2\Theta(1) \\ \kappa\Theta(1) & \nu_i^+ + \kappa\Theta(1) & \partial w/\partial w' \end{pmatrix}.$$

The last $2n-4$ columns of $\mathscr{D}_4$ can be transformed to the form

$$\begin{pmatrix} 0 & & E+\kappa\Theta(1) \\ & \nu_i^+ + \kappa\Theta(1) & \\ \kappa\Theta(1) & & \kappa\Theta(1) \end{pmatrix},$$

where E is the $(n-2)\times(n-2)$ identity matrix. To verify the condition $\operatorname{rk}\mathscr{D} = n$ it now remains to multiply the matrices $\mathscr{D}_1$ and $\mathscr{D}_4$ and discard the terms of higher order of smallness.

This proves that each of the two smooth pieces of Σ^* projects regularly on the space of x for $\kappa \neq 0$. In view of (2.36), for $\kappa \neq 0$ the phase velocity vectors $\partial(z, w)/\partial t$ of the system (2.4) at the points $(z, w) \in \Gamma^*$ together with the tangent vectors to Γ^* form a basis for the tangent planes to each of the two smooth pieces of Σ^*. Since both these planes project regularly on the space of x, it follows from this that the trajectories of (2.1) intersect the switching surface $\Gamma^*\backslash\mathscr{N}_2$ transversally.

The surface $\pi\Gamma^*$ divides the space of x into two open domains. By (2.30), the component $(1-2A_i\lambda_i^+)\kappa$ of the matrix $\mathscr{D}_4$ is of single sign both for $i=0$ and for $i=1$. Therefore, at points of the surface $\pi\Gamma^*$ both phase velocity vectors are directed into the same half-space, and the trajectories themselves of the maximum principle intersect the surface $\pi\Gamma^*$, passing from one open domain to the other. Thus, the trajectories of the maximum principle filling $\pi\Sigma^*$ are disjoint. The theorem is proved.

We remark that, as a rule, $\operatorname{codim}\pi S_0 = 2$, and the conditions of Theorem 4 are satisfied in the following form:

$$\operatorname{rk}\mathscr{D}\pi|_\sigma(\partial/\partial z_3, \partial/\partial z_4, \partial/\partial w) = n.$$

The problem of Lawden, in which $\operatorname{codim}\pi S_0 = 1$, constitutes an exception. For this problem it must be proved directly that the matrix $\mathscr{D}x/\mathscr{D}(z, w)$ can be reduced to the form (2.42).

Chapter 3. Chattering trajectories in a neighborhood of the manifold of singular trajectories of order three and higher

1. Generalizations and conjectures

In Chapter 2 it was shown that the asymptotics of chattering trajectories of a Hamiltonian system (2.4) in a neighborhood of the manifold of singular trajectories of second order is given by the solution of the Fuller problem. It is natural to assume that the asymptotics of the solutions lying in a neighborhood of the singular manifold of order k is given by the solution of the following Problem 5, which is a k-dimensional analogue of the Fuller problem.

Problem 5. *Minimize the functional*

$$\int_0^\infty x^2\,dt \to \inf$$

on the solutions of the control system

$$x^{(k)} = u$$

with initial conditions

$$x^{(m)}(0) = x_m\,, \qquad m = 0,\ 1, \dots, k-1\,,$$

where $u \in [\beta\,, 1]$, $\beta < 0$.

In connection with this, the question of a complete description of the optimal trajectories of Problem 5 acquires exceptional importance.

It should be noted that Problem 5 with $k = 3$ and $\beta = -1$ was treated in [2], [3], and [7]. In these papers self-similar chattering trajectories and optimal trajectories tending to the origin without switchings are found. The behavior of the remaining trajectories and the structure of the switching surface in this problem were investigated by Borisov in [46]. These results will be presented in the present chapter.

Before presenting the results and the conjectures on the structure of solutions of Problem 5, we consider the following trivial case of the "one-dimensional Fuller problem", when $k = 1$ and $\beta = -1$. The system of equations of the Pontryagin maximum principle in this case has the form

$$\dot\psi = x\,, \qquad \dot x = \operatorname{sgn}\psi. \tag{3.1}$$

The phase portrait of the system (3.1), which is like a "flattened saddle", is pictured in Figure 9.

The curve consisting of the two half-parabolas AO and CO is the stable tendril of the saddle, and the curve consisting of the half-parabolas OB and OD is its unstable tendril. The stable and unstable tendrils are tangent at the point O. In Figure 9 it is possible to illustrate graphically the results in Chapter 1 concerning the solution of the Fuller problem for $k = 2$. As shown in Chapter 2, the synthesis of optimal trajectories in Problem 5 with $k = 2$ and with an arbitrary $\beta < 0$ is trajectorywise

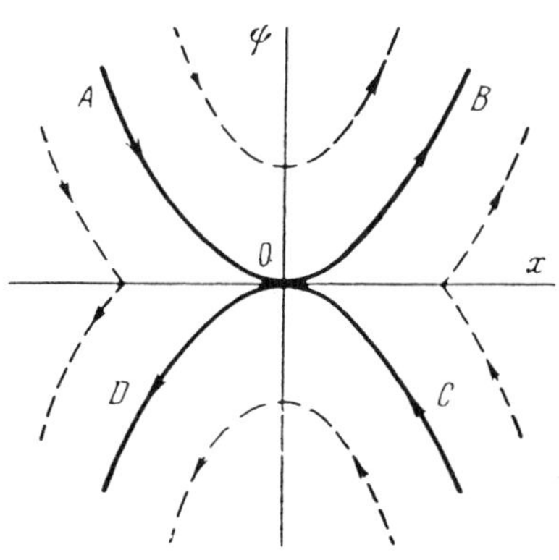

FIGURE 9

equivalent to the synthesis in Problem 1 with $\beta = -1$. The Hamiltonian of the system in Problem 1 has the form

$$\dot{z}_1 = z_2, \quad \dot{z}_2 = z_3, \quad \dot{z}_3 = z_4, \quad \dot{z}_4 = -\operatorname{sgn} z_1. \tag{3.2}$$

We will represent two-dimensional manifolds composed of trajectories of (3.2) in the form of one-dimensional curves. Then Figure 9 describes the behavior of such two-dimensional invariant manifolds. We have seen that the system (3.2) has exactly two two-dimensional invariant manifolds $\mathfrak{N}^+$ and $\mathfrak{N}^-$, each consisting of trajectories tending to the origin with increasing and decreasing t, respectively. These two manifolds are tangent at the origin and correspond to the stable and unstable tendrils in Figure 9. As shown in Lemmas 5 and 6, the remaining trajectories of (3.2) tend to the tendril $\mathfrak{N}^-$ as $t \to +\infty$ and to $\mathfrak{N}^+$ as $t \to -\infty$ in the "angle" metric generated by the orbits of the group $\widetilde{\mathfrak{G}}$. Figure 9 does not reflect the structure of the trajectories inside each two-dimensional invariant manifold, and this structure must be pictured separately (see Figures 2 and 4).

In this chapter we prove that in Problem 5 with $k = 3$ and $\beta = -1$ the system of equations

$$\dot{z}_1 = z_2, \quad \dot{z}_2 = z_3, \quad \dot{z}_3 = z_4, \quad \dot{z}_4 = z_5, \quad \dot{z}_5 = z_6, \quad \dot{z}_6 = \operatorname{sgn} z_1, \tag{3.3}$$

of the Pontryagin maximum principle has two three-dimensional manifolds $\mathfrak{N}^+$ and $\mathfrak{N}^-$. The manifold $\mathfrak{N}^+$ corresponds to the stable tendril of the saddle point O, while $\mathfrak{N}^-$ corresponds to the unstable tendril. In $\mathfrak{N}^+$ there is a one-dimensional manifold consisting of two branches, on one of which the phase point with control $u = 1$ hits the origin without switchings, while on the other the same thing happens for $u = -1$. Moreover, $\mathfrak{N}^+$ contains the two-dimensional manifold formed from the trajectories of (3.3) that are self-similar with respect to the symmetry group $\widetilde{\mathfrak{G}}$. All the remaining trajectories of the family $\mathfrak{N}^+$ tend to the origin as t increases, performing infinitely many switchings in a finite time. However, as $t \to -\infty$ the trajectories of $\mathfrak{N}^+$ that are not self-similar have only finitely many switching points. The switching surface in $\mathfrak{N}^+$ is piecewise smooth. The same structure of trajectories is valid inside $\mathfrak{N}^-$, but with reverse time.

For the four-dimensional Fuller problem

$$\dot{z}_i = z_{i+1} \quad (i = 1, 2, \dots, 7), \qquad \dot{z}_8 = -\operatorname{sgn} z_1$$

we find two families of self-similar solutions lying inside the stable four-dimensional tendril $\mathfrak{N}^+$.

The results obtained in this section lead us to the following conjectures on the structure of a solution of the Hamiltonian system of the Pontryagin maximum principle in Problem 5:

$$\begin{aligned} &\dot{z}_i = z_{i+1} \qquad (i = 1, \dots, 2k-1), \\ &\dot{z}_{2k} = (-1)^{k+1} \begin{cases} 1 & \text{for } z_1 > 0, \\ \beta & \text{for } z_1 < 0. \end{cases} \end{aligned} \tag{3.4}$$

First of all we note that the system (3.4) is homogeneous with respect to the action of the group of diffeomorphisms

$$g_\lambda \colon z_i \to \lambda^{2k+1-i} z_i, \qquad i = 1, \dots, 2k,$$

and we are interested in self-similar solutions of (3.4).

Conjecture 1. *In the phase space $\mathbf{R}^{2k}$ there are two k-dimensional manifolds $\mathfrak{N}^+$ and $\mathfrak{N}^-$ that are invariant under the action of the group of g_λ and consist of trajectories* *of the system* (3.4) *going to the origin as* t *increases and decreases, respectively. The trajectories of* (3.4) *belonging to neither* $\mathfrak{N}^+$ *nor* $\mathfrak{N}^-$ *tend to* $\mathfrak{N}^-$ *as* $t \to +\infty$ *and to* $\mathfrak{N}^+$ *as* $t \to -\infty$ *in the "angle" metric generated by the orbits of the group of* g_λ.

Conjecture 2. *In the case of even* $k = 2m$ *there are* m *one-parameter families of self-similar chattering trajectories inside the manifold* $\mathfrak{N}^+$. *In the case of odd* $k = 2m + 1$ *there are* m *one-parameter families of self-similar trajectories and two one-dimensional trajectories in* $\mathfrak{N}^+$, *each of which goes to the origin without switchings with* $u = 1$ *and* $u = \beta$, *respectively. The same structure, but with reversed time, holds inside* $\mathfrak{N}^-$.

Conjecture 3. *The manifolds* $\mathfrak{N}^+$ *and* $\mathfrak{N}^-$ *are imbedded in* $\mathbf{R}^{2k}$ *as piecewise analytic manifolds intersecting the switching surface* $z_1 = 0$ *in a piecewise analytic surface homeomorphic to* $\mathbf{R}^{k-1}$.

Suppose that the system of equations of the Pontryagin maximum principle for Problem 5 can be reduced to the form

$$
\begin{aligned}
\dot{z}_i &= z_{i+1} + f_i(z, w, u) \qquad (i = 1, \dots, 2k-1), \\
\dot{z}_{2k} &= \alpha(w) + u\beta(w) + f_{2k}(z, w, u), \\
\dot{w} &= F(z, w, u), \qquad u = \operatorname{sgn} z_1, \\
w &\in \mathbf{R}^m, \qquad z = (z_1, \dots, z_{2k}) \in \mathbf{R}^{2k},
\end{aligned}
\tag{3.5}
$$

where

$$\overline{\lim_{\lambda \to +0}} \lambda^{-(2k+1-i)} |f_i(g_\lambda, z, w, u)| < \infty.$$

We define the domain W on the plane $z = 0$ by the inequalities $(-1)^k \beta < 0$ and $|\alpha| < |\beta|$. The domain W is composed of singular trajectories of the system (3.5) of order k.

Conjecture 4. *In a small neighborhood of the manifold* W *there are two piecewise smooth bundles* $p^+ : \mathfrak{M}^+ \to W$ *and* $p^- : \mathfrak{M}^- \to W$ *with base* W *and* k*-dimensional piecewise smooth fibers* $\mathfrak{M}^+_\sigma$ *and* $\mathfrak{M}^-_\sigma$ *composed of trajectories of the system* (3.5). *Each manifold* $\mathfrak{M}^\pm_\sigma$ *is trajectorywise equivalent to the manifolds* $\mathfrak{N}^+$ *and* $\mathfrak{N}^-$ *described in Conjectures* 1, 2, *and* 3.

Conjecture 5. *If* $\mathcal{M} \subset W$ *is a Lagrangian submanifold of the singular manifold* W, *then* $(p^+)^{-1}\mathcal{M}$ *and* $(p^-)^{-1}\mathcal{M}$ *are Lagrangian submanifolds of* $\mathbf{R}^{2k+m}$.

As in Theorem 3, this yields a theorem on sufficient conditions for optimality of chattering trajectories when the projection of the corresponding Lagrangian manifold on the space of x is regular.

2. Problems with symmetries

Let us consider the following problem.

Problem 6.

$$
\begin{gathered}
\int_0^\infty f(x)\,dt \to \inf, \\
\dot{x} = Ax + bu, \quad x(0) = x^0, \quad x \in \mathbf{R}^n, \quad u \in [-1, 1].
\end{gathered}
\tag{3.6}
$$

The admissible trajectories are absolutely continuous, and the admissible controls are in the class $L_\infty[0, \infty)$ (see [47], p. 241, for the definition of the space $L_\infty[0, \infty)$). Let A be a constant $n \times n$ matrix and $b \in \mathbf{R}^n$ a constant vector. The function

$f\colon \mathbf{R}^n \to \mathbf{R}$ is of class C^2, $f \geq 0$, and $f(0) = 0$. The condition of complete controllability is assumed:

$$\operatorname{rk}\{b, Ab, \dots, A^{n-1}b\} = n. \tag{3.7}$$

Assumption 1. For arbitrary initial conditions there is an admissible trajectory of the system (3.6) hitting the origin in finite time.

It will be shown below that Assumption 1 always holds for the class of problems with symmetries satisfying condition (3.7).

Lemma 12. *Problem* 6 *has a solution for arbitrary initial conditions.*

Proof. Denote by $\mathfrak{X}$ the set of continuous functions $x(\cdot)\colon \mathbf{R}_+ \to \mathbf{R}^n$ with initial conditions $x(0) = x^0$. We consider the operator

$$K\colon L_\infty[0, \infty) \to \mathfrak{X}, \qquad Ku(t) = e^{At}x^0 + \int_0^\infty \rho_t(\tau)u(\tau)\,d\tau,$$

where $\rho_t(\tau) = e^{A(t-\tau)}$ for $0 \leq \tau \leq t$ and $\rho_t(\tau) = 0$ for $\tau > t$. The operator K assigns to a function $u(\cdot) \in L_\infty[0, \infty)$ the solution $x(t)$ of the system (3.6) with control $u = u(t)$ and initial conditions x^0. By the Banach-Alaoglu theorem ([47], Chapter V, §4, Theorem 2), the unit ball of the space $L_\infty[0, \infty)$ is weak* compact ($L_\infty[0, \infty) = L_1^*[0, \infty)$; see [47], Chapter IV, §8, Theorem 5). Let $\{u^{(n)}(\cdot)\}_{n=1}^\infty$ be a minimizing sequence of controls in Problem 6. In can be assumed without loss of generality that this sequence converges to some function $u^{(0)}(\cdot)$ in the weak* topology of $L_\infty[0, \infty)$ (otherwise we can pass to a convergent subsequence). By the definition of the weak* topology, this means that

$$\lim_{n\to\infty} \int_0^\infty \rho(\tau)u^{(n)}(\tau)\,d\tau = \int_0^\infty \rho(\tau)u^{(0)}(\tau)\,d\tau$$

for any function $\rho(\cdot) \in L_1[0, \infty)$. For any $t > 0$ the function $\rho_t(\cdot)$ is in $L_1[0, \infty)$; therefore, the sequence $x^{(n)}(t) = Ku^{(n)}(t)$ converges pointwise on $[0, \infty)$ to the function $x^{(0)}(t) = Ku^{(0)}(t)$. It follows from the form of the operator K that on any finite interval the sequence of functions $x^{(n)}(t)$ is uniformly Lipschitz, and so the convergence is uniform on any compact subset of $[0, \infty)$.

It follows from Assumption 1 that the minimal value of the functional in Problem 6 is finite (strictly less than $+\infty$) and nonnegative. By Fatou's theorem ([47], Chapter III, §6, Theorem 19),

$$\varliminf_{n\to\infty} \int_0^\infty f(x^{(n)}(t))\,dt \geq \int_0^\infty f(x^{(0)}(t))\,dt.$$

The existence of a solution in Problem 6 is thereby proved.

Assumption 2. The function f does not depend on the coordinates $x_{k+1}, \dots, x_n$ and is strictly convex with respect to $x_1, \dots, x_k$, $1 \leq k \leq n$.

Let $L = \{x | x_1 = \dots = x_k = 0\}$.

Assumption 3. For an arbitrary $x \in L\backslash\{0\}$

$$\min_{|u|\leq 1} \max_{1\leq j\leq k} |(Ax + bu)_j| > 0,$$

i.e., the surface L is strictly supporting for the vectorgram of admissible velocities of the system (3.6) at the points of $L\backslash\{0\}$.

Denote by ω the Bellman function of Problem 6, i.e., the minimal value of the functional as a function of the initial state.

Lemma 13. *For any* x^0 *there is a unique solution* $x(t, x^0)$ *of Problem* 6 *with initial conditions* x^0. *The Bellman function in Problem* 6 *is strictly convex.*

Proof. Let $x^i(t)$, $i = 1, 2$, be two distinct optimal trajectories of Problem 6 with initial conditions x^{i0}, and let $u^i(t)$ be the controls on them. Consider the trajectory $x^\lambda(t) = \lambda x^1(t) + (1-\lambda)x^2(t)$ with the control $u^\lambda(t) = \lambda u^1(t) + (1-\lambda)u^2(t)$. Denote by π the operator of orthogonal projection onto the plane $x_1 = \cdots = x_k = 0$. If $\pi x^1(t) = \pi x^2(t)$ for $t \in (t_0, t_1)$ and $x^i(t) \neq 0$, then the trajectory $\frac{1}{2}(x^1(t) - x^2(t))$ belongs to the surface $L \backslash \{0\}$ and satisfies (3.6) with the control $u = \frac{1}{2}(u^1 - u^2)$. This contradicts Assumption 3. Consequently, $\pi x^1(t) \neq \pi x^2(t)$ on a set of positive measure. Since f is strictly convex with respect to $x_1, \ldots, x_k$, we get

$$\int_0^\infty f(x^\lambda(t))\, dt < \lambda \int_0^\infty f(x^1(t))\, dt + (1-\lambda) \int_0^\infty f(x^2(t))\, dt.$$

If the initial conditions x^{10} and x^{20} coincide, this contradicts the optimality of the trajectories $x^i(t)$. If the initial conditions are distinct, then it follows from the above inequality that ω is strictly convex. The lemma is proved.

Corollary 1. ω *is Lipschitz.*

Indeed, ω is a proper convex function, and hence a Lipschitz function ([48], Part II, Theorem 10.4).

Corollary 2. *The mapping* $\mathbf{R}^n \to C[0, T]$, $x^0 \mapsto x(\cdot, x^0)$, *is continuous for any finite* $T > 0$.

Proof. Consider a sequence $\{x^n\}_1^\infty$ of points for which $\lim_{n\to\infty} x^n = x^0$. Denote by $u^0(t)$ the weak* limit of some subsequence $\{u^{n_k}(\cdot)\}_{k=1}^\infty$ of optimal controls. Let $x^0(t) = Ku^0(t)$. It can be proved just as in Lemma 12 that for any fixed t the pointwise convergence $\lim_{k\to\infty} x(t, x^{n_k}) = x^0(t)$ is valid, uniformly on any compact subset of $[0, \infty)$. By Fatou's theorem,

$$\int_0^\infty f(x^0(t))\, dt \le \underline{\lim}_{n\to\infty} \int_0^\infty f(x(t, x^{n_k}))\, dt.$$

Since the Bellman function ω is continuous, the right-hand side of the inequality is equal to $\omega(x^0)$. It follows from the uniqueness of the optimal trajectory in Problem 6 that $x^0(t) = x(t, x^0)$. The corollary is proved.

Definition 2. It will be said that Problem 6 admits the one-parameter symmetry group $\mathfrak{G} \cong \mathbf{R}_+$ ($\mathbf{R}_+$ is the multiplicative group of positive real numbers) if a continuous mapping

$$\Lambda: \mathbf{R}_+ \to \mathrm{Diff}(\mathbf{R}^n), \quad \lambda \mapsto g_\lambda(\cdot),$$

($\mathrm{Diff}(\mathbf{R}^n)$ is the group of diffeomorphisms of the space $\mathbf{R}^n$) is defined with the following properties:

1) $g_{\lambda\mu}(x) = g_\lambda(g_\mu(x))$ for any $\lambda > 0$ and $\mu > 0$.
2) If id is the identity mapping of $\mathbf{R}^n$, then $\Lambda^{-1}(\mathrm{id}) = 1$.
3) $g_\lambda(0) = 0$ for any $\lambda > 0$.
4) For any $u \in [-1, 1]$ and $x \in \mathbf{R}^n$

$$\mathscr{D} g_\lambda|_x(Ax + bu) = \lambda^\alpha(Ag_\lambda(x) + bu), \qquad f(g_\lambda(x)) = \lambda^\beta f(x)$$

for some constants $\alpha > 0$ and $\beta > 0$ independent of u and x.

5) The stationary subgroup of each point $x \neq 0$ is trivial: $\{\lambda > 0 : g_\lambda(x) = x\} = 1$.
6) $\lim_{\lambda\to+0} g_\lambda(x) = 0$ for any $x \in \mathbf{R}^n$.

It will be assumed below that Problem 6 admits the symmetry group $\mathfrak{G}$.

Lemma 14. 1) *$x(t, g_\lambda(x^0)) = g_\lambda(x(\lambda^{-\alpha}t, x^0))$ and $u(t, g_\lambda(x^0)) = u(\lambda^{-\alpha}t, x^0)$.*
2) *$\omega(g_\lambda(x)) = \lambda^{\alpha+\beta}\omega(x)$.*

Proof. We have

$$\begin{aligned}\frac{d}{dt}g_\lambda(x(\lambda^{-\alpha}t, x^0)) &= \mathscr{D} g_\lambda|_{x(\lambda^{-\alpha}t, x^0)}\frac{d}{dt}x(\lambda^{-\alpha}t, x^0)\\ &= \mathscr{D} g_\lambda|_{x(\lambda^{-\alpha}t, x^0)}(Ax(\lambda^{-\alpha}t, x^0) + bu(\lambda^{-\alpha}t, x^0))\lambda^{-\alpha}\\ &= Ag_\lambda(x(\lambda^{-\alpha}t, x^0)) + bu(\lambda^{-\alpha}t, x^0).\end{aligned}$$

Moreover,

$$\int_0^\infty f(g_\lambda(x(\lambda^{-\alpha}t, x^0)))\,dt = \lambda^\beta \int_0^\infty f(x(\lambda^{-\alpha}t, x^0))\,dt = \lambda^{\alpha+\beta}\int_0^\infty f(x(t, x^0))\,dt.$$

The mapping $x(t) \to g_\lambda(x(\lambda^{-\alpha}t))$ is one-to-one and obviously carries the optimal trajectory in Problem 6 with initial conditions x^0 into the optimal trajectory with initial conditions $g_\lambda(x^0)$. The lemma is proved.

Remark 6. Assumption 1 automatically holds for Problem 6, which admits the symmetry group $\mathfrak{G}$. By (3.7), for all sufficiently small initial conditions x^0 there is an admissible trajectory $x(t)$ of (3.6) with the control $u(t)$ and initial conditions x_0 that hits the origin in a finite time T. Then for an arbitrary $\lambda > 0$ the trajectory $g_\lambda(x(\lambda^{-1}t))$ with initial conditions $g_\lambda(x^0)$ also satisfies (3.6) with the control $u = u(\lambda^{-1}t)$ and hits the origin in time λT. Thus, controllability of (3.6) everywhere in $\mathbf{R}^n$ follows from local controllability of (3.6) in a neighborhood of the origin and the existence of the symmetry group $\mathfrak{G}$.

Lemma 15. *For any x^0 there is a finite $T_0 = T(x^0) > 0$ such that the optimal trajectory $x(t, x^0)$ in Problem 6 with initial conditions x^0 is optimal in the following problem with fixed time.*

Problem 7.

$$\int_0^{T_0} f(x)\,dt \to \inf,$$

$$\dot{x} = Ax + bu, \quad |u| \le 1, \quad x(0) = x^0, \quad x(T_0) = 0.$$

Proof. It suffices to show that the optimal trajectory in Problem 6 hits the origin in finite time. First of all we show that $\lim_{t\to\infty} x(t, x^0) = 0$. Assume that $\lim_{t\to\infty} x(t, x^0) \neq 0$ for some x^0. Then there exist an $\varepsilon > 0$ and an increasing sequence of times $\{t_n\}_1^\infty$, $\lim_{n\to\infty} t_n = \infty$, such that the points $y^n = x(t_n, x^0)$ lie outside the ball B_ε of radius ε about the origin. Let $\delta(\varepsilon) = \inf\{\omega(x) : x \in \partial B(\varepsilon)\}$. The function $\omega(x)$ is continuous and strictly convex, and hence is nonzero away from the origin. Consequently, $\delta(\varepsilon) > 0$. Since $x(t, x^0)$ is optimal, $\int_0^\infty f(x(t, x^0))\,dt$ converges. Consequently, for any $\delta > 0$ there exists a $T(\delta) > 0$ such that

$$\int_{T(\delta)}^\infty f(x(t, x^0))\,dt < \delta.$$

Choose a $\delta < \delta(\varepsilon)$. There is an index n with $t_n > T(\delta)$. Since $\omega(y^n) \ge \delta(\varepsilon)$ by assumption,

$$\int_{t_n}^\infty f(x(t, x^0))\,dt = \int_0^\infty f(x(t, y^n))\,dt = \omega(y^n) \ge \delta(\varepsilon).$$

On the other hand,

$$\int_{t_n}^\infty f(x(t, x^0))\,dt \le \int_{T(\delta)}^\infty f(x(t, x^0))\,dt < \delta,$$

i.e., $\delta(\varepsilon) < \delta$. Contradiction. The condition $\lim_{t\to\infty} x(t, x^0) = 0$ is thereby proved.

Let $\omega_C = \{x|\omega(x) = C\}$ be a level surface of the Bellman function. We consider the function

$$\tau(C) = \sup_{x\in\omega_C} \{t_C(x)|\omega(x(t_C, x)) = C/2\}.$$

Since $\omega(x(t, x))$ is strictly decreasing as t increases to zero, the function t_C is uniquely defined for $C \neq 0$. We show that $\tau(C) < \infty$. Assume that there exists a sequence $\{x^n\}_{n=1}^{\infty}$ of points belonging to ω_C such that $\omega(x(t_n, x^n)) = C/2$, $t_n = t_C(x^n)$, and $t_n \to \infty$ as $n \to \infty$. It can be assumed without loss of generality that $\{x^n\}$ converges to some point x^0. In this case the trajectory $x(t, x^0)$ is the pointwise limit of the sequence of trajectories $x(t, x^n)$ (Corollary 2). Then the trajectory $x(t, x^0)$ is outside the ball $\{x|\omega(x) < C/2\}$ for arbitrarily distant values of $t = t_n \to \infty$. This contradicts the condition $\lim_{t\to\infty} x(t, x^0) = 0$. We have thus shown that $\tau(C) < \infty$.

We show that $\tau(C/2) = \sigma^\alpha \tau(C)$, where $\sigma = 2^{-1/(\alpha+\beta)}$. In view of Lemma 14 the level surface ω_C satisfies the relation

$$\omega_{C_1} = g_\gamma \omega_{C_2} \tag{3.8}$$

for any nonzero constants C_1 and C_2, where $\gamma = (C_1/C_2)^{1/(\alpha+\beta)}$. Indeed, if $\omega(x) = C_2$ at the point x, then

$$\omega(g_\gamma(x)) = \gamma^{(\alpha+\beta)}\omega(x) = (C_1/C_2)C_2 = C_1.$$

For $C_1 = \frac{1}{2}C_2$ we get that $\omega_{C/2} = g_\sigma \omega_C$.

Suppose that for $x^0 \in \omega_C$ the inclusion $x(t, x^0) \in \omega_{C/2}$ holds at time t. Then for the point $g_\sigma(x^0)$ we get that at the time $\tau = \sigma^\alpha t$

$$\omega(x(\tau, g_\sigma(x^0))) = \omega(g_\sigma(x(\sigma^{-\alpha}\tau, x^0))) = \sigma^{(\alpha+\beta)}\omega(x(t, x^0)) = \tfrac{1}{4}C,$$

i.e., $x(\tau, g_\sigma(x^0)) \in \omega_{C/4}$. Since the mapping $g_\sigma\colon \omega_C \to \omega_{C/2}$ is one-to-one, this implies that $\tau(C/2) = \sigma^\alpha\tau(C)$.

Denote by $T(x)$ the time at which the trajectory $x(t, x)$ hits the origin. Let $\omega(x) = C$. Then

$$\begin{aligned} T(x) &\le \tau(C) + \tau(C/2) + \cdots + \tau(C/2^n) + \cdots \\ &= \tau(C)(1 + \sigma^\alpha + \cdots + \sigma^{n\alpha} + \cdots) = \tau(C)/(1 - \sigma^\alpha). \end{aligned} \tag{3.9}$$

The lemma is proved.

The continuous dependence of optimal trajectories on the initial data and the finiteness of the time to hit the origin imply the continuity of the function $T(x)$.

Corollary 3. *$T(x)$ is continuous.*

Proof. We show that $T(g_\lambda(x)) = \lambda^\alpha T(x)$. On the one hand,

$$x(t, g_\lambda(x))|_{t=\lambda^\alpha T(x)} = g_\lambda(x(\lambda^{-\alpha}t, x))|_{t=\lambda^\alpha T(x)} = g_\lambda(0) = 0,$$

and hence $T(g_\lambda(x)) \le \lambda^\alpha T(x)$, and on the other hand, this implies the estimate

$$T(x) = T(g_{1/\lambda}(g_\lambda(x))) \le (1/\lambda)^\alpha T(g_\lambda(x)),$$

i.e., $\lambda^\alpha T(x) \le T(g_\lambda(x))$.

The Bellman function ω is continuous (Corollary 1), and so $\lim_{x\to 0}\omega(x) = 0$. If $\omega(x) = C$, then $T(x) \le \sup_{y\in\omega_C} T(y)$. In view of (3.8)

$$\sup_{y\in\omega_C} T(y) = \sup_{z\in\omega_1} T(g_\nu(z)) = \nu^\alpha \sup_{z\in\omega_1} T(z),$$

where $\nu = C^{1/(\alpha+\beta)}$. It follows from (3.9) that

$$\overline{\lim_{x\to 0}} T(x) \le \frac{\tau(1)}{1-\sigma^\alpha} \lim_{x\to 0} \omega(x)^{\alpha/(\alpha+\beta)} = 0.$$

Since $T(x) \ge 0$, it follows that $\lim_{x\to 0} T(x) = 0$, i.e., we have proved that $T(x)$ is continuous at the origin.

We now consider an arbitrary sequence $\{x^n\}_{n=1}^\infty$ such that the limit $\lim_{n\to\infty} x^n = x^0$ exists. Let $T^+ = \overline{\lim}_{n\to\infty} T(x^n)$. By (3.9), $T^+ < \infty$. We show that $T(x^0) \ge T^+$. Assume that $T(x^0) = T^+ - \varepsilon < T^+$. Then there exists a subsequence $\{x^{n_k}\}_{k=1}^\infty$ such that $T(x^{n_k}) > T^+ - \varepsilon/2$. Since the optimal trajectories $x(t, x^{n_k})$ depend continuously on the initial data,

$$\lim_{k\to\infty} x(T^+ - \varepsilon, x^{n_k}) = x(T^+ - \varepsilon, x^0) = 0.$$

Let $y^k = x(T^+ - \varepsilon, x^{n_k})$. Then $\lim_{k\to\infty} y^k = 0$, while $T(y^k) \ge \varepsilon/2$. This contradicts the continuity of T at the origin.

Let $T^- = \underline{\lim}_{n\to\infty} T(x^n)$. We show that $T(x^0) \le T^-$. Assume that $T(x^0) = T^- + \varepsilon > T^-$. Then there is a subsequence $\{x^{n_k}\}_{k=1}^\infty$ such that $T(x^{n_k}) < T^- + \varepsilon/2$. By the continuous dependence of $x(t, x^{n_k})$ on the initial data,

$$\lim_{k\to\infty} x(T^- + \varepsilon/2, x^{n_k}) = x(T^- + \varepsilon/2, x^0).$$

But $x(T^- + \varepsilon/2, x^{n_k}) = 0$. Therefore, $x(T^- + \varepsilon/2, x^0) = 0$, while $T(x^0) > T^- + \varepsilon/2$, a contradiction. Thus,

$$\overline{\lim_{n\to\infty}} T(x^n) = \underline{\lim_{n\to\infty}} T(x^n) = T(x^0).$$

The corollary is proved.

Lemma 16. 1) *For arbitrary initial conditions* x^0 *there is a unique collection of conjugate variables* $\psi(t)$, $\psi(0) = \Psi(x^0)$, *satisfying the conditions of the Pontryagin maximum principle for Problem* 6:

$$\dot\psi = -A^*\psi + f'_x(x(t, x^0)), \tag{3.10}$$

$$\max_{u\in[-1,1]} \langle u\psi(t), b\rangle = \langle u(t, x^0)\psi(t), b\rangle; \tag{3.11}$$

2) *The Bellman function in Problem* 6 *is of class* C^1, *and* $-\Psi(x) = \operatorname{grad}\omega(x)$.

Proof. Fix an $x^0 \ne 0$, and let $T_0 > T(x^0)$. In view of Lemma 15 the trajectory $x(t, x^0)$ is optimal in Problem 7 with the given T_0. By the Pontryagin maximum principle for Problem 7, there exist an absolutely continuous vector-valued function $\psi(t)$ and a number $\lambda_0 \ge 0$, not both zero, satisfying the equation

$$\dot\psi = -A^*\psi + \lambda_0 f'_x(x(t, x^0)) \tag{3.12}$$

and the maximum condition (3.11). We show that $\lambda_0 \ne 0$. We remark that for $t \in [T(x^0), T_0]$ the control on the optimal trajectory is equal to zero: $u(t, x^0) = 0$, and hence $\langle\psi(t), b\rangle = 0$. If $\lambda_0 = 0$, then for the function $\psi(t) = e^{-A^*t}\psi(0)$ we obtain

$$\frac{d}{dt}\langle\psi(t), b\rangle = 0 = -\langle\psi(t), Ab\rangle,$$

$$\dots\dots\dots\dots\dots\dots\dots\dots\dots\dots\dots\dots$$

$$\frac{d^k}{dt^k}\langle\psi(t), b\rangle = 0 = (-1)^k\langle\psi(t), A^k b\rangle,$$

$$k = 1, \dots, n-1,$$

for $t \in [T(x^0), T_0]$. It follows from (3.7) that $\psi(t) = 0$ for $t \in [T(x^0), T_0]$. Then, since the system (3.12) is linear in ψ, $\psi(t) = 0$ also for arbitrary t. This shows that $\lambda_0 \neq 0$, and λ_0 can be set equal to 1.

Assume that for some optimal trajectory $x(t, x^0)$ there are two distinct solutions $\psi^1(t)$ and $\psi^2(t)$ of the system (3.10), (3.11). Then the function $\Delta\psi(t) = \psi^1(t) - \psi^2(t)$ satisfies (3.12) with $\lambda_0 = 0$ and $\Delta\psi(T(x^0)) = 0$. Therefore, $\Delta\psi(t) \equiv 0$. The first assertion of the lemma is proved.

We show that $-\Psi(x^0) = \operatorname{grad}\omega(x^0)$. Consider Problem 7 with $T_0 > T(x^0)$ and with the initial conditions $x(0) = x^0$ replaced by the condition $x(0) \in l$, where l is some hyperplane that supports the level surface of $\omega(x)$ at the point x^0. Since $\omega(x)$ is strictly convex, the trajectory $x(t, x^0)$ is the unique solution of this problem for any hyperplane l. By the maximum principle, for each l there is a collection of conjugate variables $\psi_l(t)$ which, besides (3.10) and (3.11), satisfies the additional condition of transversality ($\psi_l(0) \perp l$) and is directed into the half-space containing the origin. It has just been shown that precisely one function $\psi(t) = \Psi(x(t, x^0))$ satisfies (3.10) and (3.11). Thus, for any support hyperplane l we have that $\psi_l(0) = \Psi(x^0)$, and hence $\Psi(x^0) \perp l$. If $\Psi(x^0) \neq 0$, this means that the function $\omega(x)$ is differentiable at x^0, and the vector $-\Psi(x^0)$ is proportional, with coefficient $\lambda(x^0)$, to the vector grad $\omega(x^0)$.

If $\Psi(x^0) = 0$, then $0 \in \partial\omega(x^0)$, because the subdifferential of a convex function is upper semicontinuous with respect to inclusion ([48], Part V, Corollary 24.5.1). But ω is strictly convex, and its unique minimum is attained at the origin. Consequently, the condition $0 \in \partial\omega(x^0)$ is necessary and sufficient for a minimum and is equivalent to the condition that $x^0 = 0$. Thus, $\Psi(x^0) \neq 0$ for $x^0 \neq 0$, and at each point $x^0 \neq 0$ there is a unique support hyperplane of the level surface of $\omega(x)$. Consequently, $\omega \in C^1(\mathbf{R}^n \backslash \{0\})$ ([48], Part V, Corollary 25.5.1).

In this case the function ω must satisfy the Bellman equation:

$$-\langle \operatorname{grad}\omega(x), Ax + b\hat{u}\rangle + f(x) = 0.$$

To see this it suffices to differentiate the identity

$$\omega(x^0) = \omega(x(t, x^0)) + \int_0^t f(x(s, x^0))\, ds$$

with respect to t at $t = 0$.

Since the function $\Psi(x) = -\lambda(x)\operatorname{grad}\omega(x)$ also satisfies the condition

$$H(\Psi(x), x, \hat{u}) = 0,$$

this implies that $\lambda(x) = 1$ and $\Psi(x) = -\operatorname{grad}\omega(x)$ if $f(x) \neq 0$. At the points of the surface L the equality $\lambda(x) = 1$ holds by continuity.

It remains to prove the smoothness of $\omega(x)$ at the origin. Each optimal trajectory $x(t, x^0)$ hits the origin in a finite time $T(x^0)$, and after this time $x \equiv 0$; therefore, $\Psi(0) = 0$. The value $\Psi(x^0)$ is obtained by integrating the system (3.10), (3.11) on the interval $(0, T(x^0))$. Since $T(x^0)$ is continuous (Corollary 3), it follows from the linearity of (3.10) in ψ that $\lim_{x\to 0}\Psi(x) = 0$. Thus, Ψ is continuous everywhere in $\mathbf{R}^n$. Consequently, $\omega \in C^1(\mathbf{R}^n)$. The lemma is proved.

The set of points at which $\langle\Psi(x), b\rangle = 0$ is called the switching surface Π in Problem 6.

Lemma 17. *Suppose that Problem 6 admits the symmetry group $\mathfrak{G}$. Then the switching surface Π in Problem 6 is homeomorphic to $\mathbf{R}^{n-1}$.*

Proof. Consider the set of points $x \in \omega_C$, $C \neq 0$, at which $\operatorname{grad}\omega(x) \perp b$. Denote this set by Π_C. It is not hard to see that at points $x \in \Pi_C$ the vector b belongs to the tangent plane $T_x\omega_C$: $b \in T_x\omega_C$, and, conversely, if $b \in T_x\omega_C$, then $\operatorname{grad}\omega(x) \perp b$, i.e., $x \in \Pi_C$. Since ω is strictly convex, projection of Π_C on the hyperplane $\langle x, b\rangle = 0$ in the direction of b is a continuous one-to-one mapping. Under this projection the image of Π_C is the boundary of the image of the whole surface ω_C. Consequently, $\Pi_C \cong S^{n-2}$. By the definition of the switching surface,

$$\Pi = \bigcup_{\lambda>0} g_\lambda \Pi_C \cong \mathbf{R}^{n-1}.$$

The switching surface Π breaks up $\mathbf{R}^n$ into two subspaces Π^+ and Π^-, in one of which (Π^+) we have $\langle\Psi(x), b\rangle > 0$, and hence $\hat{u} = 1$, while in the other (Π^-) we have $\langle\Psi(x), b\rangle < 0$ and $\hat{u} = -1$. The lemma is proved.

We give an example of a Problem 6 admitting a one-parameter group of diffeomorphisms.

Problem 8.

$$\int_0^\infty \sum_{i=1}^n C_i|x_i|^{\sigma n/(n+1-i)}\,dt \to \inf,$$

$$\dot{x}_i = x_{i+1} \quad (i = 1, \dots, n-1), \qquad \dot{x}_n = u, \qquad x(0) = x^0,$$
$$u \in [-1, 1], \qquad C_i \geq 0, \quad i = 1, \dots, n, \qquad \sigma \geq 2.$$

The action of the group $\mathfrak{G}$ is given by the mapping

$$g_\lambda\colon \mathbf{R}^n \to \mathbf{R}^n, \quad g_\lambda(x) = (\lambda^n x_1, \lambda^{n-1}x_2, \dots, \lambda x_n),$$
$$\alpha = 1, \qquad \beta = \sigma n.$$

It follows from Lemma 17 that the switching surface Π in Problem 8 is homeomorphic to $\mathbf{R}^{n-1}$, and it follows from Lemma 16 that the Bellman function of Problem 8 is continuously differentiable.

3. The three-dimensional Fuller problem

The main result in this chapter is a construction of a complete qualitative picture of optimal synthesis in a special example of Problem 8: for $n = 3$, $C_1 = 1/2$, $C_3 = 0$, $\sigma = 2$, and sufficiently small $C_2 = \varepsilon/3$.

Problem 9.

$$J(x(\cdot)) = \int_0^\infty \frac{1}{2}x_1^2 + \frac{\varepsilon}{3}|x_2|^3\,dt \to \inf,$$
$$\dot{x}_1 = x_2, \qquad \dot{x}_2 = x_3, \qquad \dot{x}_3 = u, \tag{3.13}$$
$$x(0) = x^0, \qquad u \in [-1, 1].$$

The action of the symmetry group in Problem 9 is given by the mapping

$$g_\lambda\colon \mathbf{R}^3 \to \mathbf{R}^3, \quad g_\lambda(x) = (\lambda^3 x_1, \lambda^2 x_2, \lambda x_3).$$

We find the singular solutions of Problem 9. Consider the system of equations of the Pontryagin maximum principle in Problem 9:

$$\dot\psi_3 = -\psi_2, \qquad \dot\psi_2 = -\psi_1 + \varepsilon x_2^2 \operatorname{sgn} x_2, \qquad \dot\psi_1 = x_1,$$
$$\dot{x}_1 = x_2, \qquad \dot{x}_2 = x_3, \qquad \dot{x}_3 = \operatorname{sgn}\psi_3. \tag{3.14}$$

For brevity let $z = (\psi_3, \psi_2, \psi_1, x_1, x_2, x_3)$, i.e., $z_1 = \psi_3$, $z_2 = \psi_2, \dots, z_6 = x_3$. Denote by $z(t, z^0)$ the solution of (3.14) with initial conditions $z(0, z^0) = z^0$. Let

$$S = \{z \mid \psi_3 = 0\}, \qquad H_0 = \{z \mid \psi_1 x_2 + \psi_2 x_3 + |\psi_3| - \tfrac{1}{2}x_1^2 - \tfrac{1}{3}\varepsilon|x_2|^3 = 0\}.$$

We show that on the surface H_0 the system (3.14) does not have singular solutions.

Lemma 18. *For sufficiently small* $\varepsilon \geq 0$ *there exists for any trajectory* $z(t, z^0)$ *with initial conditions* $z^0 \in S \cap H_0 \backslash \{0\}$ *a* $t^0(z^0) > 0$ *such that* $z(t, z^0) \cap S = z^0$ *for* $t \in (-t^0, t^0)$.

Proof. Let us consider the set of points of S such that the following relations hold simultaneously:

$$\psi_1 x_2 + \psi_2 x_3 + |\psi_3| - \tfrac{1}{2}x_1^2 - \tfrac{1}{3}\varepsilon|x_2|^3 = 0, \qquad \psi_3 = 0,$$
$$d\psi_3/dt = -\psi_2 = 0, \qquad d^2\psi_3/dt^2 = \psi_1 - \varepsilon x_2^2 \operatorname{sgn} x_2 = 0,$$
$$d^3\psi_3/dt^3 = x_1 - 2\varepsilon|x_2|x_3 = 0,$$
$$d^4\psi_3/dt^4 = x_2 - 2\varepsilon x_3^2 \operatorname{sgn} x_2 - 2\varepsilon|x_2|u = 0.$$

(If at least one of the relations does not hold, then the property formulated in the lemma is obviously satisfied for the trajectory $z(t, z^0)$.) Successively, we get

$$\psi_3 = \psi_2 = 0, \qquad \psi_1 = \varepsilon x_2^2 \operatorname{sgn} x_2, \qquad x_1 = 2\varepsilon|x_2|x_3.$$

If $x_2 = 0$, then $x_1 = \psi_1 = 0$; then it follows from (3.13) that $x_3 = 0$. If $x_2 \neq 0$, then we get

$$u = \frac{1}{2\varepsilon}\operatorname{sgn} x_2 - \frac{x_3}{x_2}.$$

Substituting this into the equation for the surface H_0, we have $|x_2| = 3\varepsilon x_3^2$, and so $u = (1/6\varepsilon)\operatorname{sgn} x_2$. If $\varepsilon < 1/6$, then $|u| > 1$. The lemma is proved.

We describe the scheme of construction of an optimal synthesis in Problem 8. The system (3.14) has symmetry group $\widetilde{\mathfrak{G}}$—the lifting of the group $\mathfrak{G}$ to $T^*\mathbf{R}^3$:

$$\tilde{g}_\lambda \colon \mathbf{R}^6 \to \mathbf{R}^6, \quad \tilde{g}_\lambda(z) = (\lambda^6\psi_3, \lambda^5\psi_2, \lambda^4\psi_1, \lambda^3 x_1, \lambda^2 x_2, \lambda x_3).$$

It can be verified immediately that

$$z(t, \tilde{g}_\lambda(z^0)) = \tilde{g}_\lambda(z(\lambda^{-1}t, z^0)). \tag{3.15}$$

We consider the Poincaré first-return mapping $\Phi^\varepsilon \colon S \to S$, which assigns to a point $z \in S$ the point where the trajectory of (3.14) starting out from z first intersects the surface S. It is easy to see that the action of $\widetilde{\mathfrak{G}}$ on S commutes with the mapping Φ^ε. The quotient mapping $\Phi^\varepsilon/\widetilde{\mathfrak{G}}$ associating with an orbit of the action of $\widetilde{\mathfrak{G}}$ on S its image under the mapping Φ^ε is thereby well defined.

For Problem 9 with $\varepsilon = 0$ Fuller and Grensted [2] found a family of self-similar solutions, or, in their terminology, constant ratio (CR) trajectories (trajectories with constant ratio of the switching times). Let $x(t, x^0)$ be an optimal trajectory in Problem 9. Let x^1 and x^2 be two successive switching points on the trajectory $x(t, x^0)$. Denote by j the mapping $j(x) = -x$.

Definition 3. The trajectory $x(t, x^0)$ is called a *CR trajectory* if

$$x^2 = g_\lambda(j(x^1))$$

for some constant $\lambda = \lambda(x^0)$.

Suppose that the point x^0 lies on some CR trajectory with constant $\lambda = \lambda(x^0)$. Then the optimal trajectories of Problem 9 with initial conditions belonging to the orbit

$$\operatorname{Orb} x^0 = \bigcup_{\lambda > 0} g_\lambda(x^0)$$

of the point are clearly CR trajectories with the same constant λ. Let x^1 and x^2 be two successive control switching points on the CR trajectory $x(t, x^0)$. By the definition of a CR trajectory,

$$x(t, x^2) = j(g_{\lambda(x^0)}(x(t, x^1))).$$

The image $z(t) = (\Psi(x(t, x^0)), x(t, x^0))$ of a CR trajectory in the extended phase space will also be called a CR trajectory.

It is shown in [2] that the CR trajectories in Problem 9 with $\varepsilon = 0$ fill a two-dimensional surface in the extended phase space. Denote this surface by P^0. We show that the switching curve ρ_0 on P^0 is a fixed point of the mapping $\Phi^\varepsilon/\widetilde{\mathfrak{G}}$. In a neighborhood of this point $\Phi^\varepsilon/\widetilde{\mathfrak{G}}$ is a diffeomorphism and has a saddle structure with a single eigenvalue less than 1 in modulus, and with the remaining eigenvalues greater than 1 in modulus. By the Grobman-Hartman theorem [42], $\Phi^\varepsilon/\widetilde{\mathfrak{G}}$ is locally topologically equivalent to its linearization at the fixed point and has a one-dimensional $\Phi^\varepsilon/\widetilde{\mathfrak{G}}$-invariant smooth stable manifold. This invariant manifold determines the part of the switching surface in Problem 9 for $\varepsilon = 0$. "Stretching" it by using iterations of the mapping $(\Phi^\varepsilon)^{-1}$ enables us in finitely many iterations to get the whole switching surface in the Fuller Problem 9 with $\varepsilon = 0$. It will then be proved that for sufficiently small $\varepsilon > 0$ the structure of the solution in Problem 9 does not change.

The qualitative behavior of the optimal trajectories in Problem 9 is described by the next theorem.

Theorem 5. *For sufficiently small $\varepsilon \geq 0$ the following statements are valid for Problem 9:*

1. *At the points of the curve*

$$\rho = \{x | x_1 = \tfrac{1}{6}x_3^3, \ x_2 = -\tfrac{1}{2}x_3^2 \operatorname{sgn} x_3\}$$

the motion with control $u = -\operatorname{sgn} x_3$ is optimal, and with this control the trajectory of the system (3.13) hits the origin without switchings. For any $x^0 \in \mathbf{R}^3 \backslash \rho$ the optimal trajectory $x(t, x^0)$ hits the origin in finite time with infinitely many control switchings.

2. *The CR trajectories in Problem 9 fill a two-dimensional piecewise smooth surface P^ε homeomorphic to $\mathbf{R}^2$, with switching curve*

$$\rho_\varepsilon = \{x | x_1 = a(\varepsilon)x_3^3, \ x_2 = b(\varepsilon)x_3^2 \operatorname{sgn} x_3\}$$

and control $u = \operatorname{sgn}(-x_2 + b(\varepsilon)x_3^2 \operatorname{sgn} x_3)$. The optimal trajectories $x(t, x^0)$ with $x^0 \notin P^\varepsilon$ have at most finitely many switchings in reverse time.

3. *The switching surface Π^ε of all optimal trajectories is homeomorphic to $\mathbf{R}^2$ and is piecewise smooth.*

The proof of the theorem breaks up into several steps.

Proposition 1. *For sufficiently small $\varepsilon \geq 0$ the CR trajectories in Problem 9 have one and the same ratio of switching times and fill a two-dimensional piecewise smooth surface homeomorphic to $\mathbf{R}^2$.*

Proof. Let

$$\operatorname{Orb} z = \bigcup_{\lambda > 0} \tilde{g}_\lambda(z)$$

be an orbit of the action of the group $\widetilde{\mathfrak{G}}$. Consider the CR trajectory $z(t, z)$ with initial conditions $z \in O_1 = \operatorname{Orb} z^0$, where $z^0 \in S$. Then the trajectory $z(t, z^0)$ with constant control equal to 1 or -1 hits the orbit $O_2 = jO_1$ symmetric to O_1 with respect to the origin in $\mathbf{R}^6$ by the next switching time. Suppose for definiteness that the trajectories of (3.14) leave O_1 with the control $u = 1$. Since the system (3.14) is homogeneous with respect to the action of $\widetilde{\mathfrak{G}}$, the ratio of the coordinates of the successive switching points on $z(t, z)$ is constant for all $z \in O_1$. It follows from symmetry considerations that the trajectories $z(t, z)$ with initial conditions $z \in O_2$ hit the orbit O_1 by the next switching time by moving with the control $u = -1$. It is not hard to see that the time intervals between successive switchings on CR trajectories are terms of a geometric progression. If the ratio of the progression is less than 1, then the trajectory hits the origin in finite time with infinitely many switchings. By the sufficiency theorem ([35], §6.1, Theorem 5), the trajectories of (3.14) hitting the origin in finite time are optimal in Problem 9.

We choose a point $z^0 \in O_1$ such that the first switching on the trajectory $z(t, z^0)$ takes place at time $t = 1$. The orbit O_1 is completely determined by the parameter

$$\rho = -z_6(1, z^0)/z_6^0,$$

which can be used to express the coordinates of z^0. For brevity let $\alpha_+^2 = \alpha^2 \operatorname{sgn} \alpha$. Integrating (3.14), we get

(3.16)

$$\begin{aligned}
-\rho x_3 &= x_3 + 1, \\
-\rho^2 x_2 &= x_2 + x_3 + \tfrac{1}{2}, \\
-\rho^3 x_1 &= x_1 + x_2 + \tfrac{1}{2}x_3 + \tfrac{1}{6}, \\
-\rho^4 \psi_1 &= \psi_1 + x_1 + \tfrac{1}{2}x_2 + \tfrac{1}{6}x_3 + \tfrac{1}{24}, \\
-\rho^5 \psi_2 &= \psi_2 - \psi_1 - \tfrac{1}{2}x_1 - \tfrac{1}{6}x_2 - \tfrac{1}{24}x_3 - \tfrac{1}{120} + \varepsilon \int_0^1 (x_2 + x_3 s + \tfrac{1}{2}s^2)_+^2 \, ds, \\
0 &= -\psi_2 + \tfrac{1}{2}\psi_1 + \tfrac{1}{6}x_1 + \tfrac{1}{24}x_2 + \tfrac{1}{120}x_3 + \tfrac{1}{720} - \varepsilon \int_0^1 (1 - s)(x_2 + x_3 s + \tfrac{1}{2}s^2)_+^2 \, ds.
\end{aligned}$$

Let

$$\begin{aligned}
\chi(t, z^0) = &- \psi_2 t + \tfrac{1}{2}\psi_1 t^2 + \tfrac{1}{6}x_1 t^3 + \tfrac{1}{24}x_2 t^4 + \tfrac{1}{120}x_3 t^5 \\
&+ \tfrac{1}{720}t^6 - \varepsilon \int_0^t (t - s)(x_2 + x_3 s + \tfrac{1}{2}s^2)_+^2 \, ds.
\end{aligned}$$

For the control on the trajectory $z(t, z^0)$ with initial conditions z^0 satisfying (3.16) to equal $\operatorname{sgn} z_1$ it is necessary and sufficient that, first, there be no roots of the equation

$$\chi(t, z^0) = 0$$

on $(0, 1)$, and, second,

$$\lim_{t \to +0} \operatorname{sgn} \chi(t, z^0) = 1. \tag{3.17}$$

Lemma 19. *Let* $\varepsilon = 0$. *In this case if* (3.17) *holds, then* $z_2^0 \le 0$. *If* $z_i^0 \ne 0$ *for* $i = 3, \dots, 6$, *then the condition* $z_2^0 < 0$ *is sufficient for the absence of roots of the equation* $\chi(t, z^0) = 0$ *on* $(0, 1)$.

Proof. By the Budan-Fourier theorem ([49], Chapter 9, §41), the number of zeros of a polynomial $P(t)$ of degree n on an interval (α, β), $\alpha < \beta$, is bounded above by the difference $S(\alpha) - S(\beta)$, where $S(t)$ is the number of sign changes in the sequence $P(t), P'(t), \dots, P^{(n)}(t)$.

For $\varepsilon = 0$ the function $\chi(t, z^0)$ is a polynomial. Since the orbits O_1 and O_2 are centrally symmetric,

$$\operatorname{sgn} \chi^{(k)}(1, z^0) = -\operatorname{sgn} \chi^{(k)}(0, z^0), \qquad k = 0, 1, \dots, 5.$$

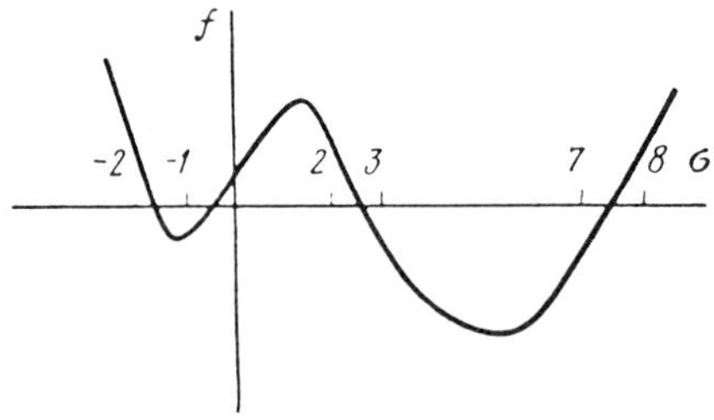

FIGURE 10

Therefore, $S(0) = S(1)$. The lemma is proved.

We rewrite (3.16) in the form

$$\begin{aligned} x_3 &= -1/(1+\rho), \\ x_2 &= (1-\rho)/(2(1+\rho)(1+\rho^2)), \\ x_1 &= (-\rho^3+2\rho^2+2\rho-1)/(6(1+\rho)(1+\rho^2)(1+\rho^3)), \\ \psi_1 &= (-\rho^6+3\rho^5+5\rho^4-5\rho^2-3\rho+1)/(24(1+\rho)\cdots(1+\rho^4)), \\ \psi_2 &= (\rho^{10}-4\rho^9-9\rho^8-3\rho^7+12\rho^6 \\ &\quad +22\rho^5+12\rho^4-3\rho^3-9\rho^2-4\rho+1)/(120(1+\rho)\cdots(1+\rho^5)) \\ &\quad -\frac{\varepsilon}{1+\rho^5}\int_0^1\left(\frac{1-\rho}{2(1+\rho)(1+\rho^2)}-\frac{s}{1+\rho}+\frac{s^2}{2}\right)_+^2 ds. \end{aligned} \tag{3.18a}$$

Here $\rho = \rho(\varepsilon) \in (0, 1)$ is a root of the equation

(3.18b)

$$\begin{aligned} &(\rho+1)^2(1-\rho)(\rho^2+1)(\rho^2+\rho+1)(\rho^8-7\rho^7-2\rho^6+8\rho^5+17\rho^4+8\rho^3-2\rho^2-7\rho+1) \\ &+720\varepsilon(1+\rho)(1+\rho^2)(1+\rho^3)(1+\rho^4) \\ &\times\int_0^1\left(\frac{1-\rho}{2(1+\rho)(1+\rho^2)}-\frac{s}{1+\rho}+\frac{s^2}{2}\right)_+^2 ds \\ &-720\varepsilon(1+\rho)(1+\rho^2)(1+\rho^3)(1+\rho^4)(1+\rho^5) \\ &\times\int_0^1(1-s)\left(\frac{1-\rho}{2(1+\rho)(1+\rho^2)}-\frac{s}{1+\rho}+\frac{s^2}{2}\right)_+^2 ds = 0. \end{aligned}$$

Suppose first that $\varepsilon = 0$. Setting $\sigma = \rho + 1/\rho$, we get that

$$(\rho+1)^2(1-\rho)(\rho^2+1)(\rho^2+\rho+1)f(\sigma) = 0,$$

where $f(\sigma) = \sigma^4 - 7\sigma^3 - 6\sigma^2 + 29\sigma + 23$.

Corresponding to the value $\rho = 1$ is $\psi_2^0 = \frac{1}{240} > 0$, and this does not satisfy the condition (3.17). The polynomial $f(\sigma)$ has four real roots, of which only two are greater than 2: $\sigma_0 = 2.3126\ldots$ and $\sigma_1 = 7.2131\ldots$ (Figure 10).

For σ_0 we get $\rho_0 = 0.5757\ldots$, and $\rho_1 = 0.1414\ldots$ for σ_1. Corresponding to the value ρ_1 is $\psi_2^0 = -\chi'(0, z^0) > 0$, and this does not satisfy (3.17). For $\rho = \rho_0$ a direct computation shows that $\chi'(0, z^0) > 0$ and $\chi^{(k)}(0, z^0) \neq 0$ for $k = 2, \ldots, 6$. By Lemma 19, this is a sufficient condition for the existence of a CR trajectory.

It follows from the implicit function theorem that for sufficiently small $\varepsilon > 0$ any solution of (3.18b) that is close to 1 and ρ_1, respectively, determines a solution of (3.18a) for which $\chi'(0, z^0) < 0$. This does not satisfy (3.17). Since for $\varepsilon = 0$ equation (3.18b) has a simple root ρ_0, for any sufficiently small $\varepsilon \geq 0$ there is a root $\rho(\varepsilon)$ of (3.18b) depending continuously on ε and satisfying $\rho(0) = \rho_0$, together with both the condition (3.17) and the condition that there be no roots of the equation $\chi(t, z^0) = 0$ on $(0, 1)$. This value of $\rho(\varepsilon)$ thereby determines a unique one-parameter family of CR trajectories in Problem 9. The switching curve of this family is an orbit of the action of the symmetry group, depends continuously on ε, and has the form $\rho_\varepsilon = \{x | x_1 = a(\varepsilon)x_3^3, x_2 = b(\varepsilon)x_3^2 \operatorname{sgn} x_3\}$. Proposition 1 is proved.

Denote by $t_\varepsilon(z^0)$ the minimal positive root of the equation $z_1(t, z^0) = 0$ for $z^0 \in S$. By the definition of the first-return mapping Φ^ε we have $\Phi^\varepsilon(z^0) = z(t_\varepsilon(z^0), z^0)$. It follows from (3.15) that $\Phi^\varepsilon \tilde{g}_\lambda = \tilde{g}_\lambda \Phi^\varepsilon$ for any $\lambda > 0$. Therefore, the corresponding quotient mapping $\Phi^\varepsilon/\widetilde{\mathfrak{G}}$ is defined. For example, this mapping $\Phi^\varepsilon/\widetilde{\mathfrak{G}}$ can be given explicitly as follows. Consider the surface $L = \{z \in S | t_0(z) = 1\}$. Let $\Phi^\varepsilon/\widetilde{\mathfrak{G}} = \Phi_L^\varepsilon = j\tilde{g}_{1/t}\Phi^\varepsilon$, $\Phi_L^\varepsilon \colon L \to L$, where $t = t(z^0) = t_0(\Phi^\varepsilon(z^0))$ and $j(z) = -z$. The point $M^\varepsilon = L \cap \rho_\varepsilon \cap \{\psi_2 < 0\}$ is a fixed point of the mapping Φ_L^ε.

Lemma 20. *The mapping Φ_L^ε is a local C^∞-diffeomorphism in some neighborhood of the point M^ε and has a saddle structure in it.*

Proof. In a small neighborhood of M^ε the surface L is given by

$$-\psi_2 + \tfrac{1}{2}\psi_1 + \tfrac{1}{6}x_1 + \tfrac{1}{24}x_2 + \tfrac{1}{120}x_3 + \tfrac{1}{720} = 0.$$

We choose the values ψ_1, x_1, x_2, x_3 as coordinates on L. In these coordinates the mapping can be written in the form

$$\begin{aligned} \psi_1 &\mapsto -\sigma^{-4}(\psi_1 + x_1\tau + \tfrac{1}{2}x_2\tau^2 + \tfrac{1}{6}x_3\tau^3 + \tfrac{1}{24}\tau^4), \\ x_1 &\mapsto -\sigma^{-3}(x_1 + x_2\tau + \tfrac{1}{2}x_3\tau^2 + \tfrac{1}{6}\tau^3), \\ x_2 &\mapsto -\sigma^{-2}(x_2 + x_3\tau + \tfrac{1}{2}\tau^2), \\ x_3 &\mapsto -\sigma^{-1}(x_3 + \tau), \end{aligned} \tag{3.19a}$$

where $\tau = \tau^\varepsilon(\psi_1, x_1, x_2, x_3)$ is the first positive root of the equation

$$\begin{aligned} 0 = &-(\tfrac{1}{2}\psi_1 + \tfrac{1}{6}x_1 + \tfrac{1}{24}x_2 + \tfrac{1}{120}x_3 + \tfrac{1}{720})\tau \\ &+ \tfrac{1}{2}\psi_1\tau^2 + \tfrac{1}{6}x_1\tau^3 + \tfrac{1}{24}x_2\tau^4 + \tfrac{1}{120}x_3\tau^5 \\ &+ \tfrac{1}{720}\tau^6 - \varepsilon\int_0^\tau (\tau - s)(x_2 + x_3 s + \tfrac{1}{2}s^2)_+^2\, ds, \end{aligned} \tag{3.19b}$$

and $\sigma = \sigma^\varepsilon(\psi_1, x_1, x_2, x_3)$ is the first positive root of the equation (3.19c)

$$\begin{aligned} 0 = &-\Big(\tfrac{1}{2}\psi_1 + \tfrac{1}{6}x_1 + \tfrac{1}{24}x_2 + \tfrac{1}{120}x_3 + \tfrac{1}{720} - \psi_1\tau - \tfrac{1}{2}x_1\tau^2 - \tfrac{1}{6}x_2\tau^3 \\ &\qquad - \tfrac{1}{24}x_3\tau^4 - \tfrac{1}{120}\tau^5 + \varepsilon\int_0^\tau (x_2 + x_1 s + \tfrac{1}{2}s^2)_+^2\, ds\Big) \\ &+ \tfrac{1}{2}\sigma(\psi_1 + x_1\tau + \tfrac{1}{2}x_2\tau^2 + \tfrac{1}{6}x_3\tau^3 + \tfrac{1}{24}\tau^4) + \tfrac{1}{6}\sigma^2(x_1 + x_2\tau + \tfrac{1}{2}x_3\tau^2 + \tfrac{1}{6}\tau^3) \\ &+ \tfrac{1}{24}\sigma^3(x_2 + x_3\tau + \tfrac{1}{2}\tau^2) + \tfrac{1}{120}\sigma^4(x_3 + \tau) - \tfrac{1}{720}\sigma^5. \end{aligned}$$

For $\varepsilon = 0$ the coordinates of the point M^0 are given by (3.18), with $\rho = \rho(0) = \rho_0$; moreover, $\tau^0 \equiv 1$ and $\sigma^0(M^0) = \rho_0$. The derivative $\mathscr{D}\Phi_L^0(M^0)$ has the form

$$\mathscr{D}\varphi_L^0(M^0) = \begin{pmatrix} -\rho_0^{-4}(1, 1, \tfrac{1}{2}, \tfrac{1}{6}) + 4\rho_0^{-5}(\psi_1^0 + x_1^0 + \tfrac{1}{2}x_2^0 + \tfrac{1}{6}x_3^0 + \tfrac{1}{24})\nabla\sigma^0 \\ -\rho_0^{-3}(0, 1, 1, \tfrac{1}{2}) + 3\rho_0^{-4}(x_1^0 + x_2^0 + \tfrac{1}{2}x_3^0 + \tfrac{1}{6})\nabla\sigma^0 \\ -\rho_0^{-2}(0, 0, 1, 1) + 2\rho_0^{-3}(x_2^0 + x_3^0 + \tfrac{1}{2})\nabla\sigma^0 \\ -\rho_0^{-1}(0, 0, 0, 1) + \rho_0^{-2}(x_3^0 + 1)\nabla\sigma^0 \end{pmatrix},$$

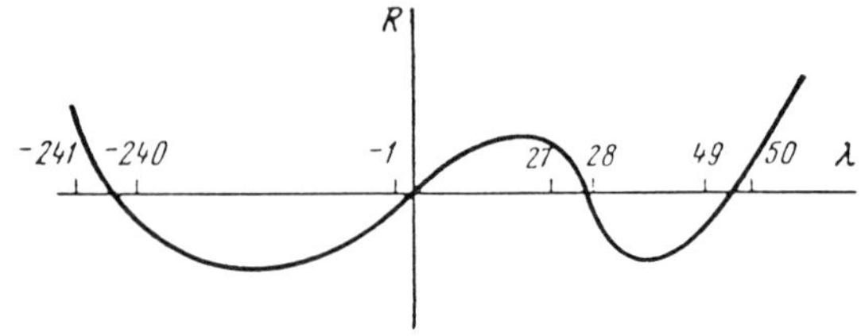

FIGURE 11

where

$$\nabla\sigma^0 = \operatorname{grad}\sigma^0(M^0) = -\left[\tfrac{1}{2}(\psi_1^0 + x_1^0 + \tfrac{1}{2}x_2^0 + \tfrac{1}{6}x_3^0 + \tfrac{1}{24}) + \tfrac{1}{3}\rho_0(x_1^0 + x_2^0 + \tfrac{1}{2}x_3^0 + \tfrac{1}{6}) + \tfrac{1}{8}\rho_0(x_2^0 + x_3^0 + \tfrac{1}{2}) + \tfrac{1}{30}\rho_0^3(x_3^0 + 1) - \tfrac{1}{144}\rho_0^4\right]^{-1}$$
$$\times\left(\tfrac{1}{2} + \tfrac{1}{2}\rho_0;\ \tfrac{1}{3} + \tfrac{1}{2}\rho_0 + \tfrac{1}{6}\rho_0^2;\ \tfrac{1}{8} + \tfrac{1}{4}\rho_0 + \tfrac{1}{6}\rho_0^2 + \tfrac{1}{24}\rho_0^3;\ \tfrac{1}{30} + \tfrac{1}{12}\rho_0 + \tfrac{1}{12}\rho_0^2 + \tfrac{1}{24}\rho_0^3 + \tfrac{1}{120}\rho_0^4\right).$$

A direct computation gives us that the eigenvalues of this matrix are the roots of a polynomial $R(\lambda)$ with coefficients dependent on ρ_0. We write out the coefficients, using the fact that ρ_0 is a root of (3.18b):

$$\begin{aligned}R(\lambda) = {} & \lambda^4(4\rho_0^7 - 29\rho_0^6 - \rho_0^5 + 35\rho_0^4 + 49\rho_0^3 + 31\rho_0^2 - \rho_0 + 10)\\ & + \lambda^3\rho_0^2(19\rho_0^7 - 57\rho_0^6 - 28\rho_0^5 + 87\rho_0^4 + 244\rho_0^3 + 213\rho_0^2 + 107\rho_0 - 18)\\ & + \lambda^2\rho_0^5(-3440\rho_0^7 - 723\rho_0^6 + 4233\rho_0^5 + 7937\rho_0^4\\ & \qquad + 3475\rho_0^3 - 1233\rho_0^2 - 3117\rho_0 - 452)\\ & + \lambda\rho_0^9(5239\rho_0^7 + 1403\rho_0^6 - 5953\rho_0^5 - 11683\rho_0^4\\ & \qquad - 5291\rho_0^3 + 1643\rho_0^2 + 4562\rho_0 - 658)\\ & + \rho_0^{14}(-71\rho_0^7 + 11\rho_0^6 + 129\rho_0^5 + 205\rho_0^4 + 79\rho_0^3 - 49\rho_0^2 - 66\rho_0 + 10).\end{aligned}$$

Computing the coefficients approximately, we get

$$R(\lambda) = 0.00001\lambda^4 + 0.0164\lambda^3 - 1.7017\lambda^2 + 31.8479\lambda + 6.8288.$$

The graph of $R(\lambda)$ is pictured in Figure 11. The polynomial $R(\lambda)$ has four real roots, of which exactly one, -0.2120, is in $[-1, 1]$. Since the mapping Φ_L^ε depends smoothly on ε as a parameter, the mapping Φ_L^ε has the same saddle structure in a neighborhood of its fixed point M^ε as the mapping Φ_L^0 has at M^0 (the eigenvalues of the derivative $\mathscr{D}\Phi_L^\varepsilon(M^\varepsilon)$ are close to the corresponding eigenvalues of $\mathscr{D}\Phi_L^0(M^0)$).

Thus, in a neighborhood of M^ε the mapping Φ_L^ε is a C^1-smooth diffeomorphism. The function $x_2^2 \operatorname{sgn} x_2$ on the right-hand side of (3.19) is C^1-smooth everywhere and C^∞-smooth at points for which $x_2 \neq 0$. It is not hard to see, however, that if the trajectory $z(t, z^0)$ with $z^0 \in S$ intersects the surface $x_2 = 0$ transversally (at a nonzero angle and with nonzero phase velocity), then it follows from the C^1-smoothness of Φ^ε at z^0 that Φ^ε is even C^∞-smooth at z^0. The solutions of (3.14) intersect the surface $x_2 = 0$, with tangency at the points at which $x_3 = 0$ simultaneously. Solving the system of equations

$$x_3^0 + t = 0, \qquad x_2^0 + x_3^0 t + \tfrac{1}{2}t^2 = 0$$

with respect to t, we get that the trajectory of (3.14) with initial conditions $z^0 \in S$ intersect the surface $x_2 = x_3 = 0$ if $x_2^0 = \frac{1}{2}(x_3^0)^2$. A direct computation gives us that the CR trajectories in Problem 9 with $\varepsilon = 0$ (and hence with an arbitrary sufficiently

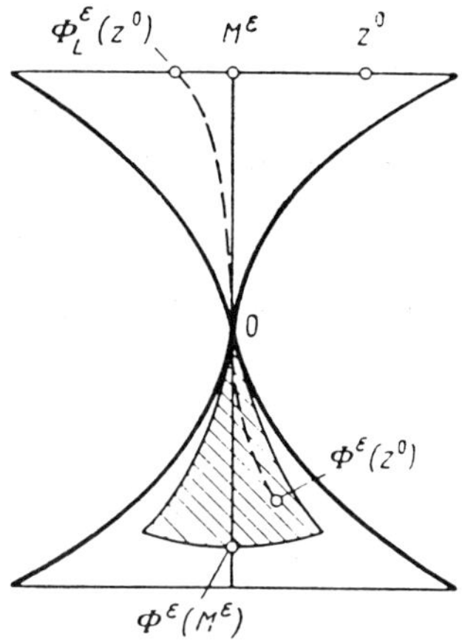

FIGURE 12

small $\varepsilon > 0$) do not satisfy this relation. Thus, the mapping Φ_L^ε is C^∞-smooth in a neighborhood of M^ε on L. The lemma is proved.

By the theorem on the stable manifold of a diffeomorphism ([42], §5.2, the Hadamard-Perron theorem), the unique eigenvalue $\lambda(\varepsilon)$ of the derivative $\mathscr{D}\Phi_L^\varepsilon(M^\varepsilon)$ less than 1 in modulus determines a Φ_L^ε-invariant C^∞-smooth curve $\gamma^\varepsilon \subset L$.

Let $\Gamma_0^\varepsilon = \operatorname{Orb} \gamma^\varepsilon$ and $\Gamma_n^\varepsilon = \Phi^\varepsilon \Gamma_{n-1}^\varepsilon$, $n = 1, 2, \dots$.

Lemma 21. 1) $\Gamma_{n+2}^\varepsilon \subset \Gamma_n^\varepsilon$.

2) *A trajectory* $z(t, z^0)$ *with* $z^0 \in \Gamma_0^\varepsilon$ *hits the origin in finite time.*

Proof. We consider a neighborhood U of the point M^ε on γ^ε and an arbitrary point $z^0 \in U$. By the definition of the mapping Φ_L^ε,

$$j\Phi_L^\varepsilon(z^0) = \tilde{g}_{\sigma^{-1}(z^0)}(\Phi^\varepsilon(z^0)),$$

where the function $\sigma(z^0)$ is defined by (3.19c). Let $\Delta = \operatorname{Orb} U$. The tangent vector to γ^ε at M^ε is an eigenvector of $\mathscr{D}\Phi_L^\varepsilon(M^\varepsilon)$ corresponding to the eigenvalue $\lambda(\varepsilon) \in (-1, 0)$. Therefore, by choosing U to be sufficiently small, we have $j\Phi^\varepsilon(\Delta) \subset \Delta$.

The functions $\sigma^\varepsilon(z^0)$ and $\tau^\varepsilon(z^0)$ defined in (3.19) are continuous in a neighborhood of M^ε, and $\sigma(M^\varepsilon) = \rho(\varepsilon) \in (0, 1)$. Choosing U so that $\sigma^\varepsilon|_U \in (0, \sigma_0)$, $0 < \sigma_0 < 1$, we get

$$j\Phi^\varepsilon(\Delta) \subset \tilde{g}_{\sigma_0^{-1}}(\Delta) \tag{3.20}$$

(see Figure 12).

Let $\tau(z^0)$ be the time along the trajectory $z(t, z^0)$ from the point $z^0 \in \Delta$ to the point of first switching of the control. Since the system (3.14) is homogeneous with respect to the action of $\widetilde{\mathfrak{G}}$ (see (3.15)), we have $\tau(\tilde{g}_\lambda(z^0)) = \lambda\tau(z^0)$. Then it follows from (3.20) that the time intervals between successive switchings on a trajectory $z(t, z^0)$ with $z^0 \in \Gamma_0^\varepsilon$ can be estimated from above by the terms of a decreasing geometric progression with ratio less than σ_0. Therefore, the trajectories $z(t, z^0)$ with $z^0 \in \Gamma_0^\varepsilon$ hit the origin in finite time. As mentioned earlier, this implies the optimality of $z(t, z^0)$ in Problem 9. The lemma is proved.

Denote by $v(x) = (x_2, x_3, \hat{u}(x))$ the section of the space $\Gamma(T\mathbf{R}^3)$ of all bounded vector fields corresponding to the optimal synthesis $\hat{u}(x)$ in Problem 9. In view of Lemma 14, under the mapping

$$x(t) \to g_\lambda x(t/\lambda)$$

the optimal trajectories in Problem 9 are carried into optimal trajectories with different initial conditions. Therefore, a system with discontinuous right-hand side $v(x)$ can be factored by the action of $\mathfrak{G}$ in $\mathbf{R}^3$ (up to trajectorywise equivalence). As the quotient space $\mathbf{R}^3/\mathfrak{G}$ we take the manifold S^2 of the form

$$S^2 = \{x | x_1^4 + x_2^6 + x_3^{12} = 1\}.$$

Consider the mapping $P\colon \mathbf{R}^3\backslash\{0\} \to S^2$ given by

$$P(x) = g_{\Lambda(x)}(x) = (x_1\Lambda^3(x), x_2\Lambda^2(x), x_3\Lambda(x)),$$

where $\Lambda(x) = (x_1^4 + x_2^6 + x_3^{12})^{-1/12}$. We define the vector field v^* on S^2 by

$$v^*(s) = \Lambda^{-1}(x)\mathscr{D}P|_x v(x),$$

where $s = P(x) \in S^2$, and $\mathscr{D}P|_x$ is the differential of P at the point x:

$$\mathscr{D}P|_x = \begin{pmatrix} \Lambda^3(x) & 0 & 0 \\ 0 & \Lambda^2(x) & 0 \\ 0 & 0 & \Lambda(x) \end{pmatrix} + \begin{pmatrix} -\frac{1}{4}\Lambda^{15}(x)(4x_1^4, & 6x_1x_2^5, & 12x_1x_3^{11}) \\ -\frac{1}{6}\Lambda^{14}(x)(4x_1^3x_2, & 6x_2^6, & 12x_2x_3^{11}) \\ -\frac{1}{12}\Lambda^{13}(x)(4x_1^3x_3, & 6x_2x_3, & 12x_3^{12}) \end{pmatrix}.$$

For any fixed $\lambda_0 > 0$ we have $\Lambda(g_{\lambda_0}(x)) = \lambda_0^{-1}\Lambda(x)$; therefore, this gives us

$$\begin{aligned}
&\mathscr{D}P|_{g_{\lambda_0}(x)}v(g_{\lambda_0}(x)) \\
&= \left[\begin{pmatrix} \lambda_0^{-3}\Lambda^3(x) & 0 & 0 \\ 0 & \lambda_0^{-2}\Lambda^2(x) & 0 \\ 0 & 0 & \lambda_0^{-1}\Lambda(x) \end{pmatrix}\right. \\
&\quad \left. + \begin{pmatrix} -\frac{1}{4}\Lambda^{15}(x)(4\lambda_0^{-3}x_1^4, & 6\lambda_0^{-2}x_1x_2^5, & 12\lambda_0^{-1}x_1x_3^{11}) \\ -\frac{1}{6}\Lambda^{14}(x)(4\lambda_0^{-3}x_1^3x_2, & 6\lambda_0^{-2}x_2^6, & 12\lambda_0^{-1}x_2x_3^{11}) \\ -\frac{1}{12}\Lambda^{13}(x)(4\lambda_0^{-3}x_1^3x_3, & 6\lambda_0^{-2}x_2^5x_3, & 12\lambda_0^{-1}x_3^{12}) \end{pmatrix}\right] \begin{pmatrix} \lambda_0^2x_2 \\ \lambda_0x_3 \\ \hat{u} \end{pmatrix} \\
&= \lambda_0^{-1}\mathscr{D}P|_x v(x).
\end{aligned}$$

We have shown that the definition of $v^*(s)$ does not depend on the choice of the point x, $s = P(x)$.

Let $T = \Pi^\varepsilon \cap S^2$ be the intersection with S^2 of the switching surface Π^ε of the optimal trajectories in Problem 9. The field v^* is discontinuous at the points of the curve T and analytic at the remaining points of S^2.

Let S_0^2 be either of the two "hemispheres" separated out on S^2 by the curve $\mathrm{Э} = S^2 \cap P^\varepsilon$ ($\mathrm{Э}$ is the trace of CR trajectories on S^2). Let $Q = \rho \cap S_0^2$ be the point of intersection with S_0^2 of the curve

$$\rho = \{x | x_1 = \tfrac{1}{6}x_3^3, \ x_2 = -\tfrac{1}{2}x_3^2 \operatorname{sgn} x_3\}.$$

It is not hard to see that Q is the only fixed point of the vector field v^* on S_0^2, since among the trajectories of (3.14) only the trajectory

$$\psi_1 = -\tfrac{1}{24}x_3^4 \operatorname{sgn} x_3, \quad \psi_2 = -\tfrac{1}{120}x_3^5, \quad \psi_3 = -\tfrac{1}{720}x_3^6 \operatorname{sgn} x_3,$$
$$x_1 = \tfrac{1}{6}x_3^3, \quad x_2 = -\tfrac{1}{2}x_3^2 \operatorname{sgn} x_3$$

is invariant under the action of $\widetilde{\mathfrak{G}}$.

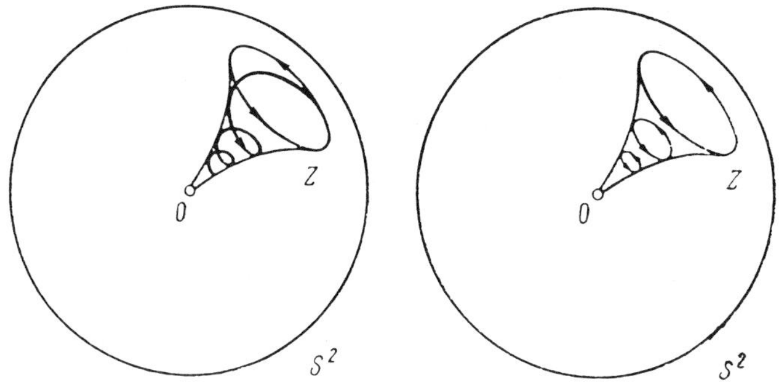

FIGURE 13

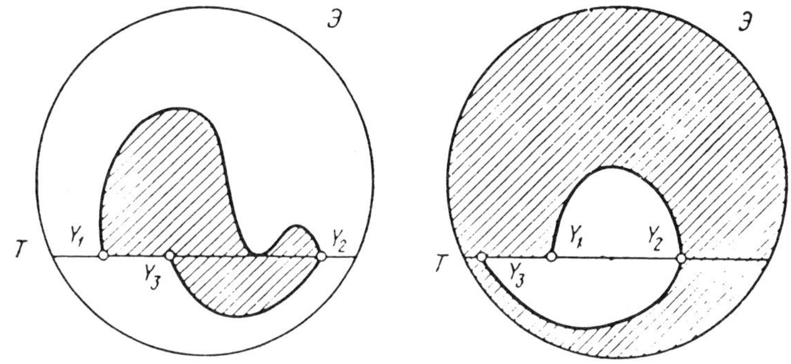

FIGURE 14

Proposition 2. *For sufficiently small* $\varepsilon \geq 0$ *the curve* Э *is the unique closed cycle of the field* v^* *on* S^2.

Proof. We show first that S^2 does not contain cycles corresponding to trajectories of (3.13) with constant control $u = 1$ or $u = -1$. Each such cycle Z of v^* would have to determine infinitely many trajectories of (3.13) with constant control $u = \pm 1$ that belong to Orb Z (Figure 13). This contradicts the fact that the origin is a nonsingular point of the system (3.13) with constant control $u \neq 0$.

Assume now that there is a closed trajectory Z of the field v^* on S^2 that does not coincide with Э. We represent S_0^2 as a two-dimensional disk with boundary Э. The segment T divides S_0^2 into two parts: S^+ and S^-, in which v^* corresponds to the controls $u = 1$ and $u = -1$. By Lemma 18, the intersection of the trajectories of v^* with the curve T is a discrete set. Therefore, it follows from the theorem on the uniqueness of solutions of differential equations that through each point of S^2 there is a unique trajectory of v^* both for $t \geq 0$ and for $t \leq 0$.

We show that the cycle Z of v^* must be a "two-component" cycle, i.e., must consist of two arcs $Z^+ \subset S^+$ and $Z^- \subset S^-$. Let $Y_1, Y_2, Y_3 \in T$ be three successive control switching points on the trajectory Z. We claim that $Y_1 = Y_3$.

Assume that $Y_1 \neq Y_3$. Then either $Y_3 \in (Y_1, Y_2) \subset T$ or $Y_3 \in T \backslash [Y_1, Y_2]$. In both cases the trajectory Z must lie in an open domain not containing Y_1 after passing through Y_3 (this domain is shaded in Figure 14), and hence Z is not a cycle.

This implies that there are two orbits $\Sigma_i \subset S$, $i = 0, 1$, of the action of $\tilde{\mathfrak{G}}$ in $\mathbf{R}^6$ interchanged by the first-return mapping Φ^ε (the trajectories $z(t, z^0)$ with $z^0 \in \Sigma_i$ hit the orbit Σ_{1-i} by the next switching time). Following Ryan [7], we call such trajectories *bi-constant ratio* (BCR) *trajectories* of Problem 9. We show that for sufficiently small $\varepsilon \geq 0$ the BCR trajectories of Problem 9 coincide with the CR trajectories found earlier. We remark that for $\varepsilon = 1/2$ a family of BCR trajectories in Problem 9 was found in [7] that does not coincide with the family of CR trajectories.

Suppose that the trajectory $z(t, z^0)$ starting from a point $z^0 \in \Sigma_0$ with the control $u = 1$ during the time $t = 1$ hits the curve Σ_1, and then with the control $u = -1$ during a time τ hits some point $z^1 = \tilde{g}_\mu z^0 \in \Sigma_0$. For such a trajectory to hit the origin it is necessary that $0 < \mu < 1$. This implies that $0 > x_3(1+\tau) > x_3(0)$, i.e., $x_3 + 1 - \tau > x_3$ and $0 < \tau < 1$. Integrating (3.14), we find that the coordinates of z^0 and the parameters μ and τ are connected by the relations

$$\begin{aligned}
\mu x_3 &= x_3 + 1 - \tau,\\
\mu^2 x_2 &= x_2 + x_3 + \tfrac{1}{2} + (x_3+1)\tau - \tfrac{1}{2}\tau^2,\\
\mu^3 x_1 &= x_1 + x_2 + \tfrac{1}{2}x_3 + \tfrac{1}{6} + (x_2 + x_3 + \tfrac{1}{2})\tau + \tfrac{1}{2}(x_3+1)\tau^2 - \tfrac{1}{6}\tau^3,\\
\mu^4 \psi_1 &= \psi_1 + x_1 + \tfrac{1}{2}x_2 + \tfrac{1}{6}x_3 + \tfrac{1}{24} + (x_1 + x_2 + \tfrac{1}{2}x_3 + \tfrac{1}{6})\tau\\
&\quad + \tfrac{1}{2}(x_2 + x_3 + \tfrac{1}{2})\tau^2 + \tfrac{1}{6}(x_3+1)\tau^3 - \tfrac{1}{24}\tau^4,\\
\mu^5 \psi_2 &= \psi_2 - \psi_1 - \tfrac{1}{2}x_1 - \tfrac{1}{6}x_2 - \tfrac{1}{24}x_3 - \tfrac{1}{120}\\
&\quad - (\psi_1 + x_1 + \tfrac{1}{2}x_2 + \tfrac{1}{6}x_3 + \tfrac{1}{24})\tau - \tfrac{1}{2}(x_1 + x_2 + \tfrac{1}{2}x_3 + \tfrac{1}{2}x_3 + \tfrac{1}{6})\tau^2\\
&\quad - \tfrac{1}{6}(x_2 + x_3 + \tfrac{1}{2})\tau^3 - \tfrac{1}{24}(x_3+1)\tau^4 + \tfrac{1}{120}\tau^5 + O(\varepsilon),\\
0 &= -(\psi_2 - \psi_1 - \tfrac{1}{2}x_1 - \tfrac{1}{6}x_2 - \tfrac{1}{24}x_3 - \tfrac{1}{120})\\
&\quad + \tfrac{1}{2}(\psi_1 + x_1 + \tfrac{1}{2}x_2 + \tfrac{1}{6}x_3 + \tfrac{1}{24})\tau + \tfrac{1}{6}(x_1 + x_2 + \tfrac{1}{2}x_3 + \tfrac{1}{6})\tau^2\\
&\quad + \tfrac{1}{24}(x_2 + x_3 + \tfrac{1}{2})\tau^3 + \tfrac{1}{120}(x_3+1)\tau^4 - \tfrac{1}{720}\tau^5 + O(\varepsilon),\\
0 &= -\psi_2 + \tfrac{1}{2}\psi_1 + \tfrac{1}{6}x_1 + \tfrac{1}{24}x_2 + \tfrac{1}{120}x_3 + \tfrac{1}{720} + O(\varepsilon).
\end{aligned} \tag{3.21}$$

For the maximum condition $u = \operatorname{sgn}\psi_3$ to be satisfied on such a trajectory it is necessary that

1) $d\psi_3/dt|_{t=+0} \geq 0$, whence $\psi_2 \leq 0$; and
2) $d\psi_3/dt|_{t=1+0} \leq 0$, whence

$$\psi_2 - \psi_1 - \tfrac{1}{2}x_1 - \tfrac{1}{6}x_2 - \tfrac{1}{24}x_3 - \tfrac{1}{120} + O(\varepsilon) \geq 0.$$

For $\mu = 1$ and $\tau = 1$ it follows from (3.21) that $\psi_2 = \frac{1}{240} + O(\varepsilon)$. This does not satisfy the condition $\psi_2 \leq 0$. Thus, for all sufficiently small $\varepsilon \geq 0$ there is a $\delta \geq 0$ such that the solution of (3.21) with $(\mu, \tau) \in K_\delta = \{(\mu, \tau) | |\mu - 1| < \delta, |\tau - 1| < \delta\}$ does not determine a BCR trajectory of (3.14).

We show that for $(\mu, \tau) \in (0, 1) \times (0, 1) \setminus K_\delta$ there is a unique solution of (3.21) satisfying the maximum condition, and this solution coincides with one of the CR trajectories of Problem 9 found earlier.

We transform the system (3.21) as follows. From the last equation we express ψ_2 in terms of the remaining coordinates of the point z^0:

$$\psi_2 = \tfrac{1}{2}\psi_1 + \tfrac{1}{6}x_1 + \tfrac{1}{24}x_2 + \tfrac{1}{120}x_3 + \tfrac{1}{720} + O(\varepsilon).$$

Let us substitute this in the remaining equations of (3.21). We get

$$x_3 = (1-\tau)/(\mu-1),$$
$$x_2 = (\mu(-\tau^2+2\tau+1)-\tau^2-2\tau+1)/(2(\mu^2-1)(\mu-1)),$$

$$x_1 = (\mu^3(-\tau^3+3\tau^2+3\tau+1)+\mu^2(-2\tau^3-6\tau^2+2)$$
$$+\mu(-2\tau^3+6\tau+2)-\tau^3-3\tau^2-3\tau+1)/(6(\mu^3-1)(\mu^2-1)(\mu-1)),$$

$$\psi_1 = (\mu^6(-\tau^4+4\tau^3+6\tau^2+4\tau+1)+\mu^5(-3\tau^4-12\tau^3-6\tau^2+4\tau+3)$$
$$+\mu^4(-5\tau^4-4\tau^3+18\tau^2+20\tau+5)+\mu^3(-6\tau^4-8\tau^3+8\tau+6)$$
$$+\mu^2(-5\tau^4-20\tau^3-18\tau^2+4\tau+5)+\mu(-3\tau^4-4\tau^3+6\tau^2+12\tau+3)$$
$$-\tau^4-4\tau^3-6\tau^2-4\tau+1)/(24(\mu^4-1)(\mu^3-1)(\mu^2-1)(\mu-1)).$$

Substituting the relations found into the sixth and fifth equations of the system (3.21), we get equations connecting the parameters μ and τ:

$$\begin{aligned}
&\mu^{10}(-\tau^5+6\tau^4+15\tau^3+20\tau^2+15\tau+5)\\
&\quad+\mu^9(-5\tau^5-30\tau^4-45\tau^3-20\tau^2+21\tau+19)\\
&\quad+\mu^8(-14\tau^5-36\tau^4+30\tau^3+175\tau^2+165\tau+40)\\
&\quad+\mu^7(-29\tau^5-66\tau^4+35\tau^3+155\tau^2+144\tau+61)\\
&\quad+\mu^6(-49\tau^5-171\tau^4-260\tau^3-50\tau^2+159\tau+71)\\
&\quad+\mu^5(-66\tau^5-279\tau^4-215\tau^3+215\tau^2+279\tau+66)\\
&\quad+\mu^4(-71\tau^5-159\tau^4+50\tau^3+260\tau^2+171\tau+49)\\
&\quad+\mu^3(-61\tau^5-144\tau^4-155\tau^3-35\tau^2+66\tau+29)\\
&\quad+\mu^2(-40\tau^5-165\tau^4-175\tau^3 30\tau^2+36\tau+14)\\
&\quad+\mu(-19\tau^5-21\tau^4+20\tau^3+45\tau^2+30\tau+5)\\
&\quad-5\tau^5-15\tau^4-20\tau^3-15\tau^2-6\tau+1+O(\varepsilon)=0;
\end{aligned} \tag{3.22a}$$

$$\begin{aligned}
&\mu^{15}+\mu^{14}(-6\tau+8)+\mu^{13}(-15\tau^2+30\tau+14)\\
&\quad+\mu^{12}(-20\tau^3+45\tau^2+36\tau+29)\\
&\quad+\mu^{11}(-15\tau^4+20\tau^3-30\tau^2+66\tau+49)\\
&\quad+\mu^{10}(-5\tau^5-21\tau^4-175\tau^3-35\tau^2+171\tau+66)\\
&\quad+\mu^9(-19\tau^5-165\tau^4-155\tau^3+260\tau^2+279\tau+71)\\
&\quad+\mu^8(-40\tau^5-144\tau^4+50\tau^3+215\tau^2+159\tau+61)\\
&\quad+\mu^7(-61\tau^5-159\tau^4-215\tau^3-50\tau^2+144\tau+40)\\
&\quad+\mu^6(-71\tau^5-279\tau^4-260\tau^3+155\tau^2+165\tau+19)\\
&\quad+\mu^5(-66\tau^5-171\tau^4+35\tau^3+175\tau^2+21\tau+5)\\
&\quad+\mu^4(-49\tau^5-66\tau^4+30\tau^3-20\tau^2+15\tau)\\
&\quad+\mu^3(-29\tau^5-36\tau^4-45\tau^3+20\tau^2)\\
&\quad+\mu^2(-14\tau^5-30\tau^4+15\tau^3)\\
&\quad+\mu(-5\tau^5+6\tau^4)-\tau^5+O(\varepsilon)=0.
\end{aligned} \tag{3.22b}$$

We first consider the case $\varepsilon = 0$. After cancelling $\mu + 1$, we rewrite (3.22) in the form

$$Q(\tau, \mu) = 0, \tag{3.23a}$$

$$\widetilde{Q}(\tau, \mu) = -\mu^{14} Q(\tau/\mu, 1/\mu), \tag{3.23b}$$

where $Q(\tau, \mu) = \sum_{i=1}^{6} \langle \overline{q}_i, \overline{M} \rangle \tau^{6-i}$, with

$$\begin{aligned}
\overline{M} &= (\mu^9, \mu^8, \dots, \mu, 1);\\
\overline{q}_1 &= (-1, -4, -10, -19, -30, -36, -35, -26, -14, -5);\\
\overline{q}_2 &= (6, -36, 0, -66, -105, -174, 15, -159, -6, -15);\\
\overline{q}_3 &= (15, -60, 90, -55, -205, -10, 60, -215, 40, -20);\\
\overline{q}_4 &= (20, -40, 215, -60, 10, 205, 55, -90, 60, -15);\\
\overline{q}_5 &= (15, 6, 159, -15, 174, 105, 66, 0, 36, -6);\\
\overline{q}_6 &= (5, 14, 26, 35, 36, 30, 19, 10, 4, 1).
\end{aligned}$$

Multiplying the first equation by τ^5 and subtracting it from the second, we get

$$\begin{gathered}
\langle \overline{q}_1, \overline{M} \rangle (\mu^5 - \tau^{10}) + \langle \overline{q}_2, M \rangle \tau (\mu^4 - \tau^8) + \langle \overline{q}_3, \overline{M} \rangle \tau^2 (\mu^3 - \tau^6)\\
+ \langle \overline{q}_4, \overline{M} \rangle \tau^3 (\mu^2 - \tau^4) + \langle \overline{q}_5, \overline{M} \rangle \tau^4 (\mu - \tau^2) = 0.
\end{gathered} \tag{3.24}$$

The solutions of (3.22) with $\mu = \tau^2$ correspond to the CR trajectories of Problem 9 found earlier. In this case the switching curves Σ_0 and Σ_1 turn out to be centrally symmetric. To see this it suffices to show that the solutions of (3.23) with $\mu = \tau^2$ determine trajectories of (3.14) that carry the projection of the switching curve Σ_0 on the space of x into a centrally symmetric curve. It follows from (3.23) that

$$\begin{aligned}
x_3 &= -1/(1+\tau),\\
x_2 &= (1-\tau)/(2(1+\tau)(1+\tau^2)),\\
x_1 &= (-\tau^3 + 2\tau^2 + 2\tau - 1)/(6(1+\tau)(1+\tau^2)(1+\tau^3)).
\end{aligned}$$

These relations coincide with (3.18) for $\tau = \rho$. As a corollary we get that (3.23) has at most two solutions with $\mu = \tau^2 - (\rho_0^2, \rho_0)$ and (ρ_1^2, ρ_1). Of them, only the one pair (ρ_0^2, ρ_0) satisfies the condition $\psi_2 \le 0$. A direct computation gives us

$$\det \begin{pmatrix} \partial Q/\partial \tau & \partial Q/\partial \mu \\ \partial \widetilde{Q}/\partial \tau & \partial \widetilde{Q}/\partial \mu \end{pmatrix}\Bigg|_{(\rho_0^2, \rho_0)} \neq 0.$$

In view of the implicit function theorem, for any sufficiently small $\varepsilon \ge 0$ there is in a small neighborhood of the curve $\mu = \tau^2$ exactly one solution of (3.22) determining BCR trajectories of Problem 9. This solution is continuously dependent on ε and coincides with the family of CR trajectories of Problem 9 found earlier in Proposition 1.

We show that there are no other (i.e., with $\mu \neq \tau^2$) solutions of (3.23). Dividing (3.24) by $\mu - \tau^2$ and regrouping the terms, we get

$$A_1(\tau^8 + \mu^4) + A_2(\tau^7 + \mu^3\tau) + A_3(\tau^6 + \mu^2\tau^2) + A_4(\tau^5 + \mu\tau^3) + A_5\tau^4 = 0, \tag{3.25}$$

where

$$\begin{aligned} A_1 &= -\mu^9 - 4\mu^8 - 10\mu^7 - 19\mu^6 - 30\mu^5 - 36\mu^4 - 35\mu^3 - 26\mu^2 - 14\mu - 5, \\ A_2 &= 6\mu^9 - 36\mu^8 - 66\mu^6 - 105\mu^5 - 174\mu^4 + 15\mu^3 - 159\mu^2 - 6\mu - 15, \\ A_3 &= -\mu^{10} + 11\mu^9 - 70\mu^8 + 71\mu^7 - 85\mu^6 - 241\mu^5 - 45\mu^4 \\ &\quad + 34\mu^3 - 229\mu^2 + 35\mu - 20, \\ A_4 &= 6\mu^{10} - 16\mu^9 - 40\mu^8 + 149\mu^7 - 165\mu^6 - 164\mu^5 \\ &\quad + 220\mu^4 - 104\mu^3 - 96\mu^2 + 45\mu - 15, \\ A_5 &= (\mu - 1)^4(-\mu^7 + 7\mu^6 - 21\mu^5 - 53\mu^4 + 17\mu^3 + 39\mu^2 - 8\mu - 6). \end{aligned}$$

We claim that $A_i < 0$, $i = 1, \dots, 5$, for $\mu \in (0, 1)$. It is clear that $A_1 < 0$ and $A_2 < 0$. For A_3 and A_4 we get the following obvious estimate. Since $\mu^k < \mu^{k-1}$ for $\mu \in (0, 1)$, it follows that

$$11\mu^9 - 70\mu^8 < -59\mu^8, \qquad 71\mu^7 - 85\mu^6 < -14\mu^6, \qquad 34\mu^3 - 229\mu^2 < -195\mu^2.$$

Consequently, $A_3 < -195\mu^2 + 35\mu - 20 < 0$, because the discriminant of this quadratic trinomial is negative. Consider the function A_4. Since $149\mu^7 - 165\mu^6 < -16\mu^5$, if follows that

$$A_4 < (-164\mu^2 + 220\mu - 104)\mu^3 + (-96\mu^2 + 45\mu - 15) < 0,$$

because the discriminants of the polynomials in parentheses are negative.

Let us show that $A_5 < 0$ for $\mu \in (0, 1)$. We get

$$-\mu^7 + 7\mu^6 - 21\mu^5 - 53\mu^4 + 17\mu^3 + 39\mu^2 - 8\mu - 6 = -\mu^5(\mu - 1)(\mu - 6) + h(\mu),$$

where $h(\mu) = -15\mu^5 - 53\mu^4 + 17\mu^3 + 39\mu^2 - 8\mu - 6$. Let us show that $h(\mu) < 0$ for $\mu \in (0, 1)$. The extremum points of the polynomial $h_1(\mu) = -53\mu^3 + 17\mu^2 + 39\mu - 8$ are roots of the quadratic equation $-159\mu^2 + 34\mu + 39 = 0$ and are equal to $\mu_1 = -0.3997\dots$ and $\mu_2 = 0.6135\dots$. Consequently, for $\mu \in (0, 1)$ the maximal value of $h_1(\mu)$ is attained for $\mu = \mu_2$ and is equal to $h_1(\mu_2) = 10.08\dots < 10.1$. Accordingly, the estimate $h(\mu) < h_2(\mu) = -15\mu^5 + 10,1\mu - 6$ holds for $\mu \in (0, 1)$. The equation $h_2'(\mu) = 0$ gives the unique local extremum point of the polynomial $h_2(\mu)$ on the interval $(0, 1)$: $\mu_0 = (10.1 : 75)^{1/4} \approx 0.605$. Consequently, $h(\mu) < h_2(\mu_0) = (-10.1/5 + 10.1)\mu_0 - 6 < -1.1 < 0$. The inequality $h(\mu) < 0$ is thereby proved for $\mu \in (0, 1)$. At the same time, it has been proved that (3.25) does not have solutions in the square $(0, 1) \times (0, 1)$. It follows from the implicit function theorem that for sufficiently small $\varepsilon \geq 0$ there are no solutions of (3.22) determining BCR solutions different from the CR trajectories of Problem 9. It has thereby been shown that the vector field v^* does not have closed trajectories different from Э. Proposition 2 is proved.

The restriction of the mapping Φ^ε to the switching surface of the optimal trajectories in the extended phase space induces the first-return mapping $\Phi^*: T \to T$ along trajectories of the field v^*. Let

$$\pi(\psi, x) = x, \qquad \tau_0 = \pi\Gamma_0 \cap T, \qquad \tau_{-n-1} = (\Phi^*)^{-1}\tau_{-n},$$
$$\Gamma_{-n-1} = (\Phi^\varepsilon)^{-1}\Gamma_{-n}, \qquad n = 0, 1, 2, \dots.$$

We show that there is a finite n_0 such that $\bigcup_{n=1}^{n_0} \tau_{-n} = T$.

It follows from the theorem on smooth dependence of the solutions of differential equations on the initial conditions that the first-return mapping Φ^ε is continuously

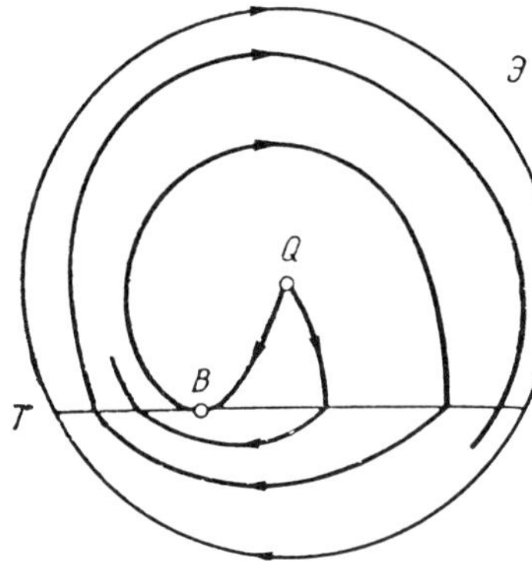

FIGURE 15

differentiable at a point $z_0 \in S$ if the trajectory $z(t, z^0)$ intersects the switching surface S at a nonzero angle and with nonzero phase velocity both at z^0 and at $\Phi^\varepsilon(z^0)$. Since $\dot{z}_1 = z_2$ for a trajectory of (3.14), the restriction $\Phi^\varepsilon|_{\Gamma_{-n+1}}$ is a diffeomorphism if the set $\Gamma^*_{-n} = \Gamma_{-n} \cap \{z|z_2 = 0\}$ is empty.

Assume that $\Gamma^*_{-n} = \varnothing$ for any n. It is asserted that the ω-limit set of the trajectories of the field v^* then gives a cycle different from Э. Take an arbitrary point $A_0 \in T$ and consider the sequence of inverse images of A_0 under the mapping Φ^*, i.e., the sequence $A_n = (\Phi^*)^{-1}A_{n-1}$, $n = 1, 2, \dots$. By assumption, for any n the restriction $(\Phi^*)^{-1}|_{[A_n, A_{n+2}]}$ is a continuous one-to-one mapping, and hence a homeomorphism. The sequences $\{A_{2k}\}_{k=1}^\infty$ and $\{A_{2k+1}\}_{k=1}^\infty$ of points are naturally ordered on the segment T, and thus have limit points A^+ and A^-, respectively. If $A^+ = A^- = A$, then the point $A \in T$ is a fixed point of v^* and hence must coincide with the point Q. However, Q does not belong to the switching surface, because $\Psi_3(Q) \neq 0$. Therefore, $A^+ \neq A^-$. In this case the trajectory of v^* leaving A^+ intersects T at A^-, and, conversely, the trajectory leaving A^- goes through A^+. This gives the required cycle.

It is thus proved that for some n_0 the set $\Gamma^*_{-n_0}$ is not empty. By Lemma 16, $\operatorname{grad}\omega(x^0) = (-\Psi_1(x^0), 0, 0)$ at the point $x^0 \in \pi\Gamma^*_{-n_0}$, and the plane $x_1 = x_1^0$ is tangent to the corresponding level surface of the Bellman function. Since ω is strictly convex, there are exactly two symmetric points with this property, and the set $\bigcup_{n=0}^\infty \Gamma^*_{-n}$ consists of exactly two centrally symmetric orbits of the action of $\widetilde{\mathfrak{G}}$. Consequently, on $T_0 = S_0^2 \cap T$ we have a unique point B such that $\Psi_2(B) = 0$. Since the function $\Psi(x) = -\operatorname{grad}\omega(x)$ is equal to zero only at the origin, $\Psi_1(B) \neq 0$. By Corollary 2, the trajectories $z(t, z^0)$ optimal in Problem 9 depend continuously on the initial data. Therefore, the mapping $(\Phi^*)^{-1}$ extends continuously to the closure of the interval Γ_{-n_0}, and $(\Phi^*)^{-1}\Gamma_{-n_0} = T \backslash \Gamma_{-n_0}$. Since the trajectories of the field v^* have the property of right- and left-sided uniqueness, the trajectories leaving the points of the interval $[B, \Phi^* B) \subset T$ in reverse time no longer intersect the switching surface T and tend with a constant control to the unique singular point Q of v^* on S_0^2. The phase portrait of the trajectories of v^* is given in Figure 15.

It remains to prove smoothness of the switching surface Π^ε. We remark that by construction the switching surface $\widetilde{\Pi}^\varepsilon = \pi^{-1}\Pi^\varepsilon$ of the optimal trajectories in Problem 9 in the extended phase space is of class C^1, with the possible exception of the points of the curve

$$\xi = \{(\psi, x)|x \in \operatorname{Orb}(B \cup j(B)), \ \psi = \Psi(x)\}.$$

Lemma 22. *The restriction of the projection π to the surface $\widetilde{\Pi}^\varepsilon$ is regular.*

Proof. Denote by $\mathscr{D}^+ \subset \mathbf{R}^3$ the domain in which the optimal control is equal to 1, and by $\mathscr{D}^- \subset \mathbf{R}^3$ the domain in which it is equal to -1. Let S^* stand for the set of points in $\mathscr{D}^+ \cup \mathscr{D}^-$ at which the projection π is not regular.

Since the trajectories of (3.14) intersect S transversally, $\pi^{-1}\mathscr{D}^+$ and $\pi^{-1}\mathscr{D}^-$ (the inverse images of $\mathscr{D}^+$ and $\mathscr{D}^-$ under the mapping π) are smooth manifolds. According to Sard's theorem, the measure of S^* is zero. Consequently, $\mathscr{D}^+\backslash S^*$ and $\mathscr{D}^-\backslash S^*$ are open dense subsets of $\mathscr{D}^+$ and $\mathscr{D}^-$. The mapping $\pi|_{\pi^{-1}(\mathbf{R}^3)}$ has a continuous inverse mapping $\Psi\colon \mathbf{R}^3 \to \mathbf{R}^6$, $x \mapsto (x, \Psi(x))$. Therefore, if $x^0 \in S^*$, then

$$\lim_{x\to x^0,\, x\in\mathscr{D}^\pm\backslash S^*} \|\Omega(x)\| = \infty,$$

where $\Omega(x) = \mathscr{D}\Psi(x)/\mathscr{D}(x)$, and $\|\Omega\|$ denotes the operator norm $\|\Omega\| = \sup_{\|y\|=1}\|\Omega y\|$. Thus, it suffices to show that the matrix $\Omega = \mathscr{D}^2\omega/\mathscr{D}x^2$ has bounded norm.

The eigenvector of the matrix $\mathscr{D}\Phi^\varepsilon_L(M^\varepsilon)$ tangent to the invariant curve γ^ε of Φ^ε has nonzero projection on the space of x. Therefore, in a small neighborhood of the curve ρ^ε (it can be assumed without loss of generality that this neighborhood coincides with Γ^ε_0) the projections of the trajectories of (3.14) on the space of x intersect the switching surface Π^ε transversally. By the theorem on the smooth dependence of the solutions of differential equations on the initial data, the vector-valued function $\Psi(x)$ is continuously differentiable at the points of $\mathscr{D}^+$ and $\mathscr{D}^-$ swept out by the trajectories of (3.13) with initial data belonging to Γ^ε_0 and with constant control equal to 1 at points of $\mathscr{D}^+$ and to -1 at points of $\mathscr{D}^-$.

For definiteness we consider the part of $\mathscr{D}^+$ swept out by the trajectories of (3.13) going in reverse time from points of Γ^ε_0 to points of the surface Γ^ε_{-1} with the control $u = 1$. In this case the function $\Psi(x)$ extends smoothly from points of $\mathscr{D}^+$ to Γ^ε_{-1}. If $\Gamma^\varepsilon_{-1} \cap \xi = \varnothing$, then the restriction $\pi|_{\Gamma^\varepsilon_{-1}}$ is regular, and the surface $\pi\Gamma^\varepsilon_{-1}$ is smooth. We show that $\Psi(x)$ also extends smoothly from the points of $\mathscr{D}^-$ to the surface Γ^ε_{-1}. Assume that there exists a point $x^0 \in \pi\Gamma^\varepsilon_{-1} \cap \operatorname{cl} S^*$. Then, by Sard's theorem, there is a point $x^1 \in \Pi^\varepsilon \cap \operatorname{cl}(\mathscr{D}^-\backslash S^*)$ arbitrarily close to x^0 such that

$$\|\Omega^-\| > \|\Omega^+\|,$$
$$\Omega^- = \lim_{x\to x^1,\, x\in\mathscr{D}^-\backslash S^*} \Omega(x), \qquad \Omega^+ = \lim_{x\to x^1,\, x\in\mathscr{D}^+\backslash S^*} \Omega(x).$$

Denote by $\tilde\omega$ a smooth extension of ω from $\mathscr{D}^+$ to $\mathscr{D}^-$ in a neighborhood of x^1 along trajectories of (3.13) with the control $u = 1$. It is easy to see that $\tilde\omega \geq \omega$. The function ω is strictly convex; therefore, the matrices Ω^+ and Ω^- are positive-definite. For the function $\Delta = \tilde\omega - \omega$ we get that $\Delta(x^1) = 0$ and $\operatorname{grad}\Delta(x^1) = 0$, while the matrix $(\mathscr{D}^2\Delta/\mathscr{D}x^2)(x^1) = \Omega^+ - \Omega^-$ has at least one negative eigenvalue. This contradicts the fact that $\Delta \geq 0$ in a neighborhood of x^1.

Thus, the restriction $\pi|_{\mathscr{D}^-}$ is also a regular mapping in a neighborhood of points of Γ^ε_{-1}. Repeating our arguments, we get that the mapping $\pi|_{\widetilde{\Pi^\varepsilon}\backslash\xi}$ is regular at points of Γ^ε_{-2}, Γ^ε_{-3}, and so on. Lemma 22 is thereby proved, and with it Theorem 5.

4. The four-dimensional Fuller problem

We present a computation of self-similar trajectories in the four-dimensional Fuller problem.

Problem 10.

$$\int_0^\infty x^2\,dt \to \inf,$$

$$x^{(\mathrm{IV})} = u, \qquad u \in [-1, 1],$$

$$x^{(k)}(0) = x_k^0, \qquad k = 0, 1, 2, 3.$$

The system of equations of the Pontryagin maximum principle can be written for Problem 10 in the form

$$\psi_4^{(8)} = -\operatorname{sgn}\psi_4 \tag{3.26a}$$

or, equivalently,

$$\begin{gathered} \dot\psi_4 = -\psi_3, \quad \dot\psi_3 = -\psi_2, \quad \dot\psi_1 = -\psi_1, \quad \dot\psi_1 = x_1, \\ \dot x_1 = x_2, \quad \dot x_2 = x_3, \quad \dot x_3 = x_4, \quad \dot x_4 = -\operatorname{sgn}\psi_4. \end{gathered} \tag{3.26b}$$

We look for a solution of (3.26) such that

$$\psi_4^{(l)}(t_k)/\psi_4^{(l)}(t_{k+1}) = \text{const}, \quad l = 1, \dots, 7, \qquad \psi_4(t_k) = 0,$$

where t_k denotes the kth control switching time on the corresponding trajectory, with t_0 taken equal to 0. The solutions of (3.26) with this property will be called the CR trajectories of Problem 10. The optimality of a CR trajectory in Problem 10 follows from the sufficiency theorem ([35], §6.1, Theorem 5). The CR trajectories of Problem 10 are determined by the following system of equations:

$$\begin{aligned} -\psi_3 + \frac{1}{2}\psi_2 - \frac{1}{3!}\psi_1 - \frac{1}{4!}x_1 - \frac{1}{5!}x_2 - \frac{1}{6!}x_3 - \frac{1}{7!}x_4 - \frac{1}{8!} &= 0, \\ -\mu^7\psi_3 = \psi_3 - \psi_2 + \frac{1}{2}\psi_1 + \frac{1}{3!}x_1 + \frac{1}{4!}x_2 + \frac{1}{5!}x_3 + \frac{1}{6!}x_4 + \frac{1}{7!}&, \\ -\mu^6\psi_2 = \psi_2 - \psi_1 - \frac{1}{2}x_1 - \frac{1}{3!}x_2 - \frac{1}{4!}x_3 - \frac{1}{5!}x_4 - \frac{1}{6!}&, \\ -\mu^5\psi_1 = \psi_1 + x_1 + \frac{1}{2}x_2 + \frac{1}{3!}x_3 + \frac{1}{4!}x_4 + \frac{1}{5!}&, \\ -\mu^4 x_1 = x_1 + x_2 + \frac{1}{2}x_3 + \frac{1}{3!}x_4 + \frac{1}{4!}&, \\ -\mu^3 x_2 = x_2 + x_3 + \frac{1}{2}x_4 + \frac{1}{3!}&, \\ -\mu^2 x_3 = x_3 + x_4 + \frac{1}{2}&, \\ -\mu x_4 = x_4 + 1&. \end{aligned} \tag{3.27}$$

The maximum condition $u = \operatorname{sgn}\psi_4$ holds on the corresponding trajectories of the system (3.26) if and only if: 1) the equation $\chi(t, \psi, x) = 0$, where

$$\chi(t, \psi, x) = -\psi_3 + \frac{1}{2}\psi_2 t - \frac{1}{3!}\psi_1 t^2 - \frac{1}{4!}x_1 t^3 - \frac{1}{5!}x_2 t^4 - \frac{1}{6!}x_2 t^5 - \frac{1}{7!}x_4 t^6 - \frac{1}{8!}t^7,$$

does not have roots in $(0, 1)$; and 2)

$$\operatorname{sgn}\chi(+0, \psi, x) = 1. \tag{3.28}$$

For these two conditions to hold it is necessary that $\psi_3 \le 0$, and sufficient that $\psi_3 < 0$, $\psi_k \ne 0$, $k = 1, 2$, and $x_k \ne 0$, $k = 1, \dots, 4$ (see Lemma 19).

From (3.27) we obtain

$$x_4 = -1/(1+\mu), \quad x_3 = (1-\mu)/(2(1+\mu)(1+\mu^2)),$$
$$x_2 = (-\mu^3 + 2\mu^2 + 2\mu - 1)/(6(1+\mu)(1+\mu^2)(1+\mu^3)),$$
$$x_1 = (-\mu^6 + 3\mu^5 + 5\mu^4 - 5\mu^2 - 3\mu + 1)/(24(1+\mu)(1+\mu^2)(1+\mu^3)(1+\mu^4)),$$
$$\psi_1 = (-\mu^{10} + 4\mu^9 + 9\mu^8 + 3\mu^7 - 12\mu^6 - 22\mu^5 - 12\mu^4 + 3\mu^3 + 9\mu^2 + 4\mu - 1)/(120(1+\mu)(1+\mu^2)\cdots(1+\mu^5)),$$
$$\psi_2 = (\mu^{15} - 5\mu^{14} - 14\mu^{13} - 9\mu^{12} + 21\mu^{11} + 61\mu^{10} + 70\mu^9 + 31\mu^8 - 31\mu^7 - 70\mu^6 - 61\mu^5 - 21\mu^4 + 9\mu^3 + 5\mu - 1)/(720(1+\mu)(1+\mu^2)\cdots(1+\mu^6)),$$
$$\psi_3 = (-\mu^{21} + 6\mu^{20} + 20\mu^{19} + 19\mu^{18} - 30\mu^{17} - 129\mu^{16} - 207\mu^{15} - 179\mu^{14} - 11\mu^{13} + 233\mu^{12} + 415\mu^{11} + 415\mu^{10} + 233\mu^9 - 11\mu^8 - 179\mu^7 - 207\mu^6 - 129\mu^5 - 30\mu^4 + 19\mu^3 + 20\mu^2 + 6\mu - 1)/(5040(1+\mu)(1+\mu^2)\cdots(1+\mu^7)).$$

Here $\mu \in (0, 1)$ is a root of the equation

$$\mu^{24} - 9\mu^{23} - 9\mu^{22} - 14\mu^{21} + 47\mu^{20} + 112\mu^{19} + 217\mu^{18} + 215\mu^{17} + 136\mu^{16} - 136\mu^{15} - 424\mu^{14} - 728\mu^{13} - 800\mu^{12} - 728\mu^{11} - 424\mu^{10} - 236\mu^9 + 136\mu^8 + 215\mu^7 + 217\mu^6 + 112\mu^5 + 47\mu^4 - 14\mu^3 - 9\mu^2 - 9\mu + 1 = 0.$$

The equation determining μ is a reciprocal equation. Setting $\sigma = \mu + 1/\mu$, we reduce its order:

$$\sigma^{11} - 9\sigma^{10} - 21\sigma^9 + 85\sigma^8 + 191\sigma^7 - 158\sigma^6 - 586\sigma^5 - 254\sigma^4 + 329\sigma^3 + 282\sigma^2 - 28\sigma + 56 = 0. \tag{3.29}$$

The equation (3.29) has exactly three roots greater than 2. Indeed, by the Budan-Fourier theorem (see Lemma 19), the number of zeros on $[2, \infty)$ can be estimated from above by the difference $S(2)-S(\infty) = 3$. It can be verified directly that there are three roots: $\sigma_1 = 2.0821\ldots$, $\sigma_2 = 2.6958\ldots$, and $\sigma_3 = 10.081\ldots$. Corresponding to the value σ_2 is $\mu_2 = 0.4445\ldots$, for which $\psi_3 > 0$. In this case the condition (3.28) does not hold. For σ_1 and σ_3 the values of μ are equal to $\mu_1 = 0.7402\ldots$ and $\mu_3 = 0.1001\ldots$. Then $\psi_3 < 0$, $\psi_1 \neq 0$, $\psi_2 \neq 0$, and $x_k \neq 0$, $k = 1, \ldots, 4$, i.e., the sufficient conditions for the existence of a CR trajectory are satisfied. In the 8-dimensional extended phase space of Problem 10 there are thus two families of self-similar solutions of the type of CR trajectories. Each of the two families fills a two-dimensional surface homeomorphic to $\mathbf{R}^2$. The projections of these surfaces on the space $\mathbf{R}^4$ are in general position, i.e., intersect in a point.

Chapter 4. Examples

1. Structural stability of a synthesis of optimal trajectories in the Fuller problem

We need the following uncomplicated generalization of Theorems 1–4. Let us consider the system (2.4) under the same assumptions about its right-hand side as in Chapter 2, except for the condition (2.5). Suppose now that $f_i(z, w, u) = \varepsilon\varphi_i(z, w, u)$, $i = 1, 2, 3, 4$, and the function φ_i satisfy the restrictions

$$\overline{\lim_{\kappa\to+0}} |\varphi_i(g_\kappa(z), w, u)|\kappa^{-(4-i)} < \infty. \tag{4.1}$$

Proposition 3. *For all sufficiently small values of the parameter* ε *the assertions of Theorems* 1–4 *remain true for the system* (2.4) *with right-hand side satisfying* (4.1).

Proof. Under the condition (4.1) the mapping $\widetilde{\Phi}^2$ constructed in Chapter 2 will depend continuously on ε as a parameter: $\widetilde{\Phi}^2 = F^\varepsilon$, and will be given by equations (2.11), with functions of the form $O(\varepsilon)$ added on the right-hand side. For $\varepsilon = 0$ the mapping F^0 has two fixed points with a neighborhood in which F^0 is a hyperbolic diffeomorphism, and so for all sufficiently small ε the mapping F^ε will also have two fixed points with coordinates continuously dependent on ε. In a neighborhood of them F^ε will have the same hyperbolic structure as F^0. The rest of the proof of Theorems 1–4 does not change.

As an example we consider the following problem.

Problem 11.

$$\begin{gathered}\int_0^T (x^2(1+\varepsilon f(x, y)) + \varepsilon y^4 g(x, y))\, dt \to \inf, \\ \dot{x} = y + \varepsilon(\varphi_1(x, y) + u\varphi_2(x, y)), \; \dot{y} = u + \varepsilon(\varphi_3(x, y) + u\varphi_4(x, y)), \\ u \in [A_0, A_1], \qquad A_0 < 0, \; A_1 > 0, \\ x(0) = x_0, \quad y(0) = y_0, \quad x(T) = y(T) = 0, \quad x_0^2 + y_0^2 < \delta^2, \\ f, g, \varphi_i \in C^2, \quad i = 1, 2, 3, 4, \qquad g \geq 0, \\ \varphi_1(0, 0) = \varphi_2(0, 0) = 0.\end{gathered}$$

Problem 11 is a perturbation of the Fuller-Marchal Problem 4. We show that for sufficiently small $\varepsilon \geq 0$ and $\delta > 0$ the optimal trajectories in Problem 11 are chattering trajectories.

Proposition 4. *For sufficiently small* $\varepsilon \geq 0$ *and* $\delta > 0$ *an optimal synthesis in Problem* 11 *is determined by the switching curve*

$$\Gamma = \begin{cases} x = \lambda_1(y)y^2, & y > 0, \\ x = \lambda_0(y)y^2, & y < 0, \end{cases}$$

where the functions $\lambda_0(\cdot)$ *and* $\lambda_1(\cdot)$ *are of class* C^1 *and satisfy the constraints* $\lambda_1 \in (1/(2A_0), 0)$ *and* $\lambda_0 \in (0, 1/(2A_1))$. *The optimal control* $\hat{u}(x, y)$ *is equal to* A_0 *to the left of and above the curve* Γ *and equal to* A_1 *to the right of and below* Γ.

Proof. We consider the system of equations of the Pontryagin maximum principle for Problem 11 with $\psi_0 = 1$:

$$\begin{aligned} \dot{\psi}_2 &= -\psi_1 + \varepsilon(x^2 f'_y + 4y^3 g + y^4 g'_y - \psi_1(\varphi'_{1y} + u\varphi'_{2y}) - \psi_2(\varphi'_{3y} + u\varphi'_{4y})), \\ \dot{\psi}_1 &= 2x + \varepsilon(2xf + x^2 f'_x + y^4 g'_x - \psi_1(\varphi'_{1x} + u\varphi'_{2x}) - \psi_2(\varphi'_{3x} + u\varphi'_{4x}), \\ \dot{x} &= y + \varepsilon(\varphi_1 + u\varphi_2), \qquad \dot{y} = u + \varepsilon(\varphi_3 + u\varphi_4), \\ u &= \tfrac{1}{2}(A_0 + A_1) + \tfrac{1}{2}\operatorname{sgn}\psi_2(A_0 - A_1). \end{aligned} \tag{4.2}$$

Let $z_1 = \psi_2$, $z_2 = -\psi_1$, $z_3 = -2x$, and $z_4 = -2y$. We must show that the right-hand side of (4.2) has the form (4.1). The one-parameter transformation group in (4.1) has the form

$$g_\kappa(\psi_2, \psi_1, x, y) = (\kappa^4\psi_2, \kappa^3\psi_1, \kappa^2 x, \kappa y)$$

for Problem 11.

We show, for example, that

$$\lim_{\kappa\to+0}\kappa^{-1}\varphi_1(\kappa^2x, \kappa y) < \infty.$$

By assumption, $\varphi_1(0, 0) = 0$. According to the mean-value theorem,

$$|\varphi_1(x, y)| < |\varphi'_{1x}(\Theta_1x, \Theta_2y)|\,|x| + |\varphi'_{1y}(\Theta_1x, \Theta_2y)|\,|y|,$$

where Θ_1 and Θ_2 are functions taking values in $[0, 1]$. For sufficiently small $\kappa > 0$ this implies that $|\varphi_1(\kappa^2x, \kappa y)| < O(1)\kappa$. The remaining relations for the functions on the right-hand sides of (4.2) are proved similarly. The assertion now follows from Theorems 1 and 3 of Chapter 2. The proposition is proved.

We remark that, since Problem 11 does not have a unique solution in the general case, other optimal trajectories can exist in principle, including nonchattering trajectories.

In order that the general problem of minimization of the integral

$$\int_0^T f(x, y)\,dt \to \inf$$

on the trajectories of a two-dimensional system

$$\dot{x} = \varphi_1 + u\varphi_2, \qquad \dot{y} = \varphi_3 + u\varphi_4$$

linearly dependent on a scalar control $u \in [-1, 1]$ and with corresponding boundary conditions have the same form as Problem 11, it is necessary to impose finitely many conditions of equality type on the integrand and the right-hand side of the system of differential relations. Namely, it is necessary that

$$\begin{aligned} f(0, 0) &= f'_x(0, 0) = f'_y(0, 0) = f''_{xy}(0, 0) = f''_{yy}(0, 0) \\ &= f'''_{xyy}(0, 0) = f'''_{yyy}(0, 0) = \varphi_1(0, 0) = \varphi_2(0, 0) = 0. \end{aligned}$$

Moreover, some conditions of inequality type must also be satisfied: $f''_{xx}(0, 0) > 0$ and $f^{\mathrm{IV}}_{yyyy}(0, 0) > 0$. Thus, the existence of chattering is a situation of finite codimension for the class of two-dimensional problems described.

2. Two model problems

We consider the following two model problems.

Problem 12 (the Fuller problem with a functional discontinued in time).

$$\begin{aligned} &\int_0^\infty e^{-\alpha t}x^2\,dt \to \inf, \qquad \alpha \geq 0, \\ &\dot{x} = y, \qquad \dot{y} = u, \qquad u \in [-1, 1], \\ &x(0) = x_0, \qquad y(0) = y_0. \end{aligned} \tag{4.3}$$

The proof of the existence and uniqueness of a solution of Problem 12 is completely analogous to that given in Lemmas 12 and 13.

Proposition 5. *Suppose that $\hat{x}(t)$, $\hat{y}(t)$, $\hat{u}(t)$ is a solution of Problem 12 with the initial conditions x^0, y^0. Then for arbitrary $\tau > 0$ the triple $\hat{x}(t+\tau)$, $\hat{y}(t+\tau)$, $\hat{u}(t+\tau)$ is a solution of Problem 12 with the initial conditions $\hat{x}(\tau)$, $\hat{y}(\tau)$.*

Proof. Assume that the trajectory $\hat{x}(t+\tau)$, $\hat{y}(t+\tau)$ is not optimal. Let $x^*(t)$, $y^*(t)$, $u^*(t)$ be a solution of Problem 12 with the initial conditions $\hat{x}(\tau)$, $\hat{y}(\tau)$. We consider the following trajectory of the system (4.3):

$$(x(t), y(t)) = \begin{cases} (\hat{x}(t), \hat{y}(t)), & t \in [0, \tau], \\ (x^*(t-\tau), y^*(t-\tau)), & t \in (\tau, \infty). \end{cases}$$

We get

$$\int_0^\infty e^{-\alpha t} x^2\,dt = \int_0^\tau e^{-\alpha t}\hat{x}^2\,dt + \int_\tau^\infty e^{-\alpha t}(x^*(t-\tau))^2\,dt;$$

$$\int_\tau^\infty e^{-\alpha t}(x^*(t-\tau))^2\,dt = e^{-\alpha\tau}\int_0^\infty e^{-\alpha s}(x^*(s))^2\,ds < e^{-\alpha\tau}\int_0^\infty e^{-\alpha s}\hat{x}^2(s+\tau)\,ds$$

$$= \int_\tau^\infty e^{-\alpha t}\hat{x}^2(t)\,dt.$$

Hence

$$\int_0^\infty e^{-\alpha t}x^2(t)\,dt < \int_0^\infty e^{-\alpha t}\hat{x}^2(t)\,dt,$$

which contradicts the assumption that the trajectory $\hat{x}(t), \hat{y}(t)$ is optimal. The proposition is proved.

Corollary 4. *Suppose that the function $u(t, x, y)$ determines an optimal synthesis in Problem 12. Then $u(t, x, y)$ can be chosen to be independent of t: $u(t, x, y) = \hat{u}(x, y)$.*

We show that for an arbitrary $\alpha \geq 0$ an optimal synthesis in Problem 12 can be given in a sufficiently small neighborhood of the origin in the form

$$u(x, y) = -\operatorname{sgn}(x + \lambda_\alpha(y)y^2 \operatorname{sgn} y),$$

where $\lambda_\alpha(\cdot) \in C^1$, $0 < \lambda_\alpha < 0.5$.

It follows from this estimate that the optimal trajectories of Problem 12 hit the origin with infinitely many switchings in a finite time.

We introduce a new phase coordinate φ corresponding to the time: $\dot{\varphi} = 1$ and $\varphi(0) = 0$. The Hamiltonian of Problem 12 can be written in the form

$$H = \psi_\varphi + \psi_x y + \psi_y u - \psi_0 e^{-\alpha\varphi}x^2, \qquad u = \operatorname{sgn}\psi_y.$$

It is easy to show that $\psi_0 \neq 0$. Let $\psi_0 = 1/2$. In this case the system of equations of the Pontryagin maximum principle for Problem 12 has the form

$$\begin{gathered}\dot{\psi}_\varphi = -\alpha e^{-\alpha\varphi}x^2, \qquad \dot{\psi}_x = xe^{-\alpha\varphi}, \qquad \dot{\psi}_y = -\psi_x, \\ \dot{x} = y, \quad \dot{y} = u, \quad \dot{\varphi} = 1, \quad u = \operatorname{sgn}\psi_y.\end{gathered} \tag{4.4}$$

If $\psi_y \equiv 0$, then $\psi_x \equiv 0$ and $x = y \equiv 0$. Consequently, the singular solutions of (4.4) fill a two-dimensional surface in the extended phase space. Setting

$$\begin{gathered}z_1 = \psi_y, \qquad z_2 = -\psi_x, \qquad z_3 = -xe^{-\alpha\varphi}, \\ z_4 = -ye^{-\alpha\varphi}, \qquad w_1 = \psi_\varphi, \qquad w_2 = \varphi,\end{gathered}$$

we rewrite (4.4) in the equivalent form

$$\begin{gathered}\dot{z}_1 = z_2, \quad \dot{z}_2 = z_3, \quad \dot{z}_3 = z_4 + \alpha z_3, \quad \dot{z}_4 = -e^{-\alpha w_2}u - z_4 - \alpha^2 z_3, \\ \dot{w}_1 = -\alpha e^{\alpha w_2}z_3^2, \qquad \dot{w}_2 = 1.\end{gathered} \tag{4.5}$$

Since the right-hand side of (4.5) does not depend on w_1, the fifth equation in (4.5) can be excluded from consideration. It follows from Theorem 1 that in the five-dimensional space of (z, w_2) there is a fiber bundle over the line $S_0 = \{(z, w) | z = 0\}$ with piecewise smooth two-dimensional fibers, each filled by chattering trajectories. By Theorem 3, the projections of these trajectories on the space of x, y, φ are

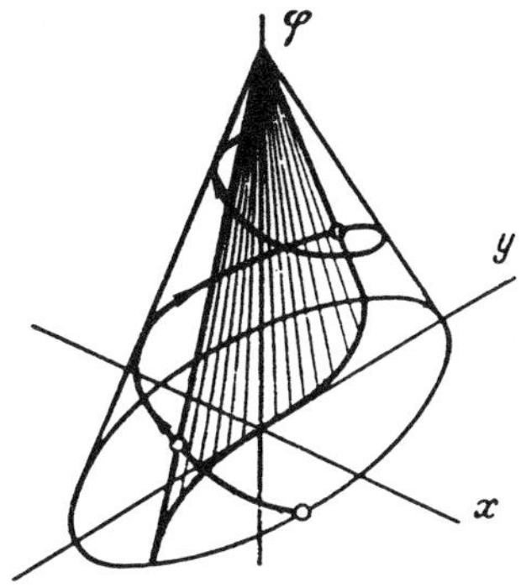

FIGURE 16

optimal in the following problem:

$$\int_0^\infty e^{-\alpha\varphi}x^2\,dt \to \inf,$$
$$\dot{x} = y, \quad \dot{y} = u, \quad \dot{\varphi} = 1, \quad u \in [-1, 1],$$
$$x(0) = x_0, \quad y(0) = y_0, \quad \varphi(0) = \varphi_0.$$

Each two-dimensional fiber of the bundle determines a switching curve of the form $x = \lambda_i y^2$, $\varphi = \varphi_0 + \nu_i y$, $i = 0$ for $y < 0$ and $i = 1$ for $y > 0$ (up to terms of higher order of smallness with respect to y). We consider an arbitrary two-dimensional fiber of the bundle in the space of (x, y, φ) (Figure 16). The trajectories of (4.4) with initial conditions belonging to the intersection of this two-dimensional fiber and the plane $\varphi = 0$ will clearly be optimal in Problem 12. Consequently, in view of Proposition 5 the projections of the switching curves of all the two-dimensional fibers on the space of (x, y) coincide and determine the switching curve of the optimal trajectories in Problem 12. The estimate $0 < \lambda < 0.5$ follows from (2.30).

Problem 13.

$$T \to \inf,$$
$$\dot{x} = y, \quad \dot{y} = u, \quad \dot{z} = e^{-\alpha x^2};$$
$$x(0) = x_0, \quad y(0) = y_0, \quad z(0) = z_0 < 1,$$
$$x(T) = y(T) = 0, \quad z(T) = 1;$$
$$u \in [-1, 1], \quad \alpha > 0.$$

The system of equations of the Pontryagin maximum principle for Problem 13 can be written in the form

$$\dot{\psi}_1 = 2\alpha x\psi_3 e^{-\alpha x^2}, \qquad \dot{\psi}_2 = -\psi_1, \qquad \dot{\psi}_3 = 0,$$
$$\dot{x} = y, \qquad \dot{y} = \operatorname{sgn}\psi_2 \qquad \dot{z} = e^{-\alpha x^2}.$$

We look for the solutions of this system such that $\psi_3 = 1/(2\alpha)$. The manifold of singular trajectories is determined by the system of equations $\psi_1 = \psi_2 = x = y = 0$. By means of the substitution $z_1 = \psi_2$, $z_2 = -\psi_1$, $z_3 = -x$, $z_4 = -y$, $w = z$ the system of equations of the Pontryagin maximum principle can be reduced to the form

$$\dot{z}_1 = z_2, \qquad \dot{z}_2 = z_3 + z_3(e^{-\alpha z_3^2} - 1), \qquad \dot{z}_3 = z_4,$$
$$\dot{z}_4 = -\operatorname{sgn} z_1, \qquad \dot{w} = 1 + (e^{-\alpha z_3^2} - 1).$$

Since the manifold of singular trajectories in Problem 13 is one-dimensional, the conditions of Theorem 1 and the condition of regular projectibility in Theorem 4 hold. A one-parameter family of chattering trajectories filling a piecewise smooth surface homeomorphic to $\mathbf{R}^2$ hits each point of the line $x = y = 0$. Since the field $(\psi_1/H, \psi_2/H, \psi_3/H)$, where $H = \psi_1 y + |\psi_2| + \psi_3 e^{-\alpha x^2} \neq 0$, is a potential field in an open neighborhood of the line $x = y = 0$, the chattering trajectories hitting the half-line $x = y = 0$, $z < 1$ are optimal in Problem 13. We remark that it follows from Lemma 11 that the switching surface of the corresponding fiber bundle is a two-dimensional C^1-smooth surface.

3. A problem in mathematical economics

Let us consider the problem of optimal allocation of resources in an economic model of controllable technological progress. Assume for simplicity that in this model there are two factors influencing the development of production. The first is measured by a quantity which we denote by x_1, and the second by a quantity x_2. The flow of producible output (the rate of production) depends on these two quantities (x_1, x_2). This flow is denoted by $Q(x_1, x_2)$ and is called the production function of the model. In the case when we do not apportion investments toward the improvement of production (technological progress) the rate of production of x_1 and x_2 stays constant. Assume that we apportion a fixed fraction $sQ(x_1, x_2)$ of the producible output for purposes of technological progress, where s is a fixed number, $0 < s < 1$. A part $usQ(x_1, x_2)$ of the fraction is directed toward the improvement of the factor x_1, and the remaining part $(1-u)sQ(x_1, x_2)$ to the improvement of x_2. Here u is a controlling parameter with the constraint $u \in [0, 1]$. Investments directed toward the improvement of one or the other factor affect the rate of growth of this factor. One usually assumes instantaneous convertability of funds and considers a model with continuous time. We are interested in the situation when the investments affect the acceleration (second derivative with respect to time) of x_1 and x_2. Thus, we get the controllable system

$$\ddot{x}_1 = usQ(x_1, x_2), \qquad \ddot{x}_2 = (1-u)sQ(x_1, x_2). \tag{4.6}$$

The coefficient s is a fixed number and is included in the function $Q(x_1, x_2)$ itself in the subsequent computations. As a result, the system (4.6) takes the form

$$\begin{aligned} \dot{x}_1 &= y_1, \qquad & \dot{y}_1 &= uQ(x_1, x_2), \\ \dot{x}_2 &= y_2, \qquad & \dot{y}_2 &= (1-u)Q(x_1, x_2), \end{aligned} \tag{4.7}$$

where $x_1, x_2, y_1, y_2 \geq 0$ and $u \in [0, 1]$.

The following natural restrictions are imposed on the production function Q:

$$Q > 0, \qquad \partial Q/\partial x_1 > 0, \qquad \partial Q/\partial x_2 > 0.$$

These conditions ensure the inequalities $x_1, x_2, y_1, y_2 \geq 0$ during the whole course of the process whenever they hold at the initial point. Thus, the problem formulated here is actually not a problem with phase constraints.

We consider the problem of hitting some terminal set $\mathscr{M} \subset \mathbf{R}^4_+$ as quickly as possible from an initial position $x_1(0) = x_{10}$, $x_2(0) = x_{20}$, $y_1(0) = y_{10}$, $y_2(0) = y_{20}$ along trajectories of (4.7).

Problem 14.

$$T \to \inf,$$
$$\dot{x}_1 = y_1, \quad \dot{y}_1 = u_1 Q(x_1, x_2), \quad \dot{x}_2 = y_2, \quad \dot{y}_2 = u_2 Q(x_1, x_2);$$
$$u_1 + u_2 = 1, \qquad u_i \geq 0, \quad i = 1, 2;$$
$$x_i(0) = x_{i0}, \qquad y_i(0) = y_{i0}, \qquad i = 1, 2;$$
$$(x_1(T), y_1(T), x_2(T), y_2(T)) \subset \mathscr{M}.$$

We first consider the special case of Problem 14 when $Q(x_1, x_2) = x_1 x_2$.

Problem 15.

$$T \to \inf,$$
$$\dot{x}_1 = y_1, \quad \dot{y}_1 = u_1 x_1 x_2, \quad \dot{x}_2 = y_2, \quad \dot{y}_2 = u_2 x_1 x_2,$$
$$x_i(0) = x_{i0}, \qquad y_i(0) = y_{i0}, \qquad i = 1, 2;$$
$$(x_1(T), y_1(T), x_2(T), y_2(T)) \in \mathscr{M} \subset \mathbf{R}^4_+;$$
$$u_1 + u_2 = 1, \quad u_i \geq 0, \quad i = 1, 2.$$

Let

$$\begin{aligned} x &= \tfrac{1}{2}(x_1 - x_2), \qquad y = \tfrac{1}{2}(y_1 - y_2), \qquad p = \tfrac{1}{2}(x_1 + x_2), \\ q &= \tfrac{1}{2}(y_1 - y_2), \qquad u = \tfrac{1}{2}(u_1 - u_2). \end{aligned} \tag{4.8}$$

In these coordinates we get

$$\begin{aligned} &\dot{x} = y, \qquad \dot{y} = u(p^2 - x^2), \qquad \dot{p} = q, \\ &\dot{q} = p^2 - x^2, \qquad u \in [-1, 1]. \end{aligned} \tag{4.9a}$$

The system of equations of the Pontryagin maximum principle has the form

$$\begin{aligned} &\dot{\psi}_1 = 2ux\psi_2 + 2x\psi_4, \qquad \dot{\psi}_2 = -\psi_1, \\ &\dot{\psi}_3 = -(2up\psi_2 + 2p\psi_4), \qquad \dot{\psi}_4 = -\psi_3, \\ &u = \operatorname{sgn} \psi_2(p^2 - x^2). \end{aligned} \tag{4.9b}$$

The manifold S_0 of singular trajectories of (4.9) is distinguished by the system of equations $\psi_2 = \psi_1 = x = y = 0$. Thus, the projection of S_0 on the space of x, y, p, q is two-dimensional. Denote this projection by L. At the points of the surface L the system (4.9a) has the form

$$\dot{p} = q, \qquad \dot{q} = p^2. \tag{4.10}$$

We consider an arbitrary curve on L of the form $L_1 = \{x = y = 0,\ p > 0,\ q = f(p) > 0\}$. The function f is chosen so that at all points $p^2 - qf'(p) > 0$. It is easily seen that the curve L_1 is transversal to the trajectories of (4.10). Let us consider a lifting $\mathscr{N}_1 \subset S_0$ of L_1 to the extended phase space such that it belongs to the level surface

$$\psi_1 y + \psi_2 u(p^2 - x^2) + \psi_3 q + \psi_4(p^2 - x^2) = 1$$

of the Hamiltonian and for which $\psi\, dx = 0$. From this we get $\psi_3 q + \psi_4 p^2 = 1$ and $\psi_3\, dp + \psi_4\, dq = 0$, and the curve $\mathscr{N}_1$ has the form

$$\mathscr{N}_1 = \left\{ x = y = \psi_1 = \psi_2 = 0;\ q = f(p);\ \psi_3 = \frac{-f'}{p^2 - qf'};\ \psi_4 = \frac{1}{p^2 - qf'} \right\}.$$

Let

$$z_1 = \psi_2, \quad z_2 = -\psi_1, \quad z_3 = -2x\psi_4, \quad z_4 = -2y\psi_4,$$
$$w_1 = \psi_4, \quad w_2 = \psi_3, \quad w_3 = p, \quad w_4 = q.$$

We get

$$\dot z_1 = z_2, \quad \dot z_2 = z_3 + uz_1z_2w_1^{-1}, \quad \dot z_3 = z_4 - w_2z_3w_1^{-1},$$
$$\dot z_4 = -2uw_1w_3^2 + uz_3^2(2w_1)^{-1} - z_4w_1^{-1}w_2.$$

The Kelley condition singles out the domain $w_1 > 0$. Thus, the system (4.9) satisfies the condition (2.5), and by Theorem 1 the manifold $\mathscr{N}_1$ determines a corresponding fiber bundle Σ^* with two-dimensional fibers filled by chattering trajectories. Since $\partial x/\partial w|_{S_0} = 0$, the condition in Theorem 4 of regular projectibility of Σ^* on the phase space is satisfied. Therefore, by Theorem 3, the manifold $\mathscr{N}_1$ determines a locally optimal synthesis in Problem 15.

A synthesis of optimal chattering trajectories for Problem 14 with an arbitrary function $Q(x_1, x_2)$ is constructed according to the same scheme. We give the corresponding computations below, though not for Problem 14 itself, but for its simplest multidimensional generalization: Problem 16.

Problem 16.

$$T \to \inf,$$

$$\begin{gathered} \dot x = y, \quad \dot y = \alpha(x, y, \omega) + u\beta(x, y, \omega), \quad \dot\omega = f(x, \omega); \\ x(0) = x_0, \quad y(0) = y_0, \quad \omega(0) = \omega_0, \\ (x(T), y(T), \omega(T)) \subset \mathscr{M}. \end{gathered} \tag{4.11}$$

Here $x, y \in \mathbf{R}^1$, $\omega \in \mathbf{R}^n$, $u \in [-1, 1]$, $\alpha, \beta: \mathbf{R}^{n+2} \to \mathbf{R}^1$, $f: \mathbf{R}^{n+1} \to \mathbf{R}^n$, and

$$\omega = \begin{pmatrix} \omega_1 \\ \vdots \\ \omega_n \end{pmatrix}, \quad f = \begin{pmatrix} f_1 \\ \vdots \\ f_n \end{pmatrix}.$$

The functions α, β, and f are assumed to be sufficiently smooth (of class C^4).

The system of equations of the maximum principle for Problem 16 has the form

$$\begin{gathered} \dot\psi_1 = -\psi_2(\alpha'_x + u\beta'_x) - \psi f'_x, \quad \dot\psi_2 = -\psi_1 - \psi_2(\alpha'_y + u\beta'_y), \\ \dot\psi = -\psi_2(\alpha'_\omega + u\beta'_\omega) - \psi f'_\omega, \quad u = \operatorname{sgn}(\beta\psi_2). \end{gathered} \tag{4.12}$$

Here $\psi_1, \psi_2 \in \mathbf{R}^1$, $\psi = (\psi^1, \dots, \psi^n) \in \mathbf{R}^n$, and

$$\alpha'_\omega = \left(\frac{\partial\alpha}{\partial\omega_1}, \dots, \frac{\partial\alpha}{\partial\omega_n}\right), \quad \beta'_\omega = \left(\frac{\partial\beta}{\partial\omega_1}, \dots, \frac{\partial\beta}{\partial\omega_n}\right),$$

$$f'_\omega = \begin{pmatrix} \partial f_1/\partial\omega_1 & \cdots & \partial f_1/\partial\omega_n \\ \dots\dots & & \dots\dots \\ \partial f_n/\partial\omega_1 & \cdots & \partial f_n/\partial\omega_n \end{pmatrix}.$$

Let us consider the domain of phase space in which $\beta \neq 0$. We find the singular solutions of the system (4.11), (4.12). Let $z_1 = \beta\psi_2$. Differentiating z_1 by virtue of the system (4.11), (4.12), we get

$$\dot z_1 = -\beta\psi_1 + (y\beta'_x + \alpha\beta'_y + f\beta'_\omega - \beta\alpha'_y)\psi_2.$$

Let $z_2 = -\beta\psi_1$. Then

$$\dot z_2 = \beta\psi f'_x - (\beta'_x y + \beta'_y(\alpha + u\beta) + \beta'_\omega f)\psi_1 + \beta(\alpha'_x + u\beta'_x)\psi_2.$$

Let $z_3 = \beta\psi f'_x$ and

$$\dot z_3 = (\beta'_x y + \beta'_y(\alpha+u\beta)+\beta'_\omega f)\psi f'_x + \beta(-\psi_2(\alpha'_\omega+u\beta'_\omega)-\psi f'_\omega)f'_x + \beta\psi(f''_{xx}y + f''_{x\omega}f),$$

where

$$f''_{x\omega} = \begin{pmatrix} \partial^2 f_1/\partial x\,\partial\omega_1 & \cdots & \partial^2 f_1/\partial x\,\partial\omega_n \\ \cdots\cdots & \cdots & \cdots\cdots \\ \partial^2 f_n/\partial x\,\partial\omega_1 & \cdots & \partial^2 f_n/\partial x\,\partial\omega_n \end{pmatrix}.$$

Let $z_4 = \beta\psi(-f'_\omega f'_x + f''_{x\omega}f + f''_{xx}y)$. Differentiating z_4 with respect to t by virtue of the system (4.11), (4.12), we see that

$$\frac{\partial}{\partial u}\frac{dz_4}{dt} = \beta^2\psi f''_{xx}\ (\operatorname{mod}(z_1 = z_2 = z_3 = z_4 = 0)).$$

Assume that $\psi f''_{xx}|_{S_0} \neq 0$, where S_0 is the surface $S_0 = \{x, y, \omega, \psi_1, \psi_2, \psi \mid z_i = 0,\ i = 1, \dots, 4\}$. Then

$$\operatorname{rk}\left(\frac{\mathscr{D}(z_1, z_2, z_3, z_4)}{\mathscr{D}(x, y, \omega, \psi_1, \psi_2, \psi)}\right)\Bigg|_{S_0} = \operatorname{rk}\begin{pmatrix} 0 & 0 & 0 & 0 & \beta & 0 \\ 0 & 0 & 0 & -\beta & 0 & 0 \\ \beta\psi f''_{xx} & 0 & \beta\psi f''_{x\omega} & 0 & 0 & (\beta f'_x)^T \\ * & \beta\psi f''_{xx} & * & 0 & 0 & * \end{pmatrix} = 4.$$

Consequently, S_0 is a smooth manifold of dimension $2n$. Supplementing $z_1, \dots, z_4$ by the functions ω and ψ to form a basis, we can rewrite the system (4.11), (4.12) in the form

$$\tag{4.13} \begin{gathered} \dot z_1 = z_2 + \Theta_4, \qquad \dot z_2 = z_3 + \Theta_3, \qquad \dot z_3 = z_4 + \Theta_2, \\ \dot z_4 = A(z, \omega, \psi) + u\beta^2\psi f''_{xx}, \\ \dot\psi = \Psi(z, \omega, \psi), \qquad \dot\omega = \Omega(z, \omega, \psi), \qquad u = \operatorname{sgn} z_1, \end{gathered}$$

where Θ_i denotes functions satisfying the relations

$$\varlimsup_{\kappa\to+0} \Theta_i(g_\kappa(z), \omega, \psi)\kappa^{-i} < \infty, \qquad g_\kappa(z) = (\kappa^4 z_1, \kappa^3 z_2, \kappa^2 z_3, \kappa z_4).$$

The conditions of Theorem 1 hold in the domain of phase space in which

$$\psi f''_{xx} < 0, \qquad |A| < -\beta^2\psi f''_{xx}.$$

Therefore, a one-parameter family of chattering trajectories hits each point of the manifold S_0. The construction of a Lagrangian manifold determining a locally optimal synthesis in Problem 16 now goes according to the same scheme as for Problem 15. We remark that the conditions of regular projectibility of the corresponding Lagrangian manifolds in the general position situation are trivially satisfied for the same reasons as in Problem 15.

4. The Lawden problem

Let us begin by sketching the history of the investigation of this problem. We make the reservation at once that this history is rather confused. Many results relating to the problem have been rediscovered again and again, both because of inadequate information and because of the feelings of dissatisfaction often left by papers falling short in clarity and mathematical rigor. Therefore, while presenting our own view of the history of the question, we are ready to consider other versions.

The problem of minimizing the fuel expenditure of a rocket moving in a Newtonian central gravity field when the engine thrust is bounded in modulus was investigated by Lawden at the beginning of the 1960s in [28]. As the most impressive result of this investigation, Lawden explicitly found extremals of the Pontryagin maximum principle on which the engine thrust took an intermediate value different both from zero and from its maximal possible value in the case of motion in a two-dimensional plane with nonfixed time to hit the terminal manifold. These trajectories are singular trajectories in the sense of the theory of optimal control, and are called *intermediate-thrust* trajectories. Incidentally, we remark that the system of equations of the Pontryagin maximum principle for the time-optimal problem turn out to be essentially the same as for the minimum-fuel problem. The difference between these two problems manifests itself only in the transversality conditions and in the investigation of the question of normality of the extremals.

Somewhat later Marchal [32] found explicit equations for the intermediate-thrust trajectories in the case of the planar problem with fixed time. The most interesting of these trajectories are the so-called reversible arcs, when the use of an intermediate thrust leads to the stopping of the rocket in space at some moment of time and then to a motion in the reverse direction along the trajectory just followed.

The question of optimality of intermediate-thrust trajectories at once attracted universal attention as something of fundamental value from the point of view of space navigation. It is not an exaggeration to assert that it was this question that gave rise to the theory of higher-order necessary conditions for optimality in the case of singular optimal control regimes.

The first result in this area is due to Kelley [50]. However, the Kelley condition became degenerate on the Lawden spirals and did not give any information about their optimality. Higher-order conditions, also inspired by the Lawden problem, were obtained by Kopp and Moyer [51] and Robbins [30]. These conditions turned out to be more informative. Robbins [30] and a number of other authors used a somewhat different approach (see, for example, [31] and [52]) and established that the Lawden spirals, i.e., the intermediate-thrust trajectories, are not optimal in the problem with nonfixed time. It was also proved [30] that sufficiently short parts of certain intermediate-thrust trajectories are optimal in the problem with fixed time.

The theory of higher-order necessary conditions for optimality later separated off from the theory of space navigation that had given rise to it and began to live an independent existence, while a number of unsolved problems remained in the Lawden problem itself. One of them is the problem of optimality of three-dimensional intermediate-thrust trajectories. Another (more important, in our view) is the problem of constructing an optimal synthesis in the Lawden problem. The first serious step in the study of this last problem was made in [33] by Breakwell and Dixon, who, in studying a neighborhood of a fixed intermediate-thrust trajectory of Marchal reversible-arc type, established that the associated problem of minimizing the second variation is the Fuller-Marchal problem (Problem 4 in Chapter 2). Therefore, [33] proposed using the synthesis in the Fuller-Marchal problem to correct small deviations from a singular trajectory in the Lawden problem.

Below we describe an optimal synthesis in the Lawden problem itself (in the time-optimal variant) for various terminal manifolds.

We consider a system of differential equations describing a control of an aircraft in a central gravity field,

$$\ddot{\bar{r}} = \frac{1}{m}\bar{T} - \mu\bar{r}/|\bar{r}|^3, \qquad \dot{m} = -c^{-1}T. \tag{4.14}$$

Here $\bar{r} = (x, y, z)$ is the radius vector, and m is the mass of the aircraft. The control is implemented by ejecting a part of the mass. The direction $\overline{T} = (T_x, T_y, T_z)$ of the ejection is arbitrary, the velocity is bounded by the maximal thrust $T = \|\overline{T}\| \in [0, 1]$, and μ and c are positive constants. We look for trajectories carrying the object from an initial state $\bar{r}(0), \dot{\bar{r}}(0), m(0)$ to a final state $\bar{r}(\tau), \dot{\bar{r}}(\tau), m(\tau)$ belonging to some terminal manifold $\mathscr{M} \subset \mathbf{R}^7$. As the functional to be minimized in this control problem we can take either the flight time $\tau \to \inf$ or the fuel expenditure

$$\int_0^\tau T\, ds \to \inf .$$

We consider the case when the motion of the aircraft takes place in a two-dimensional plane (the plane of the ecliptic) and the flight time is minimized.

Problem 17.

$$\tau \to \inf ,$$

$$\tag{4.15} \begin{gathered} \dot{X} = U, \qquad \dot{U} = \tfrac{T}{m}\cos\Theta - \mu X(X^2+Y^2)^{-3/2}, \\ \dot{Y} = V, \qquad \dot{V} = \tfrac{T}{m}\sin\Theta - \mu Y(X^2+Y^2)^{-3/2}, \qquad \dot{m} = -T/c; \\ X(0) = X_0, \quad U(0) = U_0, \quad Y(0) = Y_0, \quad V(0) = V_0, \quad m(0) = m_0; \\ (X(\tau), U(\tau), Y(\tau), V(\tau), m(\tau)) \subset \mathscr{M} \subset \mathbf{R}^5. \end{gathered}$$

The thrust $T \in [0, 1]$ and the angle Θ giving the direction of thrust serve as the controlling parameters. It will be shown below that in a neighborhood of the singular manifold of Problem 17 there is a fiber bundle with piecewise smooth fibers that is filled by chattering trajectories. We construct a whole family of different locally optimal syntheses that consists entirely of chattering trajectories adjoining singular trajectories of reversible-arc type.

Reduction of the problem. For the system (4.15) there is a one-parameter symmetry group induced by rotations about the origin of coordinates in the (X, Y)-plane. Consequently, it is possible to reduce the order of the system (4.15) by 1. Let

$$r = (X^2+Y^2)^{1/2}, \qquad x = (XU+YV)(X^2+Y^2)^{-1/2},$$
$$y = (XV - YU)(X^2+Y^2)^{-1/2}, \qquad \varphi = \Theta - \arccos X(X^2+Y^2)^{-1/2}.$$

Since the angle Θ is chosen arbitrarily, the value φ can be regarded as a new independent control parameter. Problem 17 is equivalent to the following problem.

Problem 18.

$$\tau \to \inf ,$$

$$\tag{4.16} \begin{gathered} \dot{r} = x, \qquad \dot{x} = \frac{u}{m}\cos\varphi - \frac{u}{r^2} + \frac{y^2}{r}, \\ \dot{y} = \frac{u}{m}\sin\varphi - \frac{xy}{r}, \qquad \dot{m} = -u/c, \qquad u \in [0, 1], \\ r(0) = r_0, \quad x(0) = x_0, \quad y(0) = y_0, \quad m(0) = m_0, \\ (r(\tau), x(\tau), y(\tau), m(\tau)) \in \mathscr{M} \subset \mathbf{R}^4. \end{gathered}$$

The system of equations of the Pontryagin maximum principle for Problem 18 can be written in the form

$$\tag{4.17} \begin{gathered} \dot{\psi}_r = \left(-\frac{2\mu}{r^3} + \frac{y^2}{r^2}\right)\psi_x - \frac{xy}{r^2}\psi_y, \qquad \dot{\psi}_x = -\psi_r + \frac{y}{r}\psi_y, \\ \dot{\psi}_y = -\frac{2y}{r}\psi_x + \frac{x}{r}\psi_y, \qquad \dot{\psi}_m = \frac{u}{m^2}(\cos\varphi\,\psi_x + \sin\varphi\,\psi_y). \end{gathered}$$

The control u, φ is found from the maximum condition

(4.18)
$$\max_{v\in[0,1],\xi\in[0,2\pi]} v(\psi_x\cos\xi+\psi_y\sin\xi-\tfrac{1}{c}\psi_m) = u(t)(\psi_x\cos\varphi(t)+\psi_y\sin\varphi(t)-\tfrac{1}{c}\psi_m).$$

We consider the solutions of the system (4.16)–(4.18) that are nonsingular with respect to the control component φ, i.e., for which $\psi_x^2+\psi_y^2\neq 0$. In this case it follows from (4.18) that

$$\cos\varphi=\psi_x(\psi_x^2+\psi_y^2)^{-1/2},\qquad \sin\varphi=\psi_y(\psi_x^2+\psi_y^2)^{-1/2}.$$

Let us find the solutions of (4.16)–(4.18) that are singular with respect to the control component u. We have

$$H=H_0+uH_1,\qquad H_0=x\psi_r+(-\mu/r^2+y^2/r)\psi_x-\frac{xy}{r}\psi_y,$$
$$H_1=\frac{1}{m}(\psi_x^2+\psi_y^2)^{1/2}-\frac{1}{c}\psi_m.$$

Let $z_1=H_1$. Differentiating z_1 by virtue of the system (4.16)–(4.18), we get successively that

(4.19)
$$\dot z_1=\frac{z_2}{m^2(z_1+\psi_m/c)},\qquad \dot z_2=z_3,\qquad \dot z_3=z_4,\qquad \dot z_4=A+uB,$$

where

$$z_2=-\psi_r\psi_x-\frac{y}{r}\psi_x\psi_y+\frac{x}{r}\psi_y^2,$$
$$z_3=\psi_r^2+\psi_x^2\left(\frac{2\mu}{r^3}+\frac{y^2}{r^2}\right)-\frac{2xy}{r^2}\psi_x\psi_y+\left(\frac{x^2}{r^2}-\frac{\mu}{r^3}\right)\psi_y^2,$$
$$z_4=\mu r^{-4}(-8r\psi_r\psi_x-6x\psi_x^2+10y\psi_x\psi_y-x\psi_y^2),$$
$$A=-4xr^{-1}z_4+\mu r^{-4}(4x\psi_r\psi_x-18y\psi_r\psi_y+8r\psi_r^2+(22\mu r^{-2}-34y^2r^{-1})\psi_x^2+(\mu r^{-2}+9y^2r^{-1}-2x^2r^{-1})\psi_y^2),$$
$$B=3\mu\psi_x(3\psi_y^2-2\psi_x^2)/(mr^4(\psi_x^2+\psi_y^2)^{1/2}).$$

If $\psi_x(3\psi_y^2-2\psi_x^2)\neq 0$, then the singular trajectories have second order (local order). We find a parametric representation of the surface

$$S_0=\{z_1=z_2=z_3=z_4=0\}.$$

Since $z_1=0$,

$$\psi_m=cm^{-1}(\psi_x^2+\psi_y^2)^{1/2}.$$

It follows from the equation $rz_2+\mu^{-1}r^4z_4=0$ that

$$-9r\psi_r\psi_x-6x\psi_x^2+9y\psi_x\psi_y=0.$$

For singular trajectories this implies that either $\psi_x=0$ or

(4.20)
$$\psi_r=(-2x\psi_x+3y\psi_y)/(3r).$$

Suppose that $\psi_x\neq 0$. We substitute (4.20) in the equations $z_2=0$ and $z_3=0$:

$$\frac{2}{3}\frac{x}{r}\psi_x^2-\frac{2y}{r}\psi_x\psi_y+\frac{x}{r}\psi_y^2=0,$$
$$\left(\frac{4}{9}\frac{x_2}{r^2}+\frac{y^2}{r^2}+\frac{2\mu}{r^3}\right)\psi_x^2-\frac{10xy}{3r^2}\psi_x\psi_y+\left(\frac{x^2}{r^2}+\frac{y^2}{r^2}-\frac{\mu}{r^3}\right)\psi_y^2=0,$$

or

$$2x\psi_x^2 - 6y\psi_x\psi_y + 3x\psi_y^2 = 0, \tag{4.21}$$

$$(4x^2 + 9y^2 + 18\mu/r)\psi_x^2 - 30xy\psi_x\psi_y + (9x^2 + 9y^2 - 9\mu/r)\psi_y^2 = 0. \tag{4.22}$$

Multiplying (4.21) by $-5x$ and adding to (4.22), we get

$$(-6x^2 + 9y^2 + 18\mu/r)\psi_x^2 + (-6x^2 + 9y^2 - 9\mu/r)\psi_y^2 = 0,$$

and hence

$$3y^2 - 2x^2 = 3\mu(\psi_y^2 - 2\psi_x^2)/r(\psi_x^2 + \psi_y^2). \tag{4.23}$$

It follows from (4.21) that

$$y = x(2\psi_x^2 + 3\psi_y^2)/(6\psi_x\psi_y). \tag{4.24}$$

Substituting (4.24) in (4.23), we get

$$x^2\frac{4\psi_x^4 - 12\psi_x^2\psi_y^2 + 9\psi_y^4}{12\psi_x^2\psi_y^2} = \frac{3\mu(\psi_y^2 - 2\psi_x^2)}{r(\psi_x^2 + \psi_y^2)},$$

or

$$x^2 = \frac{36\mu\psi_x^2\psi_y^2(\psi_y^2 - 2\psi_x^2)}{r(3\psi_y^2 - 2\psi_x^2)^2(\psi_x^2 + \psi_y^2)}.$$

As we see, the singular trajectories of (4.16)–(4.18) can be found in the domain $\psi_y^2 \geq 2\psi_x^2$. The points of the manifold S_0 are parametrized by the values of ψ_x, ψ_y, r, and m:

$$\begin{aligned} x &= \pm\frac{6\psi_x\psi_y(\psi_y^2 - 2\psi_x^2)^{1/2}}{(3\psi_y^2 - 2\psi_x^2)(\psi_x^2 + \psi_y^2)^{1/2}}\left(\frac{\mu}{r}\right)^{1/2}, \\ y &= \pm\frac{(2\psi_x^2 + 3\psi_y^2)(\psi_y^2 - 2\psi_x^2)^{1/2}}{(3\psi_y^2 - 2\psi_x^2)(\psi_x^2 - \psi_y^2)^{1/2}}\left(\frac{\mu}{r}\right)^{1/2}, \\ \psi_r &= \pm\frac{\psi_y(\psi_y^2 - 2\psi_x^2)^{1/2}}{(\psi_x^2 + \psi_y^2)^{1/2}}\left(\frac{\mu}{r^3}\right)^{1/2}, \\ \psi_m &= \frac{c}{m}(\psi_x^2 + \psi_y^2)^{1/2}, \\ u &= -\mu m\psi_x(120\psi_x^6 + 108\psi_x^4\psi_y^2 + 66\psi_x^2\psi_y^4 - 27\psi_y^6) \\ &\quad \times r^{-2}(\psi_x^2 + \psi_y^2)^{-1/2}(3\psi_y^2 - 2\psi_x^2)^{-3} \\ H_0 &= -3\mu\psi_x^3r^{-2}(\psi_x^2 + \psi_y^2)^{-1}. \end{aligned} \tag{4.25}$$

Kelley's necessary condition for optimality singles out the domain

$$\psi_x(3\psi_y^2 - 2\psi_x^2) < 0.$$

Since $\psi_y^2 - 2\psi_x^2 \geq 0$, we get that $\psi_x < 0$.

We find the Jacobian of the substitution $(\psi_m, \psi_r, x, y) \to (z_1, z_2, z_3, z_4)$ at the points of S_0:

$$J = \begin{vmatrix} -1/c & 0 & 0 & 0 \\ 0 & -\psi_x & \psi_y^2r^{-1} & -\psi_x\psi_yr^{-1} \\ 0 & 2\psi_r & -2yr^{-2}\psi_x\psi_y + 2xr^{-2}\psi_y^2 & 2yr^{-2}\psi_x^2 - 2xr^{-2}\psi_x\psi_y \\ 0 & -8\mu\psi_xr^{-3} & -6\mu r^{-4}\psi_x^2 - \mu r^{-4}\psi_y^2 & 10\mu r^{-4}\psi_x\psi_y \end{vmatrix}$$

$$= -\frac{2\mu}{cr^7}\begin{vmatrix} -r\psi_x & \psi_y^2 & -\psi_x\psi_y \\ r^2\psi_r & -y\psi_x\psi_y + x\psi_y^2 & y\psi_x^2 - x\psi_x\psi_y \\ -8r\psi_x & -6\psi_x^2 - \psi_y^2 & 10\psi_x\psi_y \end{vmatrix}.$$

Substituting (4.20) in place of ψ_r, we get

$$J = \frac{-2\mu}{3cr^6}\begin{vmatrix} -\psi_x & \psi_y^2 & -\psi_x\psi_y \\ -2x\psi_x + 3y\psi_y & -3y\psi_x\psi_y + 3x\psi_y^2 & 3y\psi_x^2 - 3x\psi_x\psi_y \\ -8\psi_x & -6\psi_x^2 - \psi_y^2 & 10\psi_x\psi_y \end{vmatrix}$$
$$= \frac{2\mu\psi_x}{cr^6}(3\psi_y^2 - 2\psi_x^2)(x\psi_x\psi_y + 3y(\psi_y^2 - \psi_x^2)).$$

By (4.24),

$$J = \frac{\mu}{cr^6}(3\psi_y^2 - 2\psi_x^2)^2\frac{x(\psi_x^2 + \psi_y^2)}{\psi_y}.$$

Thus, at the points of S_0

$$J = \pm 6\mu^{3/2}c^{-1}r^{-13/2}(\psi_x^2 + \psi_y^2)^{1/2}\psi_x(3\psi_y^2 - 2\psi_x^2)(\psi_y^2 - 2\psi_x^2)^{1/2} \neq 0.$$

Consequently, S_0 is four-dimensional. Replacing ψ_x and ψ_y by $\lambda\psi_x$ and $\lambda\psi_y$ does not change x, y, r, and m in (4.25); therefore, the manifold projects regularly into a three-dimensional submanifold of the space of x, y, r, m.

Consequently, construction of an optimal synthesis with chattering assumes that necessarily not all possible singular extremals lying in the four-dimensional phase space of (x, y, r, m) take part in this synthesis. Depending on the choice of the boundary conditions, a two-dimensional submanifold X is determined in the three-dimensional manifold πS_0 of singular extremals. A neighborhood of X is fibered (over the base X) into two-dimensional fibers consisting of chattering trajectories, and this gives a typical optimal synthesis.

We are interested in the part of S_0 in which $u \in [0, 1]$. We show that this subset of S_0 is nonempty for any values of the parameters μ and c. By (4.25), $u \geq 0$ if

$$P(\psi_x, \psi_y) = 40\psi_x^6 + 36\psi_x^4\psi_y^2 + 22\psi_x^2\psi_y^4 - 9\psi_y^6 \geq 0,$$

and $u \leq 1$ if

$$-3\mu r^{-2}m\psi_x(\psi_x^2 + \psi_y^2)^{-1/2}(3\psi_y^2 - 2\psi_x^2)^{-3}P(\psi_x, \psi_y) \leq 1.$$

The last inequality holds automatically if r is sufficiently large.

Let $(\psi_x/\psi_y)^2 = \alpha$. Since $\psi_y^2 - 2\psi_x^2 \geq 0$, it follows that $0 \leq \alpha \leq 0.5$ in the domain of interest to us. The inequality $P(\psi_x, \psi_y) \geq 0$ means that

$$R(\alpha) = 40\alpha^3 + 36\alpha^2 + 22\alpha - 9 \geq 0.$$

The cubic polynomial $R(\alpha)$ is monotonically increasing for $\alpha \geq 0$. Its only positive root is $\alpha_0 = 0.2629\ldots$. Thus, $P(\psi_x, \psi_y) > 0$ if $(\psi_x/\psi_y)^2 \in (\alpha_0; 0.5)$.

We consider an arbitrary curve Γ in S_0 that is transversal to the singular trajectories of the system (4.16)–(4.18) belonging to the surface $H = 1$ and satisfying the condition $\psi\, dx|_\Gamma = 0$. Suppose that such a curve Γ is given by the equations $\psi_x = \psi_x(s)$, $\psi_y = \psi_y(s)$, $r = r(s)$, $m = m(s)$, and the values of x, y, ψ_r, ψ_m are expressed in terms of s according to (4.25). The condition $H = 1$ gives us $r^2(\psi_x^2 + \psi_y^2) = -3\mu\psi_x^3$, i.e., the value of $\psi_y(s)$ is uniquely determined from the given $\psi_x(s)$. For arbitrarily specified functions $r(s)$ and $m(s)$ the condition $\psi\, dx|_\Gamma = 0$ enables us to express $d\psi_x/ds$ in terms of $\psi_x, r, m, dr/ds, dm/ds$ in the form of a differential equation

$$d\psi_x = F_1(\psi_x, r, m)\, dr + F_2(\psi_x, r, m)\, dm$$

(if $dy|_\Gamma \neq 0$). The functional arbitrariness in the choice of such curves is exhausted by specifying the initial conditions $\psi_x(0)$ and the functions $r(s)$ and $m(s)$.

We show that the conditions of Theorem 4 hold, and hence the choice of the manifold Γ determines the field of locally optimal chattering trajectories in Problem 18, with the projection $\pi\Gamma$ of Γ on the space of (x, y, r, m) chosen as the terminal manifold.

The basic condition of Theorem 4 is equivalent to the relation (2.42). Let us show that the matrix

$$\mathscr{D} = \frac{\mathscr{D}(x, y, r, m)}{\mathscr{D}(z_1, \dots, z_4, \psi_x, \psi_y, r, m)}$$

can be reduced to the form (2.42). We have

$$\mathscr{D} = \begin{pmatrix} * & * & \partial x/\partial z_3 & \partial x/\partial z_4 & \partial x/\partial \psi_x & \partial x/\partial \psi_y & * & * \\ * & * & \partial y/\partial z_3 & \partial y/\partial z_4 & \partial y/\partial \psi_x & \partial y/\partial \psi_y & * & * \\ 0 & 0 & 0 & 0 & 0 & 0 & 1 & 0 \\ 0 & 0 & 0 & 0 & 0 & 0 & 0 & 1 \end{pmatrix},$$

where the asterisks stand for quantities whose form is not essential. A direct computation gives

$$\left.\frac{\partial x}{\partial z_3}\right|_{S_0} = \frac{18r^2\psi_x^2\psi_y^2}{x(\psi_x^2+\psi_y^2)(3\psi_y^2-2\psi_x^2)^2}, \qquad \left.\frac{\partial x}{\partial z_4}\right|_{S_0} = \frac{r^4}{3\mu(3\psi_y^2-2\psi_x^2)},$$

$$\left.\frac{\partial y}{\partial z_3}\right|_{S_0} = \frac{3\psi_y^2+2\psi_x^2}{6\psi_x\psi_y}\left.\frac{\partial x}{\partial z_3}\right|_{S_0}, \qquad \left.\frac{\partial y}{\partial z_4}\right|_{S_0} = \frac{\psi_y}{\psi_x}\left.\frac{\partial x}{\partial z_4}\right|_{S_0},$$

$$\left.\frac{\partial y}{\partial \psi_x}\right|_{S_0} = \left.\frac{\partial x}{\partial \psi_x}\right|_{S_0} \frac{\psi_x\psi_y(\psi_y^2+2\psi_x^2)(\psi_y^2+6\psi_x^2)}{2(3\psi_y^6-10\psi_x^2\psi_y^4-2\psi_x^4\psi_y^2-4\psi_x^6)}.$$

At the points of the singular manifold S_0 we have $\partial x/\partial z_3 \neq 0$, and in the case of general position

$$\begin{vmatrix} \partial x/\partial z_3 & \partial x/\partial \psi_x \\ \partial y/\partial z_3 & \partial y/\partial \psi_x \end{vmatrix} \neq 0.$$

Therefore, $\mathscr{D}$ can be transformed to the form

$$\mathscr{D} = \begin{pmatrix} * & * & \partial x/\partial z_3 & * & 0 & * & 0 & 0 \\ * & * & 0 & A & B & * & 0 & 0 \\ 0 & 0 & 0 & 0 & 0 & 0 & 1 & 0 \\ 0 & 0 & 0 & 0 & 0 & 0 & 0 & 1 \end{pmatrix}$$

with the help of linear combinations of rows, where the values of $\partial x/\partial z_3$, A, and B are nonzero.

If the manifold $\mathscr{N}_2$ does not lie in the plane $\psi_y = \text{const}$ and the projection of $\operatorname{Span}(\partial/\partial z_3, T\mathscr{N}_2)$ on the space of (x, y, r, m) is regular, then the conditions of Theorem 4 are satisfied for Lawden's problem.

5. Chattering regimes in the problem of controlling a robot

The results in this section were published in [27].

The problem of controlling a robot can be posed in two variants [25]. A moving part is attached to a massive vertical cylinder rotating about its axis. In the first variant the part has the form of a rod rotating in the vertical plane, and in the second the form of a horizontal shaft moving back and forth (Figures 17 and 18; see next page). In the system there are two controlling parameters: the torque acting on the vertical cylinder, and the force acting on the moving part. Both controls are bounded in modulus.

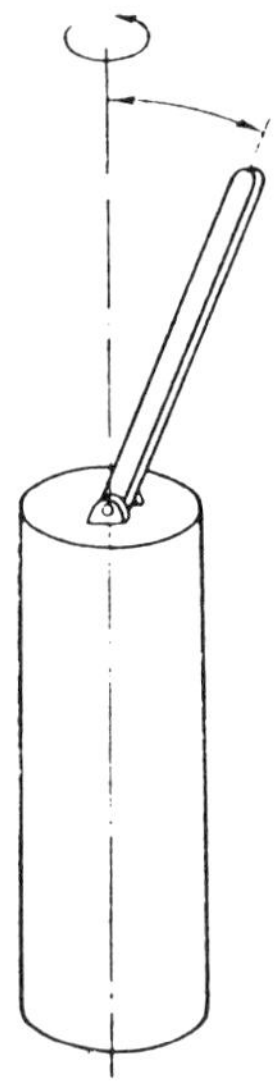

FIGURE 17

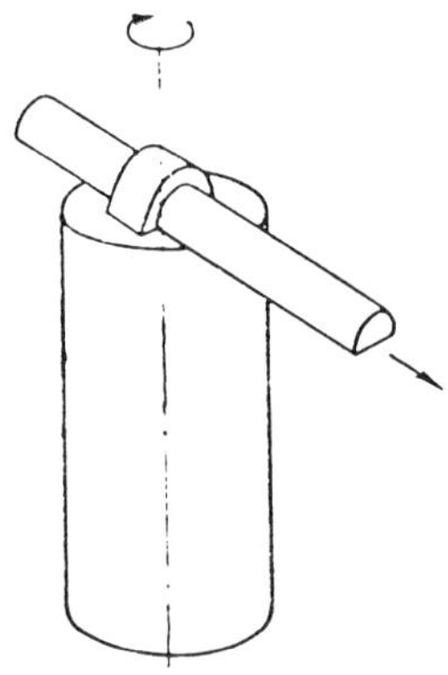

FIGURE 18

For the first variant the equations of motion have the form

$$(4.26)\qquad \begin{gathered} \dot{x}_1 = x_2, \quad \dot{x}_2 = (v - x_2 x_4 \sin 2x_3)/(1 + \sin^2 x_3), \\ \dot{x}_3 = x_4, \quad \dot{x}_4 = u + \sin x_3 + C x_2^2 \sin 2x_3. \end{gathered}$$

For the second variant,

$$(4.27)\qquad \dot{x}_1 = x_2/(1 + x_3^2), \quad \dot{x}_2 = v, \quad \dot{x}_3 = x_4, \quad \dot{x}_4 = u + x_2^2 x_3/(1 + x_3^2)^2.$$

The controls $u(\cdot)$ and $v(\cdot)$ are measurable functions, and

$$(4.28)\qquad |u| \le u_0, \qquad |v| \le 1.$$

The trajectories $X(\cdot) = (x_1(\cdot), \dots, x_4(\cdot))$ are absolutely continuous.

We consider the following problem.

Problem 19.

$$T \to \min, \quad X(0) = X_0, \quad X(T) = 0,$$

where the trajectories $X(\cdot)$ *satisfy* (4.27) *with the constraints* (4.28).

All the results obtained are valid also for the problem in which the trajectories $X(\cdot)$ satisfy (4.26), (4.28) if in this problem we consider not optimal but locally optimal trajectories.

Problems of controlling manipulators (robots) have been actively studied in recent times (see [24]–[28]). The system (4.27), in particular, describes a manipulator of "cyclone" type [24]. The derivation of (4.26) and (4.27) and the statement of the control problem are contained in [25]. As a rule, regimes for controlling a manipulator with one or two control switchings were considered in those references. Moreover, [25] contains the untrue assertion that an optimal trajectory in Problem 19 goes to a regime singular with respect to the control component u with finitely many switchings of u (see [26] about this). It will be shown below that for sufficiently small $\varepsilon > 0$ the optimal trajectories in Problem 19 contain chattering parts for arbitrary constants A and B (with $B < 0$ and $A > \frac{1}{2}B^2$) and for the initial conditions $X_0 \in \mathscr{D}_\varepsilon$, where

$$(4.29)\qquad \mathscr{D}_\varepsilon = \{X \mid \max(|x_4|^{1/2}, |x_3|, |x_1 - A|, |x_2 - B|) < \varepsilon\}.$$

In a finite time depending continuously on the initial conditions the optimal trajectories hit the singular manifold of Problem 19—the surface $\pi=\{X|x_3=x_4=0\}$—with infinitely many switchings of the control component u and with constant $v=-1$, and then hit the origin with $u=0$ and with exactly one switching of the control v. A one-parameter family of optimal trajectories filling a two-dimensional surface $\Sigma_{\alpha,\beta}$ that is smooth outside $\Sigma_{\alpha,\beta}\cap\pi$ hits each point $(\alpha,\beta,0,0)\in\mathscr{D}_\varepsilon$. The switching points of the optimal trajectories form a three-dimensional surface P of class C^1 separating the domain $\mathscr{D}_\varepsilon$ into two subdomains $\mathscr{D}_\varepsilon^+$ and $\mathscr{D}_\varepsilon^-$ in which $u=u_0$ and $u=-u_0$, respectively.

Reduction of the problem. Let us consider the system of equations

$$\dot{x}_1=\sigma x_2,\quad \dot{x}_2=v,\quad \dot{x}_3=x_4,\quad \dot{x}_4=u+x_2^2x_3/(1+x_3^2)^2, \tag{4.30}$$

with control $\sigma\in[0,1]$, u, and v, and with the conditions (4.28) satisfied. We denote by M_R the set of absolutely continuous trajectories $X(\cdot)$ satisfying (4.27) for $0\le t\le\varepsilon R$ and (4.30) with $t>\varepsilon R$. The constant R will be chosen later. We consider the following problem.

Problem 20.

$$T\to\min,$$
$$X(0)=X_0,\quad X(T)=0,\quad X(\cdot)\in M_R.$$

By a theorem of Filippov [53], Problems 19 and 20 have solutions for arbitrary initial conditions.

Remark 7. The optimal value of the time in Problem 20 does not exceed the optimal value of the time in Problem 19. Consequently, if an optimal trajectory in Problem 20 happens to be admissible in Problem 19, then it is simultaneously optimal in Problem 19. It will be shown below that Problems 19 and 20 really are equivalent.

All subsequent assertions are valid for $0<\varepsilon<\varepsilon_0$ for sufficiently small $\varepsilon_0=\varepsilon_0(A,B)$; as a rule, we do not make special mention of this.

Lemma 23. *Let X^* be an optimal trajectory in Problem* 20 *with initial conditions $X_0\in\mathscr{D}_\varepsilon$, and let σ^*, u^*, and v^* be the control on it. Then $\sigma^*=1$ for $t>\varepsilon R$, and $v^*=-1$ for $0\le t\le\varepsilon R$.*

Proof. For $t>\varepsilon R$ the motion in the (x_1,x_2)-plane does not depend on the behavior of the coordinates x_3 and x_4. By (4.30),

$$|x_2(t)-x_2(0)|<t,\qquad |x_1(t)-x_1(0)|<|x_2(0)|t+t^2/2,\qquad t>0.$$

Therefore, for $X_0\in\mathscr{D}_\varepsilon$ the trajectory of (4.30) on $[0,\alpha_1\varepsilon]$ satisfies

$$x_2^2|x_3|(1+x_3^2)^{-2}<\alpha_2\varepsilon,$$

where the constant α_2 can be chosen to depend only on α_1.

As is known ([15], Paragraph 10), in the simplest time-optimal problem

$$T\to\inf,\quad \dot{x}_3=x_4,\quad \dot{x}_4=u,\quad |u|\le\tilde{u}_0,$$
$$\tilde{u}_0=u_0-\varepsilon\alpha_2,\quad x_3(0)=x_{30},\quad x_4(0)=x_{40},$$
$$x_3(T)=x_4(T)=0,$$

the optimal time for hitting the origin is

$$T_{\text{opt}}=\kappa x_{20}+2(\kappa\tilde{u}_0x_{10}+x_{20}^2/2)^{1/2}, \tag{4.31}$$

where $\kappa=\operatorname{sgn}(x_{10}-x_{20}^2\operatorname{sgn}x_{20}/(2\tilde{u}_0))$.

Consequently, by choosing u in $[-(u_0 - \varepsilon\alpha_2), u_0 - \varepsilon\alpha_2]$ it is possible to move the projection of the trajectory $X(\cdot)$ on the (x_3, x_4)-plane to the origin of this plane in a time of order ε. Therefore, the optimal time T^* in Problem 20 is determined only by the conditions $x_1(T^*) = x_2(T^*) = 0$.

Let $x_1(\varepsilon R) = \alpha$ and $x_2(\varepsilon R) = \beta$. It follows from the conditions $A > B^2/2$ and $B < 0$ that $\alpha > \beta^2/2$ and $\beta < 0$. To find an expression for T^* in terms of α and β we consider the following problem.

Problem 21.

$$T \to \inf,$$
$$\dot{x}_1 = \sigma x_2, \quad \dot{x}_2 = v; \quad \sigma \in [0, 1], \quad v \in [-1, 1],$$
$$x_1(0) = x_{10}, \quad x_2(0) = x_{20}, \quad x_1(T) = x_2(T) = 0.$$

A solution of Problem 21 exists by a theorem of Filippov [53]. Let $x_1(\cdot)$, $x_2(\cdot)$ be an optimal trajectory in Problem 21, and let $\sigma(\cdot)$, $v(\cdot)$ be the control on it. By virtue of the Pontryagin maximum principle, there exist absolutely continuous functions $\psi_1(\cdot)$ and $\psi_2(\cdot)$ that are not simultaneously zero and that satisfy the conjugate system

$$\dot{\psi}_1 = 0, \qquad \dot{\psi}_2 = -\sigma\psi_1 \tag{4.32}$$

and the maximum condition

$$\max_{\sigma\in[0,1]} \sigma\psi_1(t)x_2(t) = \sigma(t)\psi_1(t)x_2(t),$$
$$\max_{v\in[-1,1]} v\psi_2(t) = v(t)\psi_2(t) \quad \text{(a.e.)}.$$

It follows from the first equation in (4.32) that $\psi_1(t) = \psi_{10}$. If $\psi_{10} = 0$, then from the second equation we have $\psi_2(t) = \psi_{20} \neq 0$. In this case the control component v is constant on the trajectory $x_1(\cdot), x_2(\cdot)$. All such trajectories passing through the origin fill the domain G_0 bounded by the line $x_1 = 0$ and the parabola branch $x_1 = -\frac{1}{2}x_2^3 \operatorname{sgn} x_2$. The control v is equal to -1 for $x_2 > 0$ and to 1 for $x_2 < 0$. The control component σ is not uniquely determined. Interior to the domain G_0 there is a domain of indifference with respect to σ, and on the boundary of this domain $\sigma = 0$ for $x_1 = 0$ and $\sigma = 1$ for $x_1 = -\frac{1}{2}x_2^2 \operatorname{sgn} x_2$.

If $\psi_{10} > 0$, then the control σ is equal to 1 for $x_2 > 0$ and to 0 for $x_2 < 0$. If $\psi_{10} < 0$, then, conversely, $\sigma = 1$ for $x_2 < 0$, and $\sigma = 0$ for $x_2 > 0$. On the part where $\sigma = 0$ we have $\psi_2 = \text{const}$, which implies that the control v is constant and has modulus 1. On the part where $\sigma = 1$ we have $\psi_2(t) = -\psi_{10}t + \psi_{20}$, and an optimal trajectory hits the origin with at most one switching of the control component v. Combining our arguments, we get a synthesis in Problem 21 (Figure 19).

Let us compute the optimal time T_{opt} for hitting the origin in Problem 21 for the initial conditions $x_{10} = \alpha$ and $x_{20} = \beta$. Integrating the equation $\dot{x}_2 = v$, we get that $T_{\text{opt}} = (\beta - x_2) + (0 - x_2)$, where x_2 is the ordinate of a point where the curves

$$x_1 + \tfrac{1}{2}x_2^2 = \alpha + \tfrac{1}{2}\beta^2, \qquad x_1 = \tfrac{1}{2}x_2^2$$

intersect (a switching point), i.e., $x_2 = -(\alpha + \beta^2/2)^{1/2}$. Thus, in Problem 20 we have that $\sigma \equiv 1$ for $t > \varepsilon R$, and the optimal time for hitting the origin is

$$T^* = 2(\alpha + \beta^2/2)^{1/2} + \beta + \varepsilon R. \tag{4.33}$$

Problem 20 has been reduced to the minimization of the function (4.33) on the solutions of the system (4.27), (4.28) which are defined for $0 \leq t \leq \varepsilon R$.

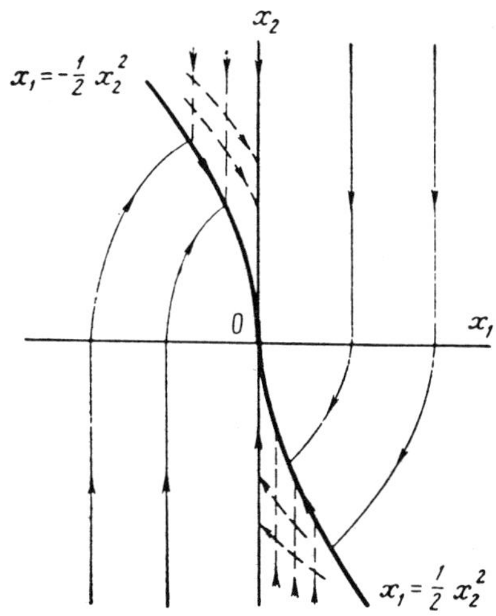

FIGURE 19

We show that if $v^* \neq -1$ on a set of positive measure for $0 \leq t \leq \varepsilon R$, then the trajectory X^* is not optimal in Problem 20. Let us consider the trajectory X^0 of (4.27) with the control $u^0 = u^*$, $v^0 = -1$. Let $y(t) = x_3^*(t) - x_3^0(t)$ and $z(t) = x_2^*(t) - x_2^0(t)$. Then

$$z(t) = \int_0^t (1 + v^*(\tau))\,d\tau \geq 0, \tag{4.34}$$

$$\ddot{y} = (x_2^*)^2 x_3^* (1 + (x_3^*)^2)^{-2} - (x_2^0)^2 x_3^0 (1 + (x_3^0)^2)^{-2}. \tag{4.35}$$

Regrouping the terms on the right-hand side of (4.35), singling out $y(t)$ and $z(t)$ as linear factors, and denoting the coefficients of $y(t)$ and $z(t)$ by $a(X^*, X^0)$ and $b(X^*, X^0)$, we get

$$\ddot{y} = a(t)y + b(t)z,$$

where

$$\begin{aligned} a &= (x_2^*)^2(1 + (x_3^*)^2)^{-2} \\ &\quad - x_3^0(x_2^0)^2(2 + (x_3^*)^2 + (x_3^0)^2)(x_3^* + x_3^0)(1 + (x_3^*)^2)^{-2}(1 + (x_3^0)^2)^{-2}, \\ b &= x_3^0(x_2^* + x_2^0)(1 + (x_3^*)^2)^{-2}. \end{aligned}$$

Since $y(0) = 0$ and $\dot{y}(0) = 0$, the system (4.35) can be rewritten as the integral equation

$$K_t(y(\cdot)) = z(\cdot),$$

where K_t is the operator $C^0[0, t] \to C^0[0, t]$ given by

$$(K_t w)(s) = w(s) - \int_0^s (s - \tau)a(\tau)w(\tau)\,d\tau,$$

and

$$z(s) = \int_0^s (s - \tau)b(\tau)z(\tau)\,d\tau.$$

Since

$$\max_{s\in[0,t]} |K_t(w)(s)| \geq \max_{s\in[0,t]} |w(s)|(1 - O(\varepsilon)),$$

it follows that K_t has a bounded inverse, and $y(\cdot)|_{[0,t]} = K_t^{-1}z(\cdot)$. Consequently, there is a $C > 0$ such that

$$|y(t)| < C\varepsilon\|z(\cdot)\|_{C[0,t]} = C\varepsilon z(t). \tag{4.36}$$

We now estimate

$$\Delta x_1 = x_1^*(\varepsilon R) - x_1^0(\varepsilon R) = \int_0^{\varepsilon R}\left(\frac{x_2^*}{1+(x_3^*)^2} - \frac{x_2^0}{1+(x_3^0)^2}\right)dt.$$

Under the integral sign we single out the terms containing $y(t)$ and $z(t)$:

$$\frac{x_2^*}{1+(x_3^*)^2} - \frac{x_2^0}{1+(x_3^0)^2} = \frac{z(t)}{1+(x_3^*)^2} + \frac{x_2^0(x_3^* + x_3^0)y(t)}{(1+(x_3^*)^2)(1+(x_3^0)^2)}.$$

By (4.36) and the fact that $|x_3^0(t)| < O(\varepsilon)$, we obtain

$$\Delta x_1 \geq (1 - O(\varepsilon))\int_0^{\varepsilon R} z(t)\,dt.$$

It now follows from (4.34) that

$$x_2^*(\varepsilon R) > x_2^0(\varepsilon R), \qquad x_1^*(\varepsilon R) > x_1^0(\varepsilon R).$$

After computing the partial derivatives of the function (4.33) with respect to α and β we see that $\partial T^*/\partial\alpha > 0$ and $\partial T^*/\partial\beta > 0$ in the domain $\beta < 0$, $\alpha > \beta^2/2$, i.e., for fixed α the function T^* is monotonically decreasing as β decreases to the intersection with the parabola $\alpha = \beta^2/2$, and for fixed β it is monotonically decreasing as α decreases to the intersection with $\alpha = \beta^2/2$. This implies the inequality $T^*(X^0(\cdot)) < T^*(X^*(\cdot))$, which contradicts the optimality of the trajectory X^*. The lemma is proved.

Let $F(X) = -x_2x_3^2(x_1 + x_2^2/2)^{-1/2}(1+x_2^2)^{-1}$. The derivative of the function $t + x_2 + 2(x_1 + x_2^2/2)^{1/2}$ with respect to t is equal to $F(X)$ by virtue of the system $\dot{x}_1 = x_2(1+x_3^2)^{-1}$, $\dot{x}_2 = -1$; therefore,

$$I = \int_0^{\varepsilon R} F(X)\,dt = (t + x_2 + 2(x_1 + x_2^2/2)^{1/2})\Big|_{t=0}^{t=\varepsilon R}.$$

It follows from Lemma 23 that Problem 20 is equivalent to the problem

$$I \to \inf,$$
$$\dot{x}_1 = x_2(1+x_3^2)^{-1}, \quad \dot{x}_2 = -1, \quad \dot{x}_3 = x_4, \quad \dot{x}_4 = u + x_2^2x_3(1+x_3^2)^{-2};$$
$$X(0) = X_0, \qquad |u| \leq u_0.$$

Theorem 6. *The solutions X^* of Problem 20 with initial conditions $X_0 \in \mathscr{D}_\varepsilon$ have a chattering part. In a finite time that is less than εR and depends continuously on X_0 the trajectory X^* goes to the singular regime $x_3 = x_4 = 0$ with infinitely many switchings of the control u. An optimal motion is carried out by the part of the singular trajectory with $u = 0$.*

Proof. 1°. *Upper estimate of the optimal value of the functional in Problem* 20. The inclusion $X(t) \subset \mathscr{D}_{\varepsilon C_1}$ is valid for $0 \leq t \leq \varepsilon R$ and for an arbitrary admissible trajectory in Problem 20. Therefore,

$$\max_{X(\cdot)\in M_R}(x_2^2x_3(1+x_3^2)^{-2};\ 0 \leq t \leq \varepsilon R) \leq \gamma_0 < C_2\varepsilon.$$

Let $x^0 = (x_3^0, x_4^0)$ be an optimal trajectory in the simplest time-optimal problem on the (x_3, x_4)-plane:

$$T \to \inf,$$
$$\dot{x}_3 = x_4, \quad \dot{x}_4 = u, \quad |u| \le u_1 = u_0 - \gamma_0,$$
$$x_i(0) = x_{i0}, \quad x_i(T) = 0, \quad i = 3, 4.$$

Consider the set

$$K(r) = \{x_3, x_4 | |x_3| + (2u_1)^{-1} x_4^2 \le r\}.$$

If $X_0 \in \mathscr{D}_\varepsilon$, then $(x_{30}, x_{40}) \in K(\lambda\varepsilon^2)$, $\lambda = 1 + 1/(2u_1)$. Let $(x_{30}, x_{40}) \in K(r)$, $r \le \lambda\varepsilon^2$. It follows from (4.31) that the time τ^0 at which the trajectory x^0 hits the point $(0, 0)$ can be estimated from above by $C_3\sqrt{r}$. Moreover, $x^0(t) \in K(r)$ for $t \in [0, \tau^0]$. Denote by $X^0(t)$ an admissible trajectory in Problem 20 whose projection on the (x_3, x_4)-space coincides with x^0. For it we get the estimate

$$\inf_{X(\cdot)\in M_R} I \le (r^2 + O(\varepsilon))(A + B^2/2)^{-1/2} \int_0^{\tau^0} (-x_2^0)\, dt \le s r^{5/2} \tag{4.37}$$

for some $s > 0$ that can be chosen to be independent of R.

2°. *Lower estimate of the optimal value of the functional in Problem* 20. In the domain $\Omega_1 = \{x_3, x_4 | |x_3| \le r/2\} \backslash K(r)$ the component $\dot{x}_3 = x_4$ of the phase velocity is bounded away from zero: $|\dot{x}_3| \ge (ru_1)^{1/2}$. Therefore, the time spent by an arbitrary admissible trajectory in the domain Ω_1 can be estimated as follows:

$$\sup_{X(\cdot)\in M_R} (\tau > 0\colon (x_3(t), x_4(t)) \subset \Omega_1, \ t \in [0, \tau]) \le (r/u_1)^{1/2} < \tfrac{5}{4}(r/u_0)^{1/2}.$$

Let $X(\cdot)$ be an arbitrary admissible trajectory in Problem 20 with $x_3(0) = \frac{1}{2} r \operatorname{sgn} x_4(0)$, and let $\Omega_2 = \{x_3, x_4 | |x_3| \ge r/2\} \backslash K(r)$. Then

$$\inf_{X(\cdot)\in M_R} (\tau > 0 | (x_3(\tau), x_4(\tau)) \notin \Omega_2) \ge 2(r/u_1)^{-1}(u_0 + \gamma_0)^{-1/2} > \tfrac{3}{2}(r/u_0)^{1/2}. \tag{4.38}$$

Let X^* be an optimal trajectory in Problem 20 with $X^*(0) \in \mathscr{D}_\varepsilon$, $(x_3^*, x_4^*)(0) \in K(r)$. We estimate the time τ when X^* hits the domain $K(r/2)$. Assume that $(x_3^*(t), x_4^*(t)) \notin K(r/2)$ for $t \in [0, \tau]$, $\tau \le \varepsilon R$. Let n be the number of intersections of the trajectory X^* with the planes $x_3 = \frac{1}{2} r \operatorname{sgn} x_4$ for $t \in [0, \tau]$. Let

$$\mu = \{t \in [0, \tau] | |x_3^*(t)| \le r/2\}, \qquad \nu = [0, \tau] \backslash \mu.$$

By (4.38), $\tau > \frac{3}{2}(n-1)(r/u_0)^{1/2}$. Consequently,

$$n < 1 + \tfrac{2}{3}\tau(u_0/r)^{1/2}, \qquad \operatorname{meas} \mu \le \tfrac{5}{4} n (r/u_0)^{1/2} \le \tfrac{5}{4}(r/u_0)^{1/2} + \tfrac{5}{6}\tau;$$
$$\operatorname{meas} \nu \ge \tfrac{1}{6}\tau - \tfrac{5}{4}(r/u_0)^{1/2}.$$

In view of (4.37),

$$\int_0^\tau F^*\, dt \le s r^{5/2}, \qquad F^* = F(X^*). \tag{4.39}$$

On the other hand,

$$\int_0^\tau F^*\, dt \ge \int_\nu F^*\, dt \ge Q_1 r^2 \operatorname{meas} \nu \ge Q_2 r^2 \tau - Q_3 r^{5/2} \tag{4.40}$$

for some $Q_i > 0$ independent of R. It follows from (4.39) and (4.40) that $\tau < Q\sqrt{r}$.

Accordingly, the value of r is reduced by a factor of two in a time no greater than $Q\sqrt{r}$; then it is reduced by another factor of two in a time no greater than $Q\sqrt{r/2}$, and so on. For $r < \lambda\varepsilon^2$ the time for an optimal trajectory X^* with initial conditions $X_0^* \in \mathscr{D}_\varepsilon$ to hit the plane $x_3 = x_4 = 0$ can thereby be estimated from above by the sum of a decreasing geometric progression:

$$\tau < Q(\sqrt{r} + \sqrt{r/2} + \cdots + \sqrt{r/2^n} + \cdots) = Q\sqrt{r}/(1 - \sqrt{1/2}) \leq Q^*\varepsilon,$$

where $Q^* = Q\sqrt{\lambda}/(1 - \sqrt{1/2})$. Choose $R > Q^*$. Then $\tau < \varepsilon R$, and from Remark 7 it follows that an optimal trajectory in Problem 20 turns out to be an optimal trajectory in Problem 19. It has thus been shown that the optimal trajectories in Problem 19 hit the plane π in a time that is less than εR and depends continuously on the initial conditions $X_0 \in \mathscr{D}_\varepsilon$.

3°. *Order of the singular trajectories.* We show that the optimal trajectories in Problem 19 with $X_0 \in \mathscr{D}_\varepsilon$ have chattering parts. To do this we show that the optimal trajectories in Problem 19 that are singular with respect to the control component u and belong to the domain $\mathscr{D}_\varepsilon$ have even order.

Consider the system of equations of the Pontryagin maximum principle for Problem 19:

$$(4.41)\qquad \begin{gathered} \dot\psi_1 = 0, \qquad \dot\psi_2 = -\psi_1(1 + x_3^2)^{-1} - 2x_2x_3\psi_4(1 + x_3^2)^{-2}, \\ \dot\psi_3 = 2x_2x_3\psi_1(1 + x_3^2)^{-2} - x_2^2(1 - 3x_3^2)\psi_4(1 + x_3^2)^{-3}, \\ \dot\psi_4 = -\psi_3, \qquad \dot x_1 = x_2(1 + x_3^2)^{-1}, \qquad \dot x_2 = v^*, \\ \dot x_3 = x_4, \qquad \dot x_4 = u^* + x_2^2x_3(1 + x_3^2)^{-2}, \\ \max_{|u| \leq u_0} u\psi_4(t) = u^*(t)\psi_4(t), \\ \max_{|v| \leq 1} v\psi_2(t) = v^*(t)\psi_2(t) \quad \text{(a.e.)}. \end{gathered}$$

We compute the trajectories of (4.41) that are singular with respect to u and lie in $\mathscr{D}_\varepsilon$. Let $\psi_4(t) = 0$ on some interval (τ_0, τ_1). Then $\psi_3(t) = 0$. Differentiating the relation $\psi_3 = 0$, we get by virtue of (4.41) that $x_2x_3\psi_1 = 0$. It follows from Lemma 23 that $x_2 \neq 0$ in $\mathscr{D}_\varepsilon$. If $\psi_1 = 0$, then, $\psi_2 = \text{const} \neq 0$ and the optimal control u^* is constant, contrary to Lemma 23. Therefore, $\psi_1 \neq 0$, and then $x_3 = x_4 = u^* = 0$. Differentiating the function $H_1 = \psi_4$, we see by virtue of (4.41) that on the singular control part

$$\frac{\partial}{\partial u}\frac{d^k}{dt^k}H_1 = 0, \quad k = 0, 1, 2, 3, \qquad \frac{\partial}{\partial u}\frac{d^4}{dt^4}H_1 = 2x_2\psi_1 \neq 0.$$

Consequently, the order (intrinsic order) of the singular trajectory is equal to two. The original assertion now follows from the fact that an optimal trajectory of even order cannot be joined with a piecewise smooth nonsingular optimal trajectory if the control is discontinuous at the joining point (Theorem A in Chapter 2). The theorem is proved.

Theorem 7. *For Problem* 19 *there is a two-parameter collection* $\mathfrak{N}_{\alpha,\beta}$ *of families of extremals of the following form. For fixed values of* α *and* β *the trajectories in the family* $\mathfrak{N}_{\alpha,\beta}$ *pass through the point* $X_{\alpha,\beta} = (\alpha, \beta, 0, 0) \in \mathscr{D}_\varepsilon$ *and contain a chattering part. The switching points on the trajectories of the family* $\mathfrak{N}_{\alpha,\beta}$ *for* $X_{\alpha,\beta} \in \mathscr{D}_\varepsilon$ *form a three-dimensional surface of class* C^1. *The trajectories of the families* $\mathfrak{N}_{\alpha,\beta}$ *are optimal in Problem* 19.

Proof. For brevity let $u_0 = 1$. We find a family $\mathfrak{N}_{\alpha,\beta}$ of trajectories of (4.41) passing through the point

$$(4.42)\qquad \begin{gathered} x_1 = \alpha, \qquad x_2 = \beta, \qquad x_3 = x_4 = \psi_3 = \psi_4 = 0, \\ \psi_1 = -1, \qquad \psi_2 = -\beta - (\alpha + \beta^2/2)^{1/2}. \end{gathered}$$

It follows from (4.42) that at points of the surface π the field ψ_1, ψ_2 coincides up to a normalizing factor with the field of conjugate variables in the simplest time-optimal problem on the plane x_3, x_4. Indeed, integrating the system

$$\dot\psi_1 = 0, \qquad \dot\psi_2 = -\psi_1, \qquad \dot x_1 = x_2, \qquad \dot x_2 = \operatorname{sgn}\psi_2, \tag{4.43}$$

we get that a trajectory of (4.43) with initial conditions (4.42) first hits the switching curve $x_1 = x_2^2/2$ with the control $v = -1$ in the time $\tau = (\alpha + \beta^2/2)^{1/2} + \beta$, and $\psi_2(\tau) = 0$ at this time, and then the trajectory hits the origin with the control $v = 1$ and without switchings in the time $(\alpha + \beta^2/2)^{1/2}$.

In (4.41) let $v^* = -1$. It is asserted that in this case the system (4.41) has the form (2.4), (4.1). It is easy to see from (4.41) that

$$\dot\psi_4 = -\psi_3, \qquad \dot\psi_3 = -2\beta x_2 + O(\varepsilon)\Theta_1,$$
$$\dot x_3 = x_4, \qquad \dot x_4 = \operatorname{sgn}\psi_4 + \Theta_2,$$

where the functions Θ_i satisfy

$$\varlimsup_{\kappa\to+0} \Theta_i(g_\kappa(\psi_4, \psi_3, x_3, x_4))\kappa^{-2} < \infty,$$
$$g_\kappa(\psi_4, \psi_3, x_3, x_4) = (\kappa^4\psi_4, \kappa^3\psi_3, \kappa^2 x_3, \kappa x_4).$$

Letting $z_1 = \psi_4$, $z_2 = -\psi_3$, $z_3 = 2\beta x_3$, $z_4 = 2\beta x_4$, we get that (4.41) satisfies the conditions of Theorem 1 with Proposition 3 taken into account. The existence of a family $\mathfrak{N}_{\alpha,\beta}$ of trajectories with the required smoothness properties is thereby proved. We prove the optimality of the trajectories in $\mathfrak{N}_{\alpha,\beta}$. It follows from (4.42) that $H = (\alpha + \beta^2/2)^{1/2} \neq 0$ on the extremals we have constructed. Theorem 2 gives us that the field of conjugate variables

$$\psi/H = (\psi_1/H, \psi_2/H, \psi_3/H, \psi_4/H)$$

corresponding to $\mathfrak{N}_{\alpha,\beta}$ is a potential field in the domain $\mathscr{D}_\varepsilon$.

Since

$$\det \left.\frac{\mathscr{D}(z_3, z_4)}{\mathscr{D}(x_3, x_4)}\right|_\pi = 4\beta^2 \neq 0$$

and

$$\frac{\mathscr{D}x_3}{\mathscr{D}(\psi_1, \psi_2, x_1, x_2)} = 0,$$

it follows from Theorem 4 that the projections of the trajectories in $\mathfrak{N}_{\alpha,\beta}$ on the space of X fill without self-intersections an open neighborhood O of the surface $\mathscr{D}_\varepsilon \cap \pi$.

However, to prove the optimality of the trajectories that go to the origin, we cannot make direct use of Theorem 3, because the endpoints of such trajectories do not lie inside the domain O covered by the field of extremals. Therefore, the two-dimensional surface π should be added to the open domain O enclosing $\pi \cap \mathscr{D}_\varepsilon$. In Theorem 6 it was proved that the optimal trajectories of Problem 19 exist and lie in the domain $\pi \cup O$. The field ψ/H, restricted to this set, is a potential field. The rest of the proof of sufficiency goes just as the proof of Theorem 3.

Let X^* be an optimal trajectory in Problem 19 with $X^*(0) \in \mathscr{D}_\varepsilon$, and let X^0 be a trajectory from $\mathfrak{N}_{\alpha,\beta}$ with $X^0(0) = X^*(0)$. We consider the contour in $O \cup \pi$ made up of these trajectories. Since ψ/H is a potential field,

$$\oint_{X^*} \frac{\psi}{H}\,dx = \oint_{X^0} \frac{\psi}{H}\,dx.$$

It follows from the Pontryagin maximum principle that the right-hand side of this relation is equal to T^0, while the left-hand side does not exceed T^*, where T^0 and T^* are the times along the trajectories X^0 and X^*, respectively, to the origin. This implies that $T^0 \leq T^*$, i.e., X^0 is an optimal trajectory. The theorem is proved.

BIBLIOGRAPHY

1. A. T. Fuller, *Relay control systems optimized for various performance criteria*, Automation and Remote Control (Proc. First Internat. Congr. IFAC, Moscow, 1960; J. F. Coales et al., editors), Vol. 1, Butterworths, London, 1961, pp. 510–519.
2. P. E. W. Grensted and A. T. Fuller, *Minimization of integral-square-error for non-linear control systems of third and higher order*, Internat. J. Control **2** (1965), 33–73.
3. Christian Marchal, *Chattering arcs and chattering controls*, J. Optim. Theory Appl. **11** (1973), 441–468.
4. ———, *Theoretical research in deterministic optimization*, ONÉRA Publ. No. 139, ONÉRA [Office National d'Études et de Recherches Aérospatiales], Paris, 1971.
5. A. T. Fuller, *Minimization of various performance indices for a system with bounded control*, Internat. J. Control **41** (1985), 1–37.
6. E. P. Ryan, *Singular optimal controls for second-order saturating conditions*, Internat. J. Control **30** (1979), 549–564.
7. E. P. Ryan and C. M. Dorling, *Minimization of a nonquadratic cost function for third-order saturating control systems*, Internat. J. Control **34** (1981), 231–258.
8. Ya. M. Bershchanskiĭ, *Fusing of singular and nonsingular parts of an optimal control*, Avtomat. i Telemekh. **1979**, no. 3, 5–12; English transl. in Automat. Remote Control **40** (1979).
9. ———, *The behavior of optimal trajectories close to the terminal part*, Studies in the Theory of Multiply Connected Systems (B. N. Petrov and M. V. Meerov, editors), "Nauka", Moscow, 1982, pp. 76–89. (Russian)
10. ———, *Trajectories of linear systems with a relay-type nonlinearity*, Avtomat. i Telemekh. **1982**, no. 7, pp. 19–27; English transl. in Automat. Remote Control **43** (1982).
11. V. R. Telesnin, *On a problem of optimizing transition processes*, Trudy Mat. Inst. Steklov. **166** (1984), 235–244; English transl. in Proc. Steklov Inst. Math. **1986**, no. 1 (166).
12. G. G. Magaril-Il'yaev, *On the Kolmogorov inequalities on a half-line*, Vestnik Moskov. Univ. Ser. I Mat. Mekh. **1976**, no. 5, 33–41; English transl. in Moscow Univ. Math. Bull. **31** (1976).
13. Pavol Brunovský and John Mallet-Paret, *Switchings of optimal controls and the equation* $y^{(4)} + |y|^\alpha \operatorname{sign} y = 0$, $0 < \alpha < 1$, Časopis Pěst. Mat. **110** (1985), 302–313.
14. M. Z. Borshchevskiĭ and I. V. Ioslovich, *On the problem of the optimal rapid braking of an axisymmetric solid rotation around its center of mass*, Prikl. Mat. Mekh. **49** (1985), 35–42; English transl. in J. Appl. Math. Mech. **49** (1985).
15. V. G. Boltyanskiĭ, *Mathematical methods of optimal control,* 2nd. rev. aug. ed., "Nauka", Moscow, 1969; English transl. of 1st ed., Holt, Rinehart and Winston, New York, 1971; German transl. of 2nd ed., Akademische Verlag, Geest & Portig, Leipzig, 1971.
16. Pavol Brunovský, *Existence of regular synthesis for general control problems*, J. Differential Equations **38** (1980), 317–343.
17. ———, *Regular synthesis for the linear-quadratic optimal control problem with linear control constraints*, J. Differential Equations **38** (1980), 344–360.
18. L. F. Zelikina, *On the question of regular synthesis*, Dokl. Akad. Nauk SSSR **267** (1982), 532–535; English transl. in Soviet Math. Dokl. **26** (1982).
19. I. A. K. Kupka, *Geometric theory of extremals—Fuller phenomenon*, Proc. Twenty-fourth IEEE Conf. Decision and Control (Fort Lauderdale, FL, 1985), IEEE, New York, 1985, pp. 711–713.
20. ———, *The ubiquity of the Fuller phenomenon*, Report No. 52, Inst. Fourier, Grenoble, 1986.
21. V. F. Borisov, *Structural stability of the synthesis of optimal trajectories in the Fuller problem*, Vestnik Moskov. Univ. Ser. I. Mat. Mekh. **1987**, no. 4, 64–66; English transl. in Moscow Univ. Math. Bull. **42** (1987).
22. H. J. Kelley, R. E. Kopp, and H. G. Moyer, *Singular extremals*, Topics in Optimization (G. Lietmann, editor), Academic Press, New York, 1967, pp. 63–103.

23. J. P. McDannel and W. F. Powers, *Necessary conditions for joining optimal singular and nonsingular subarcs*, SIAM J. Control **9** (1971), 161–173.

24. L. D. Akulenko, N. N. Bolotnik, and A. A. Kaplunov, *Optimization of regimes for controlling manipulated robots*, Preprint No. 218, Institute of Problems of Mechanics, Moscow, 1983. (Russian)

25. Hans P. Geering et al., *Time-optimal motions of robots in assembly tasks*, Trans. IEEE Automat. Control **AC-31** (1986), 512–518.

26. S. N. Osipov and A. M. Formal′skiĭ, *The problem of the fastest turn of a robot*, Prikl. Mat. Mekh. **52** (1988), 929–938; English transl. in J. Appl. Math. Mech. **52** (1988).

27. V. F. Borisov and M. I. Zelikin, *Chattering regimes in the problem of controlling a robot*, Prikl. Math. Mekh. **52** (1988), 939–946; English transl. in J. Appl. Math. Mech. **52** (1988).

28. D. F. Lawden, *Optimal trajectories for space navigation*, Butterworths, London, 1963.

29. Henry J. Kelley, *Singular extremals in Lawden's problem of optimal rocket flight*, AIAA J. **1** (1963), 1578–1580.

30. H. M. Robbins, *Optimality of intermediate-thrust arcs of rocket trajectories*, AIAA J. **3** (1965), 1094–1098.

31. C. Marchal, *Second-order tests in optimization theories*, J. Optim. Theory Appl. **15** (1975), 633–666.

32. ______, *Généralisation tridimensionnelle et étude de l'optimalité des arcs à poussée intermédiaire de Lawden (dans un champ newtonian)*, Recherche Aérospatiale No. 123 (1968), 3–13.

32. J. V. Breakwell and J. F. Dixon, *Minimum-fuel rocket trajectories involving intermediate-thrust arcs*, J. Optim. Theory Appl. **17** (1975), 465–479.

34. L. S. Pontryagin et al., *The mathematical theory of optimal processes*, 3rd ed., "Nauka", Moscow, 1976; English transls. of 1st ed., Wiley, New York, 1962, and Macmillan, New York, 1964.

35. E. B. Lee and L. Markus, *Foundations of optimal control theory*, Wiley, New York, 1967.

36. Earl A. Coddington and Norman Levinson, *Theory of ordinary differential equations*, McGraw-Hill, New York, 1955.

37. M. I. Zelikin and V. F. Borisov, *Synthesis in optimal control problems containing chattering trajectories and singular trajectories of second order*, Mat. Zametki **47** (1990), no. 1, 62–73; English transl. in Math Notes **47** (1990).

38. ______, *Synthesis in optimal control problems containing chattering trajectories and singular trajectories of second order*, Manuscript No. 4226–B89, deposited at VINITI, June 26, 1989. (Russian)

39. ______, *Fields of optimal trajectories containing singular extremals of second order and chattering extremals*, Dokl. Akad. Nauk SSSR **304** (1989), 1050–1053; English transl. in Soviet Math. Dokl. **39** (1989).

40. A. F. Filippov, *Differential equations with discontinuous right-hand sides*, "Nauka", Moscow, 1984; English transl., Reidel, Dordrecht, 1988.

41. R. M. Lewis, *Definitions of order and junction conditions in singular optimal control problems*, SIAM J. Control Optim. **18** (1980), 21–32.

42. Zbigniew Nitecki, *Differentiable dynamics*, M.I.T. Press, Cambridge, MA, 1971.

43. A. N. Shoshitaĭshvili, *On topological-type bifurcation of singular points of vector fields depending on parameters*, Trudy. Sem. Petrovsk. Vyp. **1** (1975), 279–309; English transl. in Amer. Math. Soc. Transl. (2) **118** (1982).

44. V. M. Alekseev V. M. Tikhomirov and S. V. Fomin, *Optimal control*, "Nauka", Moscow, 1979; English transl., Plenum Press, New York, 1987.

45. René Thom and Harold I. Levine, *Singularities of differentiable mappings*, Lecture Notes, Bonn. Univ., Bonn, 1959.

46. V. F. Borisov, *Construction of optimal synthesis with a chattering regime*, Dokl. Akad. Nauk SSSR **302** (1988), 785–789; English transl. in Soviet Math. Dokl. **38** (1989).

47. Nelson Dunford and Jacob T. Schwartz, *Linear operators*. Vol. I, Interscience, New York, 1958.

48. R. Tyrrell Rockafellar, *Convex analysis*, Princeton Univ. Press, Princeton, NJ, 1970.

49. A. G. Kurosh, *A course of higher algebra*, 11th ed., "Nauka", Moscow, 1975; English transl. of 10th ed., "Mir", Moscow, 1975.

50. Henry J. Kelley, *A second variation test for singular extremals*, AIAA J. **2** (1964), 1380–1382.

51. Richard E. Kopp and H. Gardner Moyer, *Necessary conditions for singular extremals*, AIAA J. **3** (1965), 1439–1444.

52. V. M. Fraeijs de Veubeke and J. Geerts, *Optimization of multiple impulse orbital transfers by the maximum principle*, XVth International Astronautical Congress (Warsaw, 1964), Vol. I, PWN, Warsaw, and Gauthier-Villars, Paris, 1965, pp. 589–616.

53. A. F. Filippov, *On certain questions in the theory of optimal control*, Vestnik Moskov. Univ. Ser. Mat. Mekh. Astr. Fiz. Khim. **1959**, no. 2, 25–32; English transl., J. Soc. Indust. Appl. Math. Ser. A: Control **1** (1962), 76–84.

Translated by H. H. McFADEN